Revolutionizing Reproduction

Revolutionizing Reproduction

The Ethical Landscape of In Vitro Gametogenesis

Edited by
CHRISTOPHER GYNGELL
JULIAN SAVULESCU
TSUTOMU SAWAI

OXFORD
UNIVERSITY PRESS

OXFORD
UNIVERSITY PRESS

Oxford University Press is a department of the University of Oxford.
It furthers the University's objective of excellence in research, scholarship,
and education by publishing worldwide. Oxford is a registered trade mark of
Oxford University Press in the UK and in certain other countries.

Published in the United States of America by Oxford University Press
198 Madison Avenue, New York, NY 10016, United States of America.

CIP data is on file at the Library of Congress

ISBN 9780197810385 (pbk.)
ISBN 9780197810378 (hbk.)

DOI: 10.1093/9780197810415.001.0001

Paperback printed by Marquis Book Printing, Canada
Hardback printed by Lightning Source, Inc., United States of America

The manufacturer's authorized representative in the EU for product safety is
Oxford University Press España S.A. of Parque Empresarial San Fernando de Henares,
Avenida de Castilla, 2 – 28830 Madrid (www.oup.es/en or product.safety@oup.com).
OUP España S.A. also acts as importer into Spain of products made by the manufacturer.

MIX
Paper | Supporting responsible forestry
FSC® C103567
FSC
www.fsc.org

Contents

Preface

The idea for this collection took shape in Kyoto in 2023, when Julian Savulescu and Chris Gyngell were invited by Tsutomu Sawai to participate in a workshop on brain organoids. During our time in Japan, we had the privilege of meeting Dr. Mitinori Saitou of Kyoto University, a pioneer in *in vitro* gametogenesis (IVG). We were struck by the remarkable progress in IVG research, much of which seemed to be unfolding quietly and without the broader public awareness it deserves. Although many scholars and commentators had previously entertained the possibility of IVG in largely theoretical terms, the concrete advances we witnessed underscored an urgent need for a closer, more up-to-date ethical and social examination of this emerging technology.

It became clear to us that IVG's promise, while inspiring new avenues for overcoming infertility and expanding reproductive choice, brings with it profound challenges for our moral, legal, and social frameworks. To date, much of the conversation about IVG has been conducted in a somewhat abstract manner, often overlooking the tangible breakthroughs now underway. This collection aims to bridge that gap by grounding ethical and social analysis in recent scientific developments, providing a timely reappraisal of what IVG might mean for individuals, families, and societies.

By bringing together diverse perspectives in one volume, we hope to spark nuanced deliberation on IVG's role in shaping reproductive medicine and broader societal norms. While the technology itself is still under development, its potential ramifications are already pressing. Our hope is that this collection will serve both as a resource and a catalyst for thoughtful dialogue on the evolving boundaries of reproductive autonomy and responsibility.

Finally, we would like to extend our sincere thanks to the Uehiro Foundation on Ethics and Education, for a generation of support for scholars (many of whom contributed to this volume), conferences, and workshops. This generous support has been integral to the preparation of this collection. We trust that the insights gathered here will encourage further reflection and, ultimately, contribute to a responsible and forward-looking approach to IVG in the years to come.

Contributors

Dorian Accoe is PhD researcher, Department of Philosophy and Moral Science, Bioethics Institute Ghent (Ghent University), Belgium. https://orcid.org/0000-0002-8873-2593

Kyoko Akatsuka, PhD, is Assistant Professor, Uehiro Research Division for iPS Cell Ethics, Center for iPS Cell Research and Application, Kyoto University, Japan. https://orcid.org/0000-0003-0254-0943

Hilary Bowman-Smart, MPH, PhD, is Research Fellow, Australian Centre for Precision Health, Adelaide University, Australia, Murdoch Children's Research Institute, Melbourne, Australia. https://orcid.org/0000-0002-2142-9696

Misao Fujita, MS, MPH, PhD, is Professor and Head, Uehiro Research Division for iPS Cell Ethics, Center for iPS Cell Research and Application (CiRA), Kyoto University, Japan. https://orcid.org/0000-0001-7919-5362

Christopher Gyngell is Team Leader, Biomedical Ethics, Murdoch Children's Research Institute, Melbourne, Australia, University of Melbourne, Australia. https://orcid.org/0000-0002-1340-3947

Katsuhiko Hayashi is Professor, Graduate School of Medicine, University of Osaka, Japan. https://orcid.org/0000-0002-2479-2941

Heidi Mertes, PhD, is Associate Professor of Medical Ethics, Department of Philosophy and Moral Sciences, Department of Public Health and Primary Care, Ghent University, Belgium. https://orcid.org/0000-0003-3029-2158

Anantharaman Muralidharan, PhD, is Research Fellow, Centre for Biomedical Ethics, Yong Loo Lin School of Medicine, National University of Singapore, Singapore. https://orcid.org/0000-0003-3380-0402

Timothy F. Murphy, PhD, is Professor of Philosophy in the Biomedical Sciences, Department of Medical Education, University of Illinois College of Medicine, Chicago, US. https://orcid.org/0000-0001-8003-550X

César Palacios-González, PhD, is Senior Research Fellow, Uehiro Oxford Institute, University of Oxford, UK. https://orcid.org/0000-0002-3825-5311

Julian Savulescu is Chen Su Lan Centennial Professor in Medical Ethics, Yong Loo Lin School of Medicine, National University of Singapore, National University of Singapore, Singapore, University of Oxford, UK. https://orcid.org/0000-0003-1691-6403

Tsutomu Sawai, PhD, is Professor (Special Recognition), Graduate School of Humanities and Social Sciences, Hiroshima University, Higashi-Hiroshima, Japan, National University of Singapore, Singapore, Singapore. https://orcid.org/0000-0002-3806-0573

Seppe Segers, PhD, is Professor of Ethics and Moral Science, Department of Philosophy and Moral Science, Bioethics Institute Ghent (Ghent University), Ghent, Belgium. https://orcid.org/0000-0001-8231-6487

Anna Smajdor is Professor of Practical Philosophy, Department of Philosophy, Classics, History of Art and Ideas, University of Oslo, Norway. https://orcid.org/0000-0002-9752-6302

Robert Sparrow, PhD, is Professor of Philosophy, Monash University, Melbourne, Australia. https://orcid.org/0000-0001-6093-9456

Sonia M. Suter, JD, MS, is Professor of Law and Kahan Family Research Professor, Law School, George Washington University Law School, Washington DC, US. https://orcid.org/0000-0001-9191-5828

Helen Watt, PhD, is Senior Research Fellow, Bios Centre, London, UK, Blackfriars Hall, University of Oxford, London, UK. https://orcid.org/0000-0002-4605-339X

Introduction—Are We Ready for a Revolution?

Over the past decade, several groundbreaking technologies have captured both academic and public attention. The controversy surrounding human genome editing—especially the creation of the so-called CRISPR twins, the first children born with edited genomes—shook the world and brought the prospect of human genetic engineering into the public imagination, prompting numerous international and national commissions, reports, and dedicated research calls. Meanwhile, as of early 2025, the rise of artificial intelligence—particularly large language models—has attracted unprecedented investment and sparked debates about an impending new era of automation. Other advances, such as the development of brain organoids—three-dimensional structures grown in vitro that partially recapitulate aspects of the human brain—and mitochondrial replacement therapy, have likewise captured headlines and ignited vigorous public debates.

Amid these high-profile developments, it is perhaps understandable that some emerging technologies have received comparatively little attention. One such technology is *in vitro* gametogenesis (IVG), which enables the generation of gametes (eggs and sperm) in the lab from virtually any cell in the body. In IVG, somatic cells such as skin or blood cells are first reprogrammed into more immature cell types called induced pluripotent stem cells, or alternatively, embryonic stem cell may be used as a starting point. These pluripotent stem cells are then directed to differentiate into cells which produce gametes. Although currently in its early stage, IVG has already been used to produce healthy, fertile offspring in mice, leading many experts to predict that human applications could emerge in the not-too-distant future.

If realized, IVG could immediately benefit the millions of couples who undergo *in vitro* fertilization (IVF) each year by eliminating the need for ovarian stimulation and egg retrieval, while enabling the generation of countless embryos in a single cycle. It could also empower same-sex couples to have genetically related children. Even more radically, IVG might pave the way for 'solo' reproduction or 'multiplex' parenting—new forms of family building previously impossible. Beyond these reproductive possibilities, IVG could vastly increase parents' ability to shape the genomes of their future children through advanced

Revolutionizing Reproduction. Edited by: Christopher Gyngell, Julian Savulescu, and Tsutomu Sawai, Oxford University Press. DOI: 10.1093/9780197810415.001.0001

embryo selection, screening not only for polygenic disease risks but also for traits like intelligence or height.

How to respond to such a transformative technology is a pressing question as IVG nears clinical viability. While bioethicists have examined its ethical implications for more than a decade, IVG has received comparatively little attention when measured against more high-profile developments.

This volume aims to bring considerations of IVG from the shadows into the spotlight. It highlights IVG as a truly novel technology with far-reaching consequences for how we understand reproduction, family, and parenthood. Drawing on scientific, philosophical, legal, and sociological perspectives, the chapters begin by outlining the foundational science behind IVG, detailing how researchers aspire to replicate complex biological processes outside the body. The focus then shifts to deeper ethical and social questions: Which longstanding assumptions about sex, parenting, and genetic relatedness should IVG prompt us to revisit? How can we balance individual autonomy and equality against potential risks, such as eugenic selection, commodification of embryos, and radical redefinitions of family? Finally, the volume asks what role bioethics should play in shaping research and governance, pushing us to consider how best to structure bioethical inquiry in a rapidly evolving technological landscape.

Chapter 1 sets the stage with Katsuhiko Hayashi, a pioneer of IVG in the lab, charting the technology's historical and technical background. He explains how early twentieth-century organ culture systems enabled researchers to cultivate gonadal tissue (producing gametes) outside the body, culminating in modern, stem cell-based approaches for deriving functional eggs and sperm in mice. Although these achievements are awe inspiring, Hayashi is careful to show just how challenging it is to replicate the intricate genetic and epigenetic signals that guide gamete maturation. While offering glimpses of remarkable possibilities—such as switching gametes between sexes or bypassing age-related fertility constraints—he foreshadows the deeper ethical questions that the remaining chapters address.

In Chapter 2, editors Christopher Gyngell, Tsutomu Sawai, and Julian Savulescu discuss the possibilities of IVG in an evolutionary context. They explain how the evolution of sexual reproduction places certain constraints on human biology that IVG promises to overcome. They survey a broad range of potential IVG applications, including restoring fertility, solo reproduction, multi-parent reproduction, posthumous reproduction, and advanced screening of embryos, and shows how each overcomes an ancient constraint of evolution. While IVG makes existing goals—like treating infertility and selecting against disease—more efficient, it also opens up entirely new reproductive possibilities.

A common argument in favour of pursuing IVG is that it promotes reproductive liberty. In Chapter 3, Sonia Suter considers this issue in detail. She argues that simply offering more reproductive options via IVG will not always translate into genuine autonomy, especially in contexts where systemic pressures, family expectations, or cultural biases exert a strong influence. Suter develops a relational approach to procreative liberty that emphasizes the importance of relationships, social norms, and cultural contexts in shaping individual choices and self-definition. Suter also notes how that shifting legal climates (e.g., post-Dobbs in the United States) threaten to restrict or reshape the availability of reproductive technologies. Whether IVG ultimately promotes or undermines relational procreative autonomy depends greatly on why people seek it and the socio-legal environment in which it becomes available.

In Chapter 4, Timothy Murphy explores IVG's potential to enable same-sex reproduction. Infertility is typically framed in physiological terms, which can exclude same-sex couples or trans individuals from mainstream fertility support. Murphy argues that if fertility interventions are embraced as legitimate medical services for opposite-sex couples, they should be made equally available to LGBTQ+ individuals. Murphy argues that in a context where opposite-sex couples access ARTs freely, excluding same-sex couples from IVG entrenches inequality. Nonetheless, cultural biases and restrictive laws often lag behind scientific progress, risking a two-tier system where the populations who might benefit most remain unable to access IVG.

Robert Sparrow, in Chapter 5, confronts us with an unsettling application of IVG: the possibility of repeated cycles of embryo creation, selection, and gamete derivation. This 'rapid-fire' laboratory-based selective breeding might further expand reproductive opportunities. However, Sparrow shows it parallels historical eugenics, reinforcing power imbalances and further commodifying offspring. Sparrow warns that ethical intentions alone do not insulate a technology from misuse, emphasizing the need to consider underlying power imbalances and the commodification of offspring.

Further developing the ethical objections to IVG, in Chapter 6 Helen Watt argues that certain uses of IVG—and in particular the 'mass-production' of embryos—undermine our recognition of embryos as morally significant offspring. She draws a distinction between interventions that restore fertility (like autologous transplantation) and large-scale embryo creation and selection. Concerned with commodification and the dignity of the child, Watt argues that parenthood should involve unconditional acceptance rather than a 'manufacturer's' mentality of testing and discarding.

In Chapter 7, César Palacios-González considers the issue of 'multiplex parenting,' a use of IVG with the capacity to radically reconfigure family structures. By deriving gametes from multiple embryonic lines in a lab, more than two individuals could be genetically related to a single child. Although such scenarios remain hypothetical, Palacios-González examines why some polyamorous or co-parenting groups might embrace the technology and what legal and ethical complexities would inevitably follow. Are all 'genetic contributors' equally responsible for childrearing? Could a child have 'grandparents' who existed only as *in vitro* embryos? 'Multiplex' arrangements reveal how profoundly IVG can blur genealogies and upend established ideas of parental responsibility.

Chapter 8 examines one of the most controversial applications of IVG: solo reproduction. Anantharaman Muralidharan explores how IVG enables the creation of gametes from an individual's somatic cells, allowing a single person to produce both sperm and egg. After comparing solo reproduction to cloning, which likewise circumvents the need for a second genetic parent—Muralidharan considers common objections such as physical risks (e.g., recessive genetic disorders), negative social consequences (e.g., stigmatization and overbearing expectations), and broader concerns about dignity, ethics, and the 'telos' of human sexuality. Most of these objections, the chapter suggests, either overextend, rely on debatable assumptions (including religious or 'dignitarian' views), or fail to distinguish genuinely problematic motives for solo reproduction from permissible ones (e.g., an abuse survivor wishing to avoid male involvement). Ultimately, Muralidharan concludes that only physiological risks to future children present a potentially decisive moral concern. Accordingly, there appear to be no intrinsic reasons to regard solo reproduction as morally objectionable.

The question of how IVG might disrupt conventional ideas of genetic parenthood is taken up directly in Chapter 9 by Hilary Bowman-Smart. Because IVG can generate sperm or eggs from embryonic tissue—or from cells never part of a fully formed adult—the usual notion that 'whoever supplies half the nuclear DNA is a genetic parent' is insufficient. In response, Bowman-Smart proposes the term 'genetic progenitor', a more neutral, less prescriptive label that decouples biological origin from moral or legal parenthood. This revised vocabulary, she argues, could help us navigate the confusing genealogies IVG might produce.

Continuing the theme of how IVG challenges traditional definitions, Chapter 10 features Anna Smajdor's exploration of how the definitions of 'gametes' and 'reproduction' also need rethinking once IVG is clinically viable. Classic morphological or meiotic criteria for defining gametes may no longer apply if cells are grown and recombined in new ways. Meanwhile, the concept of reproduction might shift: Is an embryo still the product of 'natural' sexual reproduction if its components are rearranged in a lab? Smajdor urges us to question the boundaries between biology and technology, and to ask whether 'naturalness' remains meaningful in an age of increasingly sophisticated interventions.

Chapter 11 highlights work by Heidi Mertes. She analyses how entrenched gender and age norms shape public attitudes toward IVG. Polling data show that many people welcome new reproductive tools for younger, 'traditionally infertile' couples yet become uneasy when it involves older mothers or single reproducers. Mertes argues that these reactions reflect deeper cultural biases—particularly around age and gender—and that the acceptance and integration of IVG into clinical settings will hinge on challenging these biases, not just on demonstrating safety or efficacy.

In Chapter 12, Misao Fujita and colleagues present further empirical data on how the public view IVG, this time in a Japanese context, which has one of the lowest fertility rates in the world. They show that while many Japanese respondents are open to at least some IVG applications, a significant minority feel unease about overturning traditional reproductive norms. This tension underscores the importance of understanding cultural and societal perspectives in charting IVG's future. Even if the technology matures rapidly, debates over social acceptability will be decisive in determining whether—and how—it is ultimately used.

Finally, Chapter 13 addresses how the very way we speak about and debate IVG will shape its trajectory. This theme is taken up by Seppe Segers and Dorian Accoe, who analyse the blurry boundary between normative and empirical dimensions of IVG. They emphasize that cultural, economic, and political forces influence what becomes scientifically feasible and socially permissible. Whether IVG is embraced by clinics or heavily restricted by legislation depends not only on technical progress but on which moral and societal values gain primacy.

Taken together, the chapters form a multi-layered portrait of IVG as more than just another new reproductive technology, but also a technology with truly revolutionary implications for human society. By bringing together scientific, philosophical, legal, and sociological perspectives, it hopes to bring the challenges presented by IVG into focus. In doing so, the volume addresses issues that far broader implications, such as the limits of procreative autonomy, and our obligations to future generations and how we define parents, families, and reproduction itself. IVG may still be on the horizon of clinical practice, but the chapters here underscore how pressing it is to anticipate its ethical, legal, and social challenges.

More broadly, this volume serves as a model for how we can—and should—critically engage with profoundly transformative biotechnologies as they evolve from theoretical possibilities into near-term realities. Rather than promoting a single ethical or political stance, we have brought together contributors from diverse backgrounds to present varied perspectives on IVG. By recognizing the dual nature of emerging technologies—capable of both beneficial and harmful ends—we can foster an evidence-based and pragmatic approach to their development, ensuring that bioethical debates remain nuanced and responsive to the complexities of real-world implementation.

By illuminating both the scientific possibilities and the moral complexities, the volume calls on policymakers, practitioners, and the public to engage in proactive debate. In so doing, it ensures that decisions about the future of reproductive medicine are guided not merely by technical feasibility but by thoughtful ethical and social considerations.

Ultimately, *this* volume is intended not just for specialists in bioethics or reproductive medicine, but for all those interested in how society navigates the intersection of human agency, scientific innovation, and moral responsibility. As IVG transitions from cutting-edge experiment to tangible clinical application, the insights captured here will help shape a more thoughtful conversation about the families of the future—and the shared values we hope will guide them.

1

In Vitro Gametogenesis

Creation of Gametes in a Dish

Katsuhiko Hayashi

Outline of Germ Cell Development in Mammals

In early mammalian development, germ cells are specified as primordial germ cells (PGCs) in the posterior region of pluripotent cells, typically known as the post-implantation epiblast, which is just about to start gastrulation (Figure 1.1).[1,2] Nascent PGCs form a cluster in extragonadal embryonic tissue and eventually migrate through the developing hindgut endoderm and mesentery to reach the embryonic gonads. In the gonads, PGCs continue to proliferate as oogonia in females and prospermatogonia in males and then initiate sex-specific developmental processes. During these processes, PGCs undergo epigenetic reprogramming, including genome-wide DNA demethylation and alterations in histone modifications, which are characterized by an increase in histone H3 lysine 27 trimethylation (H3K27me3) and a decrease in histone H3 lysine 9 dimethylation (H3K9me2).[3,4,5] Disruptions in this process can result in developmental disorders, such as germ cell loss, germ cell tumours, and imprinting diseases in subsequent generations, highlighting the critical need to understand this process.

In female embryos, oogonia are exposed to signals provided by surrounding ovarian somatic cells including pregranulosa cells and interstitial cells, and

[1] Mitinori Saitou and Hidetaka Miyauchi, 'Gametogenesis from Pluripotent Stem Cells', *Cell Stem Cell*, 18/6 (2016), 721–735.

[2] Walfred W. C. Tang et al., 'Specification and Epigenetic Programming of the Human Germ Line', *Nature Reviews Genetics*, 17/10 (2016), 585–600.

[3] Heather J. Lee, Timothy A. Hore, and Wolf Reik, 'Reprogramming the Methylome: Erasing Memory and Creating Diversity', *Cell Stem Cell*, 14/6 (2014), 710–719.

[4] Yoshiyuki Seki et al., 'Extensive and Orderly Reprogramming of Genome-Wide Chromatin Modifications Associated with Specification and Early Development of Germ Cells in Mice', *Developmental Biology*, 278/2 (2005), 440–458.

[5] Yoshiyuki Seki et al., 'Cellular Dynamics Associated with the Genome-Wide Epigenetic Reprogramming in Migrating Primordial Germ Cells in Mice', *Development*, 134/14 (2007), 2627–2638.

Katsuhiko Hayashi, In Vitro *Gametogenesis* In: *Revolutionizing Reproduction*. Edited by: Christopher Gyngell, Julian Savulescu, and Tsutomu Sawai, Oxford University Press.
DOI: 10.1093/9780197810415.003.0001

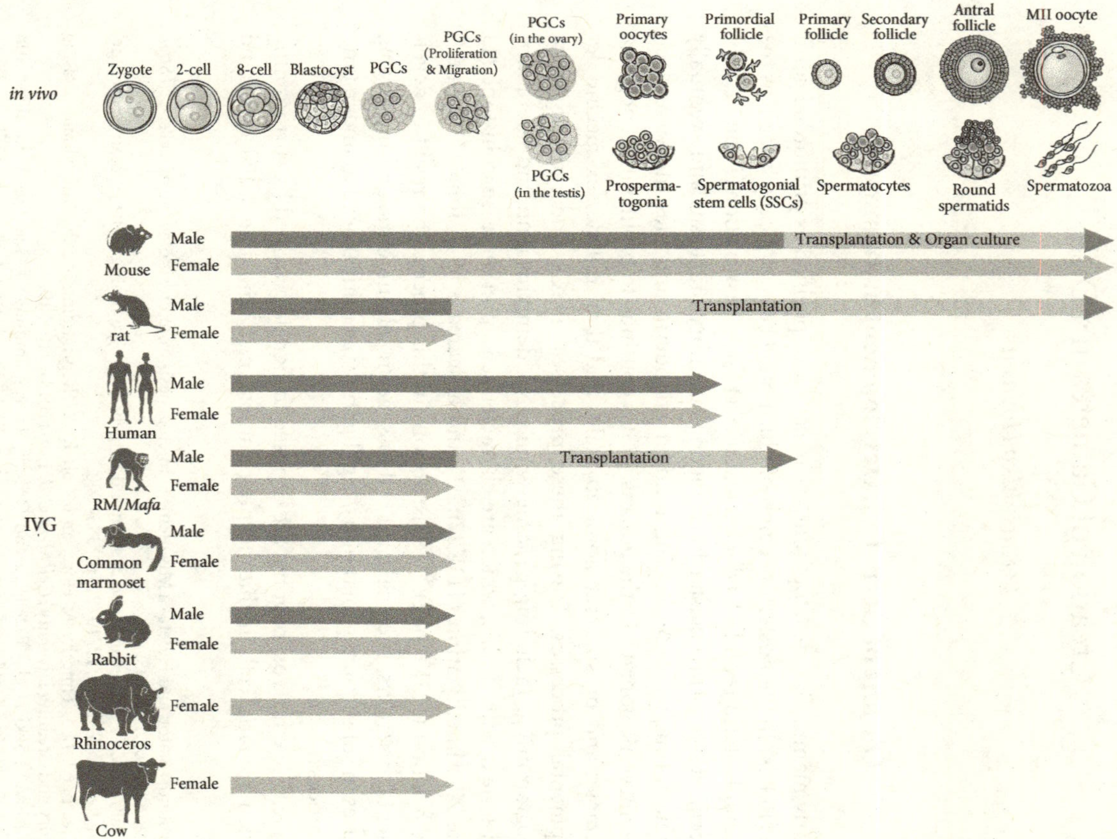

Figure 1.1 Germ-cell development *in vivo* and the current status of IVG in various mammalian species. Arrows indicate the reconstitution of germ cell development *in vitro* from pluripotent stem cells.

then enter meiosis and differentiate into primary oocytes (Figure 1.1).[6,7] Mechanistically, bone morphogenetic protein (BMP)—and retinoic acid (RA)—signalling transactivate downstream genes required for initiating meiosis.[8,9,10] These primary oocytes progress to the diplotene stage of meiotic prophase, where they form follicle structures via a close association with surrounding pregranulosa cells, and thereby establish primordial follicles. Around puberty, follicle growth starts in response to hormonal signals; this stage of follicle growth involves oocyte enlargement and the expansion of granulosa cell layers, and it culminates in the formation of mature antral follicles. The mature oocytes carry both the maternal genome and epigenome.[11,12] Following ovulation and fertilization by sperm, the oocytes complete their first and second meiotic divisions, resulting in the formation of a zygote.

In male embryos, prospermatogonia associate with Sertoli cells, and this association protects them from meiosis-initiating signals. These cells undergo mitotic arrest, acquire the paternal epigenome (which includes genomic imprints), and protect against transposon activity.[3,6,7,13,14] After birth, a partial population of prospermatogonia differentiate into spermatogonial stem cells (SSCs) (Figure 1.1). The SSCs eventually begin the process of spermatogenesis, a sequence in which mitotic stem cells give rise to haploid gametes through successive mitotic amplifications, meiotic divisions, and spermiogenesis, ultimately forming haploid spermatozoa. Testicular somatic cells, particularly Sertoli cells, play crucial roles in facilitating the development of male germ cells.[6,13,15]

Prototype of IVG: Organ Culture of Gonads

Inasmuch as *in vitro* gametogenesis (IVG) is a culture system for producing gametes, the organ culture of gonadal tissue, which has been refined over the

[6] Cassy Spiller, Peter Koopman, and Josephine Bowles, 'Sex Determination in the Mammalian Germline', *Annual Review of Genetics*, 51 (2017), 265–285.

[7] Lu Wen and Fuchou Tang, 'Human Germline Cell Development: From the Perspective of Single-Cell Sequencing', *Molecular Cell*, 76/2 (2019), 320–328.

[8] Jana Koubova et al., 'Retinoic Acid Regulates Sex-Specific Timing of Meiotic Initiation in Mice', *Proceedings of the National Academy of Sciences*, 103/8 (2006), 2474–2479.

[9] Kei-Ichiro Ishiguro et al., 'MEIOSIN Directs the Switch from Mitosis to Meiosis in Mammalian Germ Cells', *Developmental Cell*, 52/4 (2020), 429–445.

[10] So I. Nagaoka et al., 'ZGLP1 Is a Determinant for the Oogenic Fate in Mice', *Science*, 367/6482 (2020), eaaw4115.

[11] Mark A. Edson, Ankur K. Nagaraja, and Martin M. Matzuk, 'The Mammalian Ovary from Genesis to Revelation', *Endocrine Reviews*, 30/6 (2009), 624–712.

[12] Elizabeth A. McGee and Aaron J. W. Hsueh, 'Initial and Cyclic Recruitment of Ovarian Follicles', *Endocrine Reviews*, 21/2 (2000), 200–214.

[13] Gurpreet Manku and Martine Culty, 'Mammalian Gonocyte and Spermatogonia Differentiation: Recent Advances and Remaining Challenges', *Reproduction*, 149/3 (2015), 139–157.

[14] Shinichiro Chuma and Toru Nakano, 'piRNA and Spermatogenesis in Mice', *Philosophical Transactions of the Royal Society B: Biological Sciences*, 368/1609 (2013), 20110338.

[15] Michael D. Griswold, 'Spermatogenesis: The Commitment to Meiosis', *Physiological Reviews*, 96/1 (2016), 1–17.

past century, serves as its prototype. With respect to spermatogenesis, a 1920 report was the first to describe that spermatogonia could be differentiated into leptotene spermatocytes through the cultivation of rabbit testicular tissue.[16] Nearly two decades later, it was reported that cultivation of newborn mouse testicular tissue on a blood clot resulted in progression to pachytene spermatocytes.[17] Subsequent studies further refined organ culture techniques for mammalian testicular tissue, including rats and humans, by adjusting culture conditions such as temperature, CO_2 levels, and medium components.[18,19,20,21,22] Other efforts at culturing immature and adult rat testis fragments advanced spermatogenesis to the pachytene stage.[23,24] The first production of functional sperm in an organ culture system was reported by Ogawa et al.[25] In their culture system, the neonatal testes could be maintained for over seventy days and produced haploid sperm with the potential to fertilize eggs and give rise to healthy offspring.

For oogenesis, cultivation of rat and mouse perinatal ovaries on blood clots yielded growing oocytes within follicular structures.[17] Although some oocytes survived for up to twenty-two days, the tissue degenerated after one month. In 1965, Blandau et al. successfully cultured mouse E16 ovaries for fifty days without degeneration, demonstrating that E16 oocytes were the largest cells with prominent nuclei entering meiotic prophase I. These growing oocytes typically measured 50 μm in diameter and displayed characteristic germinal vesicle (GV) nuclei. Subsequent studies aimed to optimize the culture conditions, including the concentrations of oxygen, nutrients, and gonadotropin.[26,27] The first production of functional and mature oocytes was realized in an organ culture system

[16] Christian Champy, 'Quelques résultats de la méthode de culture des tissues', *Archives de Zoologie Expérimentale et Générale* (1920), 461–500.

[17] Petar N. Martinovitch, 'Development In Vitro of the Mammalian Gonad', *Nature*, 139 (1937), 413.

[18] Alan Steinberger and Elisabeth Steinberger, 'Differentiation of Rat Seminiferous Epithelium in Organ Culture', *Journal of Reproduction and Fertility*, 9 (1965), 243–248.

[19] Alan Steinberger and Elisabeth Steinberger, 'Factors Affecting Spermatogenesis in Organ Cultures of Mammalian Testes', *Journal of Reproduction and Fertility*, 2 (1967), 117–124.

[20] Alan Steinberger, Elisabeth Steinberger, and William H. Perloff, 'Mammalian Testes in Organ Culture', *Experimental Cell Research*, 36 (1964), 19–27.

[21] Elisabeth Steinberger, 'Maintenance of Adult Human Testicular Tissue in Culture', *The Anatomical Record*, 157 (1967), 327.

[22] Elisabeth Steinberger, Alan Steinberger, and William H. Perloff, 'Initiation of Spermatogenesis in Vitro', *Endocrinology*, 74/5 (1964), 788–792.

[23] Satoru Aizawa and Yoshitake Nishimune, 'In-Vitro Differentiation of Type A Spermatogonia in Mouse Cryptorchid Testis', *Journal of Reproduction and Fertility*, 56 (1979), 99–104.

[24] Cristiana Boitani, Maria G. Politi, and Teresa Menna, 'Spermatogonial Cell Proliferation in Organ Culture of Immature Rat Testis', *Biology of Reproduction*, 48 (1993), 761–767.

[25] Takuya Sato et al., 'In Vitro Production of Fertile Sperm from Murine Spermatogonial Stem Cell Lines', *Nature Communications*, 2/1 (2011), 472.

[26] Tibor Fainstat, 'Organ Culture of Postnatal Rat Ovaries in Chemically Defined Medium', *Fertility and Sterility*, 19 (1968), 317–338.

[27] John J. Eppig, 'Mouse Oocyte Development In Vitro with Various Culture Systems', *Developmental Biology*, 60 (1977), 371–388.

developed by Eppig et al.[28,29] In this system, secondary follicles were isolated from neonatal ovaries after eight days of culture, then cultured for an additional fourteen days. Fully grown germinal vesicle (GV) oocytes were then treated with FSH and EGF to mature into MII oocytes. These MII oocytes produced viable offspring through *in vitro* fertilization followed by transplantation into surrogate mothers. This culture system serves as a prototype for *in vitro* oogenesis but does not fully recapitulate the entire meiotic process of oogenesis, as oocytes in neonatal ovaries are already in meiotic prophase I. The first successful reconstitution of meiosis in oogenesis was accomplished using an organ culture system with E12.5 mouse ovaries, in which PGCs remained in a premeiotic state.[30] In this culture system, PGCs enter meiotic prophase I and form primary follicles, within which meiotic prophase I is completed. The follicles grow into mature oocytes with FSH supplementation and then resume meiosis to become MII oocytes. The MII oocytes produced by this system are capable of fertilization and developing to term. This historical series of organ culture systems offers valuable insights and strategies for the future induction of functional gametes from pluripotent stem cells.

Prototype of IVG: Spontaneous Differentiation of Germ Cells from Pluripotent Stem Cells

The concept of IVG using pluripotent stem cells was initially established in early studies that mostly relied on the spontaneous differentiation of germ cell-like cells from pluripotent stem cells. Culturing mouse embryonic stem cells (mESCs) in medium containing foetal bovine serum without leukaemia inhibitory factor (LIF) induced spontaneous differentiation into cells that share several characteristics, including promoter usage of *Pou5f1* and expression of genes such as *Kit* and *Mvh*.[31] Moreover, cell aggregates that detached from the culture dish developed into follicle-like structures expressing oocyte-specific genes, including *Zp2*, *Zp3*, and *Figla*. Remarkably, germ cell development advanced beyond the sex-determination stage without requiring the addition of gonadal somatic cells, suggesting that gonadal somatic cells could also arise from mESCs

[28] John J. Eppig and Marilyn J. O'Brien, 'Development in Vitro of Mouse Oocytes from Primordial Follicles', *Biology of Reproduction*, 54/1 (1996), 197–207.

[29] Marilyn J. O'Brien, Janice K. Pendola, and John J. Eppig, 'A Revised Protocol for In Vitro Development of Mouse Oocytes from Primordial Follicles Dramatically Improves Their Developmental Competence', *Biology of Reproduction*, 68/5 (2003), 1682–1686.

[30] Kanako Morohaku et al., 'Complete In Vitro Generation of Fertile Oocytes from Mouse Primordial Germ Cells', *Proceedings of the National Academy of Sciences*, 113/32 (2016), 9021–9026.

[31] Karin Hübner et al., 'Derivation of Oocytes from Mouse Embryonic Stem Cells', *Science*, 300/5623 (2003), 1251–1256.

within the culture system. However, there was no evidence to suggest that the oocytes matured into fully developed oocytes. An early study demonstrated that germ cells could be derived from embryoid bodies (EBs) formed from ESCs, and that these EB-derived germ cells had the potential to contribute to spermatogenesis upon transplantation into the testis. This study also highlighted the positive effect of BMP4 on germ cell derivation from EBs and thereby served as a prototype of IVG in the context of directed germ cell differentiation.[32] However, the developmental potential of mESC-derived sperm in recipient testes has not yet been tested. Similarly, another early study showed that haploid cells could be derived from EBs made from ESCs after thirteen days of culture without LIF.[33] These haploid cells could be incorporated into eggs via intracytoplasmic injection; however, the full-term developmental potential of the resulting zygotes has not been tested.

Early studies demonstrated spontaneous or directed differentiation towards the germ cell lineage using human ESCs (hESCs). In a manner similar to mESCs, EBs derived from hESCs underwent germ cell differentiation, including meiotic entry.[34] The addition of BMP4 to the culture medium of hESC-derived EBs promoted germ cell differentiation,[35] indicating a differentiation process similar to that of mouse germ cells. Following the establishment of human induced pluripotent stem cells (hiPSCs),[36] it was shown that hiPSCs can serve as an alternative source for the spontaneous differentiation of germ cells,[37] supporting the concept of germ cell derivation from somatic cells. This study also demonstrated that the efficiency of germ cell derivation from hESCs and hiPSCs was improved by co-culturing with human foetal gonadal stromal cells. Subsequent studies demonstrated the potential of hESCs as a source for induction of haploid cells. For example, modulation of germ cell-specific genes, including DAZL, DAZ, and BOULE, was shown to enable the differentiation of hESCs into haploid cells.[38] Alternatively, a later study showed that haploid cells could be derived more simply by culturing hESCs and hiPSCs under conditions

[32] Yayoi Toyooka et al., 'Embryonic Stem Cells Can Form Germ Cells In Vitro', *Proceedings of the National Academy of Sciences*, 100 (2003), 11457–11462.

[33] Niels Geijsen et al., 'Derivation of Embryonic Germ Cells and Male Gametes from Embryonic Stem Cells', *Nature*, 427/6970 (2004), 148–154.

[34] Amander T. Clark et al., 'Spontaneous Differentiation of Germ Cells from Human Embryonic Stem Cells In Vitro', *Human Molecular Genetics*, 13 (2004), 727–739.

[35] Kehkooi Kee et al., 'Bone Morphogenetic Proteins Induce Germ Cell Differentiation from Human Embryonic Stem Cells', *Stem Cells and Development*, 15/6 (2006), 831–837.

[36] Kazutoshi Takahashi et al., 'Induction of Pluripotent Stem Cells from Adult Human Fibroblasts by Defined Factors', *Cell*, 131/5 (2007), 861–872.

[37] Tae S. Park et al., 'Derivation of Primordial Germ Cells from Human Embryonic and Induced Pluripotent Stem Cells Is Significantly Improved by Coculture with Human Fetal Gonadal Cells', *Stem Cells*, 27/4 (2009), 783–795.

[38] Kehkooi Kee et al., 'Human DAZL, DAZ and BOULE Genes Modulate Primordial Germ-Cell and Haploid Gamete Formation', *Nature*, 462/7270 (2009), 222–225.

optimized for spermatogonial stem cells.[39] These systems appeared to halve the genome, as evidenced by FACS analyses, haploid-specific gene expression, and the characteristic localization of acrosin protein. However, evaluating the functionality of these haploid cells as male gametes in humans remains challenging. Similar to that in mice in the early studies, the differentiation process in humans occurred much faster *in vitro* than *in vivo*, requiring only ten to fourteen days. Furthermore, it is unclear whether mature sperm with the typical morphology—a long tail and compact head—can be generated in culture. Nevertheless, these early studies laid the foundation for subsequent IVG research by demonstrating key principles of germ cell differentiation from pluripotent stem cells.

PGC Derivation from Pluripotent Stem Cells in Model Animals

In mammals, PGCs are derived from the pluripotent epiblast through autocrine and paracrine signalling. Consistently across mammalian species, PGCs are located in the posterior region of the epiblast, where signalling molecules also guide primitive mesoderm formation. Genetic studies in mice have shown that disruptions in BMP4 and WNT3—critical genes for PGC formation—simultaneously impair mesoderm development.[40,41] Additionally, *ex vivo* experiments in mice reveal that for proper PGC development, the pluripotent epiblast must attain PGC-competence, or the ability to respond to inductive signals.[42,43,44,45] This competence is only present in the epiblast from embryonic day (E) 6.0 to 6.75 in mouse embryos, highlighting the shifting pluripotent states during early embryonic stages.

The pluripotency of pluripotent stem cells is observed in three main states: 'naïve', 'formative' and 'primed'.[46] Mouse embryonic stem cells (mESCs),

[39] Charles A. Easley et al., 'Direct Differentiation of Human Pluripotent Stem Cells into Haploid Spermatogenic Cells', *Cell Reports*, 2/3 (2012), 440–446.

[40] Kirstie A. Lawson et al., 'Bmp4 Is Required for the Generation of Primordial Germ Cells in the Mouse Embryo', *Genes and Development*, 13/4 (1999), 424–436.

[41] Jeffery R. Barrow et al., 'Wnt3 Signaling in the Epiblast Is Required for Proper Orientation of the Anteroposterior Axis', *Developmental Biology*, 312 (2007), 312–320.

[42] Karen A. Lawson and Willem J. Hage, 'Clonal Analysis of the Origin of Primordial Germ Cells in the Mouse', in J. Marsh and J. Goode, eds., *Ciba Foundation Symposium 182: Germline Development* (London: John Wiley & Sons, 2007), 68–91.

[43] Patrick P. L. Tam and Sheila X. Zhou, 'The Allocation of Epiblast Cells to Ectodermal and Germ-Line Lineages Is Influenced by the Position of the Cells in the Gastrulating Mouse Embryo', *Developmental Biology*, 178/1 (1996), 124–132.

[44] Toru Yoshimizu, Masuo Obinata, and Yoshitake Matsui, 'Stage-Specific Tissue and Cell Interactions Play Key Roles in Mouse Germ Cell Specification', *Development*, 128 (2001), 481–490.

[45] Yasuhide Ohinata et al., 'A Signaling Principle for the Specification of the Germ Cell Lineage in Mice', *Cell*, 137/3 (2009), 571–584.

[46] Jennifer Nichols and Austin Smith, 'Naive and Primed Pluripotent States', *Cell Stem Cell*, 4/6 (2009), 487–492.

which originate from the inner cell mass of the blastocyst, display naïve pluripotency. These cells, which can integrate into chimeras and contribute to germ lineages, are typically cultured under a 2i + LIF condition, which includes CHIR99021 (a GSK3 inhibitor) and PD0325901 (MEK inhibitor) along with LIF.[47,48,49] This state mirrors the peri-implantation epiblast stage (~E4.5).[47,50] In contrast, mouse epiblast stem cells (mEpiSCs), derived from the post-implantation epiblast (~E5.5-E6.5), exhibit primed pluripotency. With limited differentiation potential, they generally do not contribute to chimeras under conventional conditions.[51,52] These cells self-renew in the presence of activin and FGF, resembling the post-E6.5 epiblast.[53] The formative state lies between the naïve and primed states. Cells in the formative state are ready for multi-lineage differentiation, including differentiation to PGCs, in response to external signals.[54] This formative state has been successfully maintained in culture through the modulation of activin and WNT signalling pathways.[55,56]

As described above, due to advancements in our understanding of the nature of pluripotent states and the PGC specification process, PGC specification can now be precisely re-created *in vitro* from ESCs or iPSCs. When naïve mouse ESCs or iPSCs are cultured with activin A and bFGF, they shift to a state similar to the formative phase, though only temporarily.[57] After two days in culture, these cells acquire PGC-competence and can then differentiate into mouse PGC-like cells (mPGCLCs) in response to factors such as BMP4, SCF, LIF, and EGF (Figure 1.2). mPGCLCs display a transcriptomic profile similar to that of early-stage PGCs *in vivo*. Furthermore, mPGCLC specification is coupled with

[47] Qi-Long Ying et al., 'The Ground State of Embryonic Stem Cell Self-renewal', *Nature*, 453/7194 (2008), 519–523.

[48] Mia Buehr et al., 'Capture of Authentic Embryonic Stem Cells from Rat Blastocysts', *Cell*, 35/7 (2008), 1287–1298.

[49] Ping Li et al., 'Germline Competent Embryonic Stem Cells Derived from Rat Blastocysts', *Cell*, 135/7 (2008), 1299–1310.

[50] Thorsten Boroviak et al., 'The Ability of Inner-Cell-Mass Cells to Self-renew as Embryonic Stem Cells Is Acquired Following Epiblast Specification', *Nature Cell Biology*, 16/6 (2014), 513–525.

[51] Paul J. Tesar et al., 'New Cell Lines from Mouse Epiblast Share Defining Features with Human Embryonic Stem Cells', *Nature*, 448/7150 (2007), 196–199.

[52] I. Gabrielle M. Brons et al., 'Derivation of Pluripotent Epiblast Stem Cells from Mammalian Embryos', *Nature*, 448/7150 (2007), 191–195.

[53] Yoji Kojima et al., 'The Transcriptional and Functional Properties of Mouse Epiblast Stem Cells Resemble the Anterior Primitive Streak', *Cell Stem Cell*, 14/1 (2014), 107–120.

[54] Austin Smith, 'Formative Pluripotency: The Executive Phase in a Developmental Continuum', *Development*, 144/3 (2017), 365–373.

[55] Masaki Kinoshita et al., 'Capture of Mouse and Human Stem Cells with Features of Formative Pluripotency', *Cell Stem Cell*, 28/3 (2021), 453–471.

[56] Leqian Yu et al., 'Derivation of Intermediate Pluripotent Stem Cells Amenable to Primordial Germ Cell Specification', *Cell Stem Cell*, 28/3 (2021), 550–567.

[57] Katsuhiko Hayashi et al., 'Reconstitution of the Mouse Germ Cell Specification Pathway in Culture by Pluripotent Stem Cells', *Cell*, 146/4 (2011), 519–532.

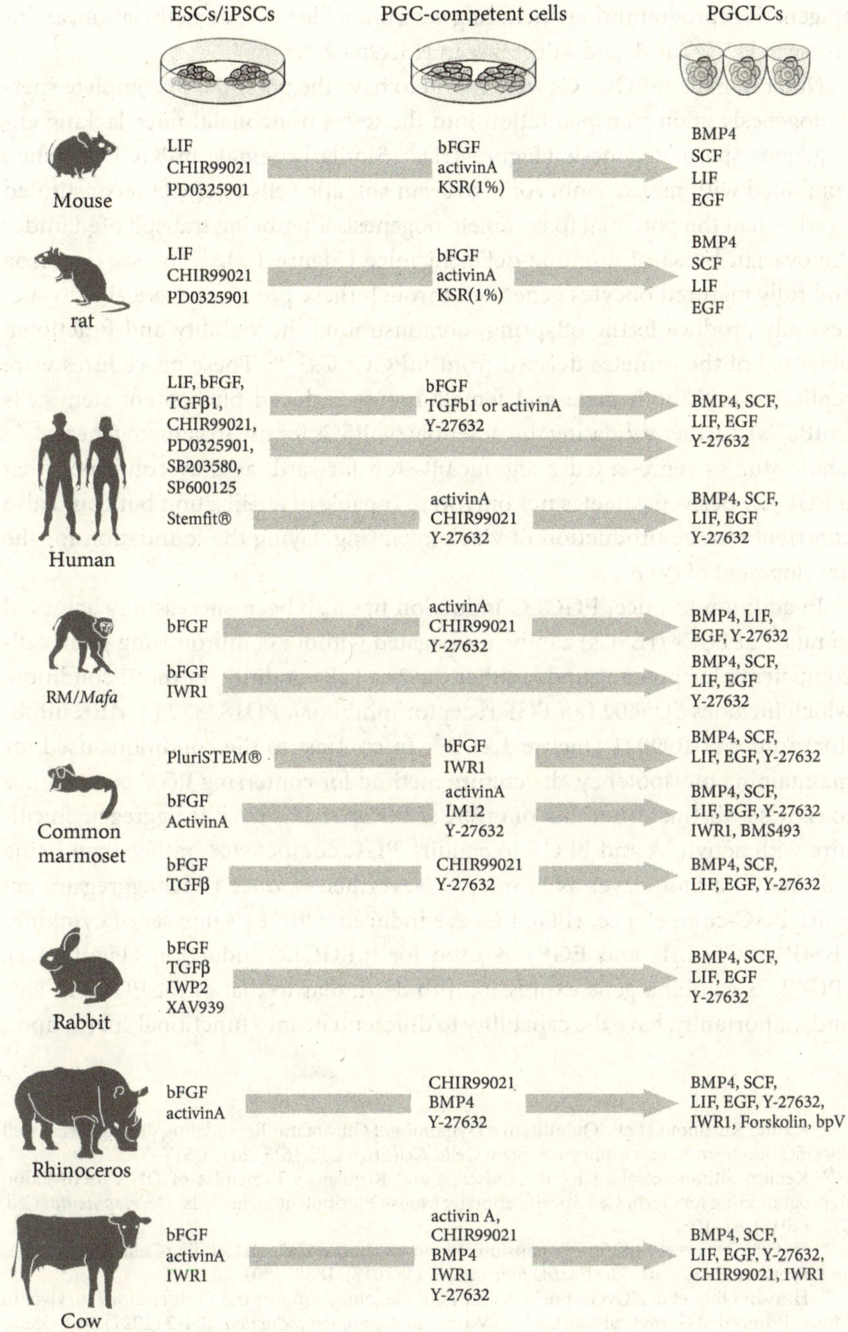

Figure 1.2 Culture conditions for PGCLCs derived via PGC-competent cells in various mammalian species.

epigenetic reprogramming, including genome-wide DNA demethylation, an increase in H3K27me3, and a decrease in H3K9me2.[58,59,60,61]

Notably, male mPGCLCs were found to have the potential to complete spermatogenesis upon transplantation into the testes of neonatal mice lacking endogenous spermatogenesis (Figure 1.1).[57] Similarly, female mPGCLCs, when combined with mouse embryonic ovarian somatic cells to create reconstituted ovaries, had the potential to complete oogenesis after being transplanted under the ovarian bursa of immune-deficient mice (Figure 1.1).[62] The spermatozoa and fully matured oocytes generated through these processes were able to successfully produce fertile offspring, demonstrating the viability and functional potential of the gametes derived from mPGCLCs.[57,62] These procedures were replicated with both male and female mouse induced pluripotent stem cells (miPSCs), further validating the potential of iPSCs for use in gametogenesis.[57,62] These studies represented a significant step forward, as they confirmed that mPGCLC-derived gametes not only were capable of fertilization but could also contribute to the production of viable offspring, laying the foundation for the development of IVG.

In addition to mice, PGCLC induction has also been successfully achieved in rats. Rat ESCs (rESCs) can be propagated without compromising germ cell-competent pluripotency under either the 2i + LIF condition or the 3i condition, which includes SU5402 (an FGF receptor inhibitor), PD184352 (a MEK inhibitor), and CHIR99021 (Figure 1.2).[48,49] In contrast to the conditions used for maintaining pluripotency, the culture method for conferring PGC competence to rESCs is distinct from that of mESCs/iPSCs: rESCs need to aggregate in culture with activin A and bFGF to acquire PGC competence, rather than being cultured in a monolayer as in mice.[63] Nevertheless, once rESC aggregates acquire PGC-competence, rPGCLCs are induced with the same set of cytokines (BMP4, SCF, LIF, and EGF) as used for mPGCLC induction (Figure 1.2). rPGCLCs exhibit a gene expression profile similar to that of rat PGCs *in vivo* and, importantly, have the capability to differentiate into functional sperm upon

[58] Kazuki Kurimoto et al., 'Quantitative Dynamics of Chromatin Remodeling during Germ Cell Specification from Mouse Embryonic Stem Cells', *Cell Stem Cell*, 16/5 (2015), 517–532.

[59] Kenjiro Shirane et al., 'Global Landscape and Regulatory Principles of DNA Methylation Reprogramming for Germ Cell Specification by Mouse Pluripotent Stem Cells', *Developmental Cell*, 39/1 (2016), 87–103.

[60] Hiroshi Ohta et al., 'In Vitro Expansion of Mouse Primordial Germ Cell-like Cells Recapitulates an Epigenetic Blank Slate', *The EMBO Journal*, 36/13 (2017), 1888–1907.

[61] Hiroshi Ohta et al., 'Cyclosporin A and FGF Signaling Support the Proliferation/Survival of Mouse Primordial Germ Cell-like Cells In Vitro†', *Biology of Reproduction*, 104/2 (2021), 344–360.

[62] Katsuhiko Hayashi et al., 'Offspring from Oocytes Derived from in Vitro Primordial Germ Cell–like Cells in Mice', *Science*, 338/6109 (2012), 971–975.

[63] Mami Oikawa et al., 'Functional Primordial Germ Cell–like Cells from Pluripotent Stem Cells in Rats', *Science*, 376/6589 (2022), 176–179.

transplantation into the testis. This is the first case to reproduce the generation of functional gametes from pluripotent stem cells, following the case in mice, supporting the idea that the induction of PGCLCs is a robust system for the initial step of IVG.

Post-implantation development of rabbit embryos differs from that in mice and rats, characterized by the formation of a disc-shaped epiblast lying on the hypoblast but lacking an amnion.[64] Spatial gene expression analyses reveal BMP expression in the posterior region of the epiblast, where PGCs emerge, as well as in the adjacent trophoblast at embryonic stage 3.[65,66,67] Expression of Wnt is observed at the rim of the epiblast during embryonic stage 2, with enrichment in the posterior region.[66] This suggests that, despite the differing configuration of post-implantation embryos, key signalling pathways are conserved in rabbits. Indeed, based on these observations, rabbit PGCLCs (rbPGCLCs) can be induced from rbESCs (Figures 1.1 and 1.2).[67] rbESCs can be propagated on mouse embryonic fibroblasts (MEF) in Essential 8 medium containing *FGF* and *TGFβ1*[68] supplemented with *XAV939* and *IWP2* (WNT inhibitors). Under these conditions, rbESCs appear to have acquired PGC-competence, as rbPGCLCs can be directly induced with the same set of cytokines (BMP4, SCF, LIF, and EGF), supplemented with Y-27632 (a ROCK inhibitor) (Figure 1.2). The transcriptome profile of rbPGCLCs is similar to that of early-stage PGCs in rabbit embryos. Although germ cell marker genes such as *PRDM1* and *NANOS3*, as well as pluripotent markers like *POU5F1* and *NANOG*, are expressed in raPGCLCs similar to those in mice and rats, they show high expression of *SOX17* and repression of *SOX2*, which play an important role in PGC specification in mice.[69] Loss of the *SOX17* gene abolishes raPGCLC induction, indicating that *SOX17* is essential for PGC specification in rabbits. The requirement of *SOX17* for PGC specification is observed in other mammalian species (see below), but not in mice.[70]

[64] Peter L. Pfeffer, 'Alternative Mammalian Strategies Leading towards Gastrulation: Losing Polar Trophoblast (Rauber's Layer) or Gaining an Epiblast Cavity', *Philosophical Transactions of the Royal Society B: Biological Sciences*, 377/1865 (2022).

[65] Clas Hopf, Christoph Viebahn, and Bernd Püschel, 'BMP Signals and the Transcriptional Repressor BLIMP1 during Germline Segregation in the Mammalian Embryo', *Development Genes and Evolution*, 221/4 (2011), 209–223.

[66] Michio Yoshida et al., 'Conserved and Divergent Expression Patterns of Markers of Axial Development in Eutherian Mammals', *Developmental Dynamics*, 245/1 (2016), 67–86.

[67] Toshihiro Kobayashi et al., 'Tracing the Emergence of Primordial Germ Cells from Bilaminar Disc Rabbit Embryos and Pluripotent Stem Cells', *Cell Reports*, 37/2, 109812 (2021).

[68] Guokai Chen et al., 'Chemically Defined Conditions for Human iPSC Derivation and Culture', *Nature Methods*, 8/5 (2011), 424–429.

[69] Federica Campolo et al., 'Essential Role of Sox2 for the Establishment and Maintenance of the Germ Cell Line', *Stem Cells*, 31/7 (2013), 1408–1321.

[70] Kenshiro Hara et al., 'Evidence for Crucial Role of Hindgut Expansion in Directing Proper Migration of Primordial Germ Cells in Mouse Early Embryogenesis', *Developmental Biology*, 330/2 (2009), 427–439.

PGC Derivation from Human and Non-Human Primate Pluripotent Stem Cells

Based on observation of post-implantation embryos of cynomolgus monkeys, the embryonic configuration and the position of PGC specification are distinct from those in mice. Primate post-implantation embryos form a disc-shaped epiblast that lies on the hypoblast and connects to the amnion. In cynomolgus monkeys, at least a portion of PGCs are specified from the nascent amnion cells, which still express pluripotent marker genes.[71] Spatial gene expression analysis reveals that the amnion cells express *BMP4*, and the cytotrophoblast cells adjacent to the nascent amnion express *WNT3A*, thereby constructing a signalling centre that induces PGCs from the pluripotent amnion.

Studies have explored various methodologies to induce human PGC-like cells (hPGCLCs) from human ESCs/iPSCs (Figures 1.1 and 1.2). One approach involves culturing ESCs under a 4i condition—using chemical inhibitors of *MEK*, *GSK3β*, *p38*, and *JNK*—creating a peri-gastrulating epiblast-like state. These ESCs/iPSCs are subsequently induced into hPGCLCs using recombinant BMP2 or BMP4 protein in combination with SCF, LIF, EGF, and Y-27632.[72] Alternatively, hPSCs cultured under conventional conditions can be first differentiated into incipient mesoderm-like cells (iMeLCs) and then induced into hPGCLCs with recombinant BMP4, SCF, LIF, EGF, and Y-27632 (Figure 1.2).[71] In both methods, hPGCLCs exhibit gene-expression profiles resembling those of early human PGCs. Further investigations have demonstrated that hPGCLCs have the potential for epigenetic reprogramming, a characteristic of germ cells, and can differentiate into early oocytes or pro-spermatogonia (see below).[73,74,75]

Using the hPGCLC induction system, the mechanisms underlying hPGCLC specification have been explored.[71,72,76,77,78,79,80] Notably, SOX17, a transcription

[71] Kotaro Sasaki et al., 'Robust In Vitro Induction of Human Germ Cell Fate from Pluripotent Stem Cells', *Cell Stem Cell*, 17/2 (2015), 178–194.

[72] Naoko Irie et al., 'SOX17 Is a Critical Specifier of Human Primordial Germ Cell Fate', *Cell*, 160/1 (2015), 253–268.

[73] Chika Yamashiro et al., 'Generation of Human Oogonia from Induced Pluripotent Stem Cells In Vitro', *Science*, 362/6412 (2018), 356–360.

[74] Young S. Hwang et al., 'Reconstitution of Prospermatogonial Specification In Vitro from Human Induced Pluripotent Stem Cells', *Nature Communications*, 11/1 (2020), 5656.

[75] Yusuke Murase et al., 'In Vitro Reconstitution of Epigenetic Reprogramming in the Human Germ Line', *Nature*, 631/8019 (2024), 170–178.

[76] Yoji Kojima et al., 'Evolutionarily Distinctive Transcriptional and Signaling Programs Drive Human Germ Cell Lineage Specification from Pluripotent Stem Cells', *Cell Stem Cell*, 21/4 (2017), 517–532.

[77] Di Chen et al., 'The TFAP2C-Regulated OCT4 Naive Enhancer Is Involved in Human Germline Formation', *Cell Reports*, 25 (2018), 3591–3602.

[78] Anastasiya Sybirna et al., 'A Critical Role of PRDM14 in Human Primordial Germ Cell Fate Revealed by Inducible Degrons', *Nature Communications*, 11/1 (2020), 1282.

[79] Merrick Pierson Smela et al., 'Testing the Role of SOX15 in Human Primordial Germ Cell Fate', *Wellcome Open Research*, 4 (2019), 122.

[80] Yoji Kojima et al., 'GATA Transcription Factors, SOX17 and TFAP2C, Drive the Human Germ-Cell Specification Program', *Life Science Alliance*, 4/5 (2021), e202000974.

factor (TF) known to play a role in endoderm differentiation in mice but to be dispensable for PGC specification in mice, has been identified as one of the most upstream TFs for hPGCLC specification. This suggests that the human PGCLC specification process is distinct from that in mice. Indeed, subsequent studies revealed that the specification of hPGCLCs involves key transcription factors, their regulatory network, and their hierarchy of actions, the last of which differs from those involved in mPGCLC specification.[76,78,79,80] Moreover, the transcriptomic profiles for hPGCLC specification diverge from those for mPGCLC specification.[71,81,82] During this process, hPGCLCs preserve the level of DNA methylation in their genome during proliferation.[83] This is presumably due to the expression of genes involved in DNA demethylation: while *DNMT3A* and *DNMT3B* are downregulated, *UHRF1* is maintained at substantial levels. This underscores the species-specific differences in the gene-expression programs that underlie epigenetic reprogramming. A portion of such reprogramming is governed by *DMRT1*, as overexpression of *SOX17* and *DMRT1* shifts the nascent state of hPGCLCs to a migratory state that precedes epigenetic reprogramming through the transactivation of *TET2* and *TET3*, which encode methylcytosine dioxygenases.[84]

Similar to the case of PGCLC induction in humans, PGCLCs can be induced from non-human primate ESCs/iPSCs, including those of common marmosets, rhesus macaques, and cynomolgus macaques (Figures 1.1 and 1.2).[85,86,87,88,89] Examination of these methods reveals that species-specific culture conditions are primarily focussed on establishing PGC competence. This likely reflects differences in the pluripotent states of ESCs/iPSCs across species. Common marmoset ESCs (cjESCs), cultured under conventional conditions on MEFs, are first induced with activin A and IM12 (a WNT agonist) into cells competent for PGCLC differentiation upon BMP signalling activation. These cells are then further differentiated into PGCLCs using the same set of cytokines (BMP4, SCF, LIF, and EGF) as used for inducing

[81] Shihori Yokobayashi et al., 'Clonal Variation of Human Induced Pluripotent Stem Cells for Induction into the Germ Cell Fate†', *Biology of Reproduction*, 96/6 (2017), 1154–1166.

[82] Yoshitake Sakai et al., 'Induction of the Germ Cell Fate from Pluripotent Stem Cells in Cynomolgus Monkeys†', *Biology of Reproduction*, 102/3 (2020), 620–638.

[83] Yusuke Murase et al., 'Long-Term Expansion with Germline Potential of Human Primordial Germ Cell-like Cells In Vitro', *The EMBO Journal*, 39/21 (2020), e104929.

[84] Naoko Irie et al., 'DMRT1 Regulates Human Germline Commitment', *Nature Cell Biology*, 25/10 (2023), 1439–1452.

[85] Enrique Sosa et al., 'Differentiation of Primate Primordial Germ Cell-like Cells Following Transplantation into the Adult Gonadal Niche', *Nature Communications*, 9/1 (2018), 5339.

[86] Sakai et al., 'Induction of the Germ Cell Fate', 620–638.

[87] Yasunari Seita et al., 'Efficient Generation of Marmoset Primordial Germ Cell-like Cells Using Induced Pluripotent Stem Cells', *eLife*, 12 (2023).

[88] Mayumi Shono et al., 'Induction of Primordial Germ Cell-like Cells from Common Marmoset Embryonic Stem Cells by Inhibition of WNT and Retinoic Acid Signaling', *Scientific Reports*, 13/1 (2023), 3186.

[89] Kurlovich et al., 'Generation of Marmoset Primordial Germ'.

mPGCLCs and hPGCLCs, supplemented with Y-27632 (a ROCK inhibitor), IWR1 (a WNT antagonist), and BMS493 (a pan-retinoic acid receptor inverse agonist) (Figure 1.2).[88] IWR1 and BMS493 play roles in the inhibition of mesoderm and neuroectoderm, respectively, thereby increasing the efficiency of PGCLC differentiation. Alternatively, common marmoset iPSCs (cjiPSCs) cultured under a feeder-free condition followed by culturing on MEF with IWR1 or on vitronectin with CHIR99021 can be differentiated into PGCLCs using the same set of cytokines and Y-27632.[87,89] In both cases, the resultant cjPGCLCs share similar characteristics with respect to gene expression and epigenetic status. cjPGCLCs are equivalent to migratory cjPGCs.[87] Rhesus macaque iPSCs (rhiPSCs), cultured under conventional conditions on MEFs, are first differentiated into iMeLCs using activin A, CHIR99021, and Y-27632. These iMeLCs are then further differentiated into rhPGCLCs with the same set of cytokines (BMP4, SCF, LIF, and EGF) supplemented with Y-27632 (Figure 1.2).[85] Upon transplantation into the testes of either mice or rhesus macaques, rhPGCLCs are further differentiated into spermatogonium-like cells that partially express spermatogonial marker genes.[85] In the case of cynomolgus monkey ESCs/iPSCs (cyESCs/cyiPSCs), the first differentiation step for iMeLCs can be skipped, as cyESCs/cyiPSCs, cultured under conventional conditions on MEFs, are directly differentiated into PGCLCs using the same set of cytokines and Y-27632 (Figure 1.2).[86] Given that the set of cytokines required for the induction of PGCLCs is the same among mice, humans, and non-human primates, species-specific differences in the culture system depend on the conditions for conferring PGC competence to ESCs/iPSCs. It appears that distinct pluripotent states influence both the acquisition of PGC competence and PGCLC differentiation, in turn suggesting that the pluripotent state should be considered when optimizing PGCLC induction.

PGC Derivation from Pluripotent Stem Cells in Large Mammals

Induction of PGCLCs from larger mammals, including livestock and wild animals, presents greater challenges compared to PGCLC induction from mice and humans. Thus far, PGCLC induction in large animals has been achieved in species such as cattle and southern/northern white rhinoceroses (Figure 1.1), with the latter achievement being driven by conservation efforts. Interestingly, even in large animals, species-specific differences in culture conditions are primarily observed in conditions required for PGC competence rather than PGC differentiation. In the case of cattle ESCs (caESCs) cultured under a conventional condition on MEFs, the PGC-competent cells are first induced with activin A, CHIR99021, a low concentration of BMP4, IWR1, and Y-27632, and then further differentiated into PGCLCs using the same set of cytokines (BMP4, SCF, LIF, and EGF) supplemented with CHIR99021 and IWR1 (Figure 1.2).[90] caPGCLCs share characteristics of gene expression and epigenetic reprogramming with hPGCLCs rather than mPGCLCs; notably, both hPGCLCs and caPGCLCs are characterized by *SOX17* expression, but not *SOX2*, during PGCLC differentiation. Southern white rhinoceros ESCs (swrESCs) and northern white rhinoceros (nwriPSCs) cultured under a conventional condition on MEFs are first induced with activin A and a low concentration of BMP4, and then further differentiated into PGCLCs using the same set of cytokines (BMP4, SCF, LIF, and EGF) supplemented with Y-27632 and IWR1 (Figure 1.2).[91] In addition to this culture condition, supplementation of forskolin and bpV (a PTEN inhibitor) further improves the number of swr/nwrPGCLCs. During this process, *SOX17*, rather than *SOX2*, is upregulated and disruption of the *SOX17* blocks swrPGCLC differentiation,[91] suggesting that the molecular network governing PGC specification in these large animals is more similar to that in primates than that in mice.

Gamete Production from Mouse PGCLCs

A gonadal somatic cell environment is required to advance PGCLCs to the sex-dependent differentiation process. For IVG, such an environment is constructed through organoid technologies using either embryonic gonadal somatic cells or mESC-derived ovarian somatic cell-like cells. In the former case, PGCLCs are reaggregated with embryonic ovarian somatic cells, followed by a series of cultures (see below). In the latter case, ovarian gonadal somatic cells are induced from mESCs by recapitulating the differentiation process of ovarian somatic cells. During embryogenesis, ovarian somatic cells originate from the nascent mesoderm and undergo a stepwise differentiation process through the intermediate mesoderm and dorsal coelomic epithelium in the genital ridge region, which eventually thickens through an epithelial–mesenchymal transition. This process can be replicated using mESCs:[92] mESCs efficiently differentiate into nascent mesoderm-like cells in the presence of BMP4, CHIR99021, and EGF, then progress to intermediate mesoderm-like cells with the addition of RA, SHH, and PD0325901, and ultimately develop into foetal ovarian somatic cell-like

[90] Atsushi Shirasawa et al., 'Efficient Derivation of Embryonic Stem Cells and Primordial Germ Cell-like Cells in Cattle', *Journal of Reproduction and Development*, 70/2 (2024), 82–95.

[91] M. Hayashi et al., 'Robust Induction of Primordial Germ Cells of White Rhinoceros on the Brink of Extinction', *Science Advances*, 8/49 (2022), eabp9683.

[92] Takashi Yoshino et al., 'Generation of Ovarian Follicles from Mouse Pluripotent Stem Cells', *Science*, 373/6552 (2021), eabe0237.

cells (FOSLCs) with BMP4 and a low concentration of FGF9. FOSLCs express *Nr5a1*, a key marker of gonadal somatic cells also known as *SF1* or *Ad4BP*, and include subsets of female gonadal somatic cells, such as supporting cells marked by *FOXL2* and stromal cells characterized by *PDGFRA* (Figure 1.3).

Based on organ culture methods for embryonic ovaries, ovary organoids composed of mPGCLCs and embryonic or mESC-derived ovarian somatic cells have been established.[92,93] The ovary organoids are cultured at the air–liquid interface for approximately three weeks in a process termed *in vitro* development, during which PGCLCs undergo meiosis, form follicular structures, and develop into immature oocytes at the secondary follicle stage (Figure 1.3). This is followed by an additional two weeks of *in vitro* growth culture, during which the oocytes reach full growth, accompanied by extensive granulosa cell proliferation (Figure 1.3). Finally, oocytes resume meiosis and acquire the potential for fertilization through *in vitro* maturation culture. Both oocytes generated from mESCs and those derived from miPSCs are fertilizable and can produce viable offspring (Figure 1.3).[92,93] This demonstrates the feasibility of IVG as a functional system. Notably, the successful production of oocytes from ovary organoids using FOSLCs underscores the potential to generate functional gametes without relying on embryonic cells. This offers a promising pathway for applications in humans and larger animals, particularly given the technical and ethical challenges associated with accessing embryonic ovarian tissues in these species.

Just as for ovary organoids, a methodology for producing testicular organoids using embryonic testicular somatic cells has been explored.[94] The testicular organoids, cultured under an air–liquid interface condition, self-organize into seminiferous tubule-like structures, in which mPGCLCs are differentiated into gonocytes, pro-spermatogonia, and eventually spermatogonia within a span of three weeks. In the testicular organoids, there are some spermatocytes that progress to meiotic prophase within eight weeks. Mouse spermatogonial stem cells can propagate indefinitely under optimal culture conditions, thus becoming germline stem cells (GSCs), while retaining their ability to differentiate into mature sperm following transplantation into the testis.[95] The potential of mPGCLC-derived spermatogonia can be demonstrated by their ability to differentiate into germline stem cell-like cells (GSCLCs) in culture. GSCLCs share

[93] Orie Hikabe et al., 'Reconstitution In Vitro of the Entire Cycle of the Mouse Female Germ Line', *Nature*, 539/7628 (2016), 299–303.

[94] Yukiko Ishikura et al., 'In Vitro Derivation and Propagation of Spermatogonial Stem Cell Activity from Mouse Pluripotent Stem Cells', *Cell Reports*, 17/10 (2016), 2789–2804.

[95] Mito Kanatsu-Shinohara et al., 'Long-Term Proliferation in Culture and Germline Transmission of Mouse Male Germline Stem Cells', *Biology of Reproduction*, 69/2 (2003), 612–616.

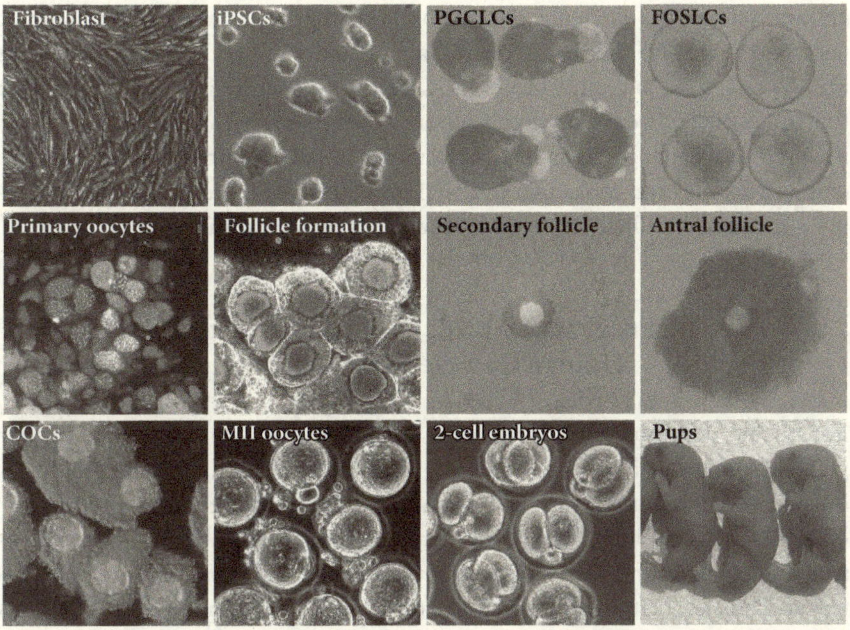

Figure 1.3 Generation of functional oocytes from a reconstituted ovary. Representative images of each step are shown: blue signals in PGCLCs indicate stella-CFP; red signals in FOSLCs indicate Foxl2-tdTomato; green, red, white, and blue signals in primary oocytes indicate stella-CFP, Foxl2-tdTomato, SYCP3, and DAPI, respectively; green/blue, red, and white signals during follicle formation indicate stella-CFP, Foxl2-tdTomato, and Nr5a1-CD271[92]; and blue signals in secondary follicles, antral follicles, and cumulus-oocyte complexes (COCs) indicate Stella-CFP.

key characteristics with bona fide GSCs, including cell morphology, expansion capacity, and transcriptomic profiles. Importantly, GSCLCs have the capacity to contribute to spermatogenesis in adult testes and produce spermatozoa capable of fertilization and full-term development.

Despite the successful establishment of GSCLCs, it is still not possible to complete spermatogenesis and produce mature spermatozoa in the testicular organoid with embryonic testis. Considering that the organ culture of testes can produce mature and functional spermatozoa *in vitro*,[25] this inability could be attributed either to the inefficiency of spermatogenesis in the testicular organoid or to substantial differences between organoid and organ culture systems. One concern is the maturity of the testicular organoid, as organoid systems are generally prone to creating embryonic rather than adult tissue environments. Indeed, even in organ culture systems, the efficiency of *in vitro* spermatogenesis decreases when infant testes are used.[96] Neither has a functional testicular organoid been created using mESCs or iPSCs, in contrast to FOSLCs. Fully reconstructing testicular tissues that are sufficiently developed to support all stages of spermatogenesis remains a significant challenge. Despite advancements in IVG in mice, replication of the complex cellular and molecular interactions within the testicular microenvironment required for efficient spermatogenesis has yet to be achieved.

Progress Towards the Production of Human Gametes Through IVG

For further development of germ cells beyond PGCLCs in humans, a proper gonadal somatic cell environment is required. Several studies have sought to induce Sertoli and granulosa cells from hESCs/iPSCs, or even directly from fibr oblasts.[97,98,99,100] While these induced cells exhibit similarities in gene expression and some morphological features, their functional capabilities remain untested. Thus far, further germ cell development in IVG using hESCs/iPSCs has been approached in part using reaggregation with mouse embryonic ovarian somatic cells, resulting in so-called xenogeneic ovarian organoids.[73] In the mouse environment, hPGCLCs progress—albeit at a slower pace and with lower efficiency than *in vivo* hPGCs—to oogonia-like cells (hOLCs) expressing genes such as *DAZL*, *DDX4*, and *PIWIL2* after approximately seventy days of culture. Notably, during transition from hPGCLCs to hOLCs, cells undergo epigenetic reprogramming including genome-wide DNA demethylation, imprint erasure, and partial X reactivation.[73] At around 120 days, some OLCs undergo meiosis, confirming that OLCs proceed along the proper differentiation program. Similar to the case of xenogeneic ovarian organoids, an aggregation culture of hPGCLCs with mouse embryonic testicular somatic cells promotes differentiation of hPGCLCs into pro-spermatogonium-like cells (pSGLCs) in a timeframe comparable to that for the differentiation of hPGCLCs into OLCs.[74] Based on

[96] Hiroyuki Sanjo et al., 'Antioxidant Vitamins and Lysophospholipids Are Critical for Inducing Mouse Spermatogenesis under Organ Culture Conditions', *The FASEB Journal*, 34/7 (2020), 9480–9497.

[97] Ingrid M. Knarston et al., 'An In Vitro Differentiation Protocol for Human Embryonic Bipotential Gonad and Testis Cell Development', *Stem Cell Reports*, 15/6 (2020), 1377–1391.

[98] Yosef Buganim et al., 'Direct Reprogramming of Fibroblasts into Embryonic Sertoli-like Cells by Defined Factors', *Cell Stem Cell*, 11/3 (2012), 373–386.

[99] Dong-Won Seol et al., 'In Vitro Derivation of Functional Sertoli-Like Cells from Mouse Embryonic Stem Cells', *Cell Transplantation*, 27/10 (2018), 1523–1534.

[100] Sabrina Piechota et al., 'Human-Induced Pluripotent Stem Cell-Derived Ovarian Support Cell Co-culture Improves Oocyte Maturation In Vitro after Abbreviated Gonadotropin Stimulation', *Human Reproduction*, 38/12 (2023), 2456–2469.

their transcriptomic profiles, the differentiation trajectory from PGCLCs to pSGLCs closely mirrors the *in vivo* process, suggesting that they retain the potential for further differentiation under appropriate culture conditions.

Aside from approaches using xenogeneic ovarian organoids or deriving gonadal somatic tissues from hESCs/iPSCs, the differentiation process from PGCLCs to OLCs and pSGLCs can be reconstituted under nearly defined conditions. This method mainly relies on the extensive proliferation ability of hPGCLCs.[75,83] hPGCLCs can undergo a more than 10^6-fold proliferation over four months when cultured with forskolin, SCF, and bFGF on m220 feeder cells.[83] Under this condition, the transcriptomic profile of proliferated hPGCLCs remains similar to that of hPGCLCs immediately after specification, suggesting that, despite extensive proliferation, their developmental state is effectively paused. The addition of BMP2 advances the developmental state, enabling hPGCLCs to differentiate into OLCs and pSGLCs while maintaining their proliferative capacity.[75] Importantly, during these transitions, genome-wide DNA demethylation occurs in hPGCLCs, indicating that BMP signalling triggers epigenetic reprogramming. However, hPGCLCs expanded with BMP2 still fail to activate the gene expression programs necessary for further progression through spermatogenesis and oogenesis, suggesting that the somatic cell environment provides unidentified signal(s) essential for advancing germ cell development.

Perspectives

IVG creates germ cell lineages as well as a gonadal somatic cell environment, thereby generating sperm and oocytes entirely in culture from pluripotent stem cells. This enables a range of innovative approaches in reproductive biology and medicine. While fully functional oocytes have thus far only been successfully produced in mice, future progress in this field promises significant advancements in our understanding of germ cell biology and the treatment of infertility.

Key areas of application include genetic modification, infertility modelling, and the practical use of IVG-derived gametes for reproduction. Genetic modification of the germline through IVG enables researchers to investigate the genetic mechanisms critical to germ cell formation and maturation. Traditionally, the investigation of gene function within germ cells has relied on genetically modified animals due to the lack of suitable cell lines. IVG reduces this dependency by allowing genetic studies to be conducted directly in cultured pluripotent stem cells. This approach is particularly valuable for understanding fertility in species that are challenging to study and offers new opportunities for exploring human reproductive biology. In this context, IVG enables the construction of infertility models using patient-specific cells, providing a unique opportunity to replicate the genetic and epigenetic characteristics associated with reproductive disorders. Such models not only enhance diagnostic capabilities for complex infertility cases but also serve as platforms for testing targeted therapies. As mouse models demonstrate the creation of functional oocytes from pluripotent stem cells, a key issue is whether such reproduction can be achieved in humans and other mammalian species. Historically, innovations in reproductive technologies, such as IVF and ESC/iPSC technologies, were developed in experimental animals before being successfully applied to humans. This precedent suggests that IVG could eventually provide mature and functional gametes in humans and larger animals. In the latter case, IVG could be particularly effective for the preservation of endangered species.

Despite its transformative potential, IVG presents notable challenges. Concerns regarding the safety of artificial gametes, such as the unknown long-term impacts of their use in reproduction, must be addressed. Ethical considerations surrounding the creation and application of IVG-derived gametes also require careful examination. Particularly when examining mouse models, the lower fidelity of *in vitro* oogenesis compared to *in vivo* oogenesis remains a concern. For example, the developmental rate of IVG-derived two-cell embryos progressing to term is reported to be 10- to 100-fold lower than that of *in vivo* embryos.[93] Similarly, the efficiency of spermatogenesis in germline stem cell-like cells (GSCLCs) is substantially lower than that in germline stem cells (GSCs) derived *in vivo*. Only three out of fifteen GSCLC lines contributed to spermatogenesis, and even these achieved spermatogenesis colonies at a rate of approximately 20%, compared to nearly 100% efficiency observed in GSCs.[94] These defects could stem from both an inadequate quality of ESCs/iPSCs and suboptimal *in vitro* culture conditions. This reminds us to take into account the critical evaluation needed for IVG-derived gametes that are intended for practical use in reproduction. In addition to these technical issues, other social impacts must be considered as reviewed in this book.

In conclusion, IVG represents a transformative tool in reproductive biology and medicine, offering unprecedented insights into gametogenesis, facilitating novel infertility treatments, and expanding reproductive options for individuals and society, while also presenting significant challenges, including concerns about safety and ethical implications.

Bibliography

Aizawa, Satoru, and Nishimune, Yoshitake, '*In Vitro* Differentiation of Type A Spermatogonia in Mouse Cryptorchid Testis', *Journal of Reproduction and Fertility*, 56 (1979), 99–104.

Barrow, Jeffery R., Howell, William D., Rule, Michael, Hayashi, Shigemi, Thomas, Kirk R., Capecchi, Mario R., et al., '*Wnt3* Signaling in the Epiblast Is Required for Proper Orientation of the Anteroposterior Axis', *Developmental Biology*, 312 (2007), 312–320, https://doi.org/10.1016/j.ydbio.2007.09.030.

Boitani, Caria, Politi, Maria G., and Menna, Tiziana, 'Spermatogonial Cell Proliferation in Organ Culture of Immature Rat Testis', *Biology of Reproduction*, 48 (1993), 761–767.

Boroviak, Thorsten, Loos, Remco, Bertone, Paul, Smith, Austin, and Nichols, Jennifer, 'The Ability of Inner-Cell-Mass Cells to Self-Renew as Embryonic Stem Cells Is Acquired Following Epiblast Specification', *Nature Cell Biology*, 16/6 (2014), 513–25, https://doi.org/10.1038/ncb2965.

Brons, I. Gabrielle M., Smithers, Lucy E., Trotter, Matthew W. B., Rugg-Gunn, Peter, Sun, Bowen, Chuva de Sousa Lopes, Susana M., et al., 'Derivation of Pluripotent Epiblast Stem Cells from Mammalian Embryos', *Nature*, 448/7150 (2007), 191–195, https://doi.org/10.1038/nature05950.

Buehr, Mia, Meek, Stephen, Blair, Kate, Yang, Jian, Ure Janice, Silva, Jose, et al., 'Capture of Authentic Embryonic Stem Cells from Rat Blastocysts', *Cell*, 135/7 (2008), 1287–1298, https://doi.org/10.1016/j.cell.2008.12.007.

Buganim, Yosef, Itskovich, Elena, Hu, Yueh-Chiang, Cheng, Albert W., Ganz, Kibibi, Sarkar, Sovan, et al., 'Direct Reprogramming of Fibroblasts into Embryonic Sertoli-like Cells by Defined Factors', *Cell Stem Cell*, 11/3 (2012), 373–386, https://doi.org/10.1016/j.stem.2012.07.019.

Campolo, Federica, Gori, Manuele, Favaro, Rebecca, Nicolis, Silvia, Pellegrini, Manuela, Botti, Flavia, et al., 'Essential Role of Sox2 for the Establishment and Maintenance of the Germ Cell Line', *Stem Cells*, 31/7 (2013), 1408–1421, https://doi.org/10.1002/stem.1392.

Champy, Christian, 'Quelques résultats de la méthode de culture des tissues', *Archives de Zoologie Expérimentale et Générale*, 60 (1920), 461–500.

Chen, Di, Liu, Wanlu, Zimmerman, Jill, Pastor, William A., Kim, Rachel, Hosohama, Linzi, et al., 'The TFAP2C-Regulated OCT4 Naive Enhancer Is Involved in Human Germline Formation', *Cell Reports*, 25 (2018), 3591–3602, https://doi.org/10.1016/j.celrep.2018.12.011.

Chen, Guokai, Gulbranson, Daniel R., Hou, Zhonggang, Bolin, Jennifer M., Ruotti, Victor, Probasco, Mitchell D., et al., 'Chemically Defined Conditions for Human iPSC Derivation and Culture', *Nature Methods*, 8/5 (2011), 424–429, https://doi.org/10.1038/nmeth.1593.

Chuma, Shinichiro, and Nakano Toru, 'piRNA and Spermatogenesis in Mice', *Philosophical Transactions of the Royal Society B: Biological Sciences*, 368/1609 (2013), 20110338, https://doi.org/10.1098/rstb.2011.0338.

Clark, Amander T., Bodnar, Megan S., Fox, Mark, Rodriguez, Ryan T., Abeyta, Michael J., Firpo, Meri T., et al., 'Spontaneous Differentiation of Germ Cells from Human Embryonic Stem Cells In Vitro', *Human Molecular Genetics*, 13 (2004), 727–739.

Easley 4th, Charles A., Phillips, Bart T., McGuire, Megan M., Barringer, Jennifer M., Valli, Hanna, Hermann, Brian P., et al., 'Direct Differentiation of Human Pluripotent Stem Cells into Haploid Spermatogenic Cells', *Cell Reports*, 2/3 (2012), 440–446, https://doi.org/10.1016/j.celrep.2012.07.015.

Edson, Mark A., Nagaraja, Ankur K., and Matzuk, Martin M., 'The Mammalian Ovary from Genesis to Revelation', *Endocrine Reviews*, 30/6 (2009), 624–712, https://doi.org/10.1210/er.2009-0012.

Eppig, John J., 'Mouse Oocyte Development In Vitro with Various Culture Systems', *Developmental Biology*, 60 (1977), 371–388.

Eppig, John J., and O'Brien, Marilyn J., 'Development in Vitro of Mouse Oocytes from Primordial Follicles1', *Biology of Reproduction*, 54/1 (1996), 197–207, https://doi.org/10.1095/biolreprod54.1.197.

Fainstat, Theodore, 'Organ Culture of Postnatal Rat Ovaries in Chemically Defined Medium', *Fertility and Sterility*, 19 (1968), 317–338.

Geijsen, Niels, Horoschak, Melissa, Kim, Kitai, Gribnau, Joost, Eggan, Kevin, and Daley, George Q., 'Derivation of Embryonic Germ Cells and Male Gametes from Embryonic Stem Cells', *Nature*, 427/6970 (2004), 148–54, https://doi.org/10.1038/nature02247.

Griswold, Michael D., 'Spermatogenesis: The Commitment to Meiosis', *Physiological Reviews*, 96/1 (2016), 1–17, https://doi.org/10.1152/physrev.00013.2015.

Hara, Kenshiro, Kanai-Azuma, Masami, Uemura, Mami, Shitara, Hiroshi, Taya, Choji, Yonekawa, Hiromichi, et al., 'Evidence for Crucial Role of Hindgut Expansion in Directing Proper Migration of Primordial Germ Cells in Mouse Early Embryogenesis', *Developmental Biology*, 330/2 (2009), 427–439, https://doi.org/10.1016/j.ydbio.2009.04.012.

Hayashi, Katsuhiko, Ogushi, Sugako, Kurimoto, Kazuki, Shimamoto, So, Ohta, Hiroshi, and Saitou, Mitinori, 'Offspring from Oocytes Derived from in Vitro Primordial Germ Cell–like Cells in Mice', *Science*, 338/6109 (2012), 971–975, https://doi.org/10.1126/science.1226889.

Hayashi, Katsuhiko, Ohta, Hiroshi, Kurimoto, Kazuki, Aramaki, Shinya, and Saitou, Mitinori, 'Reconstitution of the Mouse Germ Cell Specification Pathway in Culture by Pluripotent Stem Cells', *Cell*, 146/4 (2011), 519–532, https://doi.org/10.1016/j.cell.2011.06.052.

Hayashi, Masafumi, Zywica, Vera, Naitou, Yuki, Hamazaki, Nobuhiko, Goeritz, Frank, Hermes, Robert, et al., 'Robust Induction of Primordial Germ Cells of White Rhinoceros on the Brink of Extinction', *Science Advances*, 8/49 (2022), eabp9683, https://doi.org/10.1126/sciadv.abp9683.

Hikabe, Orie, Hamazaki, Nobuhiko, Nagamatsu, Go, Obata, Yayoi, Hirao, Yuji, Hamada, Norio, et al., 'Reconstitution In Vitro of the Entire Cycle of the Mouse Female Germ Line', *Nature*, 539/7628 (2016), 299–303, https://doi.org/10.1038/nature20104.

Hopf, Clas, Viebahn, Christoph, and Püschel, Bernd, 'BMP Signals and the Transcriptional Repressor BLIMP1 during Germline Segregation in the Mammalian Embryo', *Development Genes and Evolution*, 221/4 (2011), 209–223, https://doi.org/10.1007/s00427-011-0373-5.

Hübner, Karin, Fuhrmann, Guy, Christenson, Lane K., Kehler, James, Reinbold, Rolland, et al., 'Derivation of Oocytes from Mouse Embryonic Stem Cells', *Science*, 300/5623 (2003), 1251–1256, https://doi.org/10.1126/science.1083452.

Hwang, Young Sun, Suzuki, Shinnosuke, Seita, Yasunari, Ito, Jumpei, Sakata, Yuka, Aso, Hirofumi, et al., 'Reconstitution of Prospermatogonial Specification In Vitro from Human Induced Pluripotent Stem Cells', *Nature Communications*, 11/1 (2020), 5656, https://doi.org/10.1038/s41467-020-19350-3.

Irie, Naoko, Lee, Sun-Min, Lorenzi, Valentina, Xu, Haiqi, Chen, Jinfeng, Inoue, Masato, et al., 'DMRT1 Regulates Human Germline Commitment', *Nature Cell Biology*, 25/10 (2023), 1439–1452, https://doi.org/10.1038/s41556-023-01224-7.

Irie, Naoko, Weinberger, Leehee, Tang, Walfred W. C., Kobayashi, Toshihiro, Viukov, Sergey, Manor, Yair S., et al., 'SOX17 Is a Critical Specifier of Human Primordial Germ Cell Fate', *Cell*, 160/1 (2015), 253–268, https://doi.org/10.1016/j.cell.2014.12.013.

Ishiguro, Kei-ichiro, Matsuura, Kumi, Tani, Naoki, Takeda, Naoki, Usuki, Shingo, Yamane, Mariko, et al., 'MEIOSIN Directs the Switch from Mitosis to Meiosis in Mammalian Germ Cells', *Developmental Cell*, 52/4 (2020), 429–445.e10, https://doi.org/10.1016/j.devcel.2020.01.010.

Ishikura, Yukiko, Yabuta, Yukihiro, Ohta, Hiroshi, Hayashi, Katsuhiko, Nakamura, Tomonori, Okamoto, Ikuhiro, et al., 'In Vitro Derivation and Propagation of Spermatogonial Stem Cell Activity from Mouse Pluripotent Stem Cells', *Cell Reports*, 17/10 (2016), 2789–2804, https://doi.org/10.1016/j.celrep.2016.11.026.

Kanatsu-Shinohara, Mito, Ogonuki, Narumi, Inoue, Kimiko, Miki, Hiromi, Ogura, Atsuo, Toyokuni, Shinya, et al., 'Long-Term Proliferation in Culture and Germline Transmission of Mouse Male Germline Stem Cells', *Biology of Reproduction*, 69/2 (2003), 612–616, https://doi.org/10.1095/biolreprod.103.017012.

Kee, Kehkoi, Gonsalves, Joanna M., Clark, Amander T., and Reijo Pera, Renee A., 'Bone Morphogenetic Proteins Induce Germ Cell Differentiation from Human Embryonic Stem Cells', *Stem Cells and Development*, 15/6 (2006), 831–837, https://doi.org/10.1089/scd.2006.15.831.

Kee, Kehkoi, Angeles, Vanessa T., Flores, Martha, Nguyen, Ha Nam, and Reijo Pera, Renee A., 'Human DAZL, DAZ and BOULE Genes Modulate Primordial Germ-Cell and Haploid Gamete Formation', *Nature*, 462/7270 (2009), 222–225, https://doi.org/10.1038/nature08562.

Kinoshita, Masaki, Barber, Michael, Mansfield, William, Cui, Yingzhi, Spindlow, Daniel, Giuseppe Stirparo, Giuliano, et al., 'Capture of Mouse and Human Stem Cells with Features of Formative Pluripotency', *Cell Stem Cell*, 28/3 (2021), 453–471.e8, https://doi.org/10.1016/j.stem.2020.11.005.

Knarston, Ingrid M., Pachernegg, Svenja, Robevska, Gorjana, Ghobril, Irene, Xuan Er, Pei, Georges, Elizabeth, et al., 'An *In Vitro* Differentiation Protocol for Human Embryonic Bipotential Gonad and Testis Cell Development', *Stem Cell Reports*, 15/6 (2020), 1377–1391, https://doi.org/10.1016/j.stemcr.2020.10.009.

Kobayashi, Toshihiro, Castillo-Venzor, Aracely, Penfold, Chris A., Morgan, Michael, Mizuno, Naoaki, Tang, Walfred W. C., et al., 'Tracing the Emergence of Primordial Germ Cells from Bilaminar Disc Rabbit Embryos and Pluripotent Stem Cells', *Cell Reports*, 37/2 (2021), 109812, https://doi.org/10.1016/j.celrep.2021.109812.

Kojima, Yoji, Sasaki, Kotaro, Yokobayashi, Shihori, Sakai, Yoshitake, Nakamura, Tomonori, Yabuta, Yukihiro, et al., 'Evolutionarily Distinctive Transcriptional and Signaling Programs Drive Human Germ Cell Lineage Specification from Pluripotent Stem Cells', *Cell Stem Cell*, 21/4 (2017), 517–532, https://doi.org/10.1016/j.stem.2017.09.005.

Kojima, Yoji, Yamashiro, Chika, Murase, Yusuke, Yabuta, Yukihiro, Okamoto, Ikuhiro, Iwatani, Chizuru, et al., 'GATA Transcription Factors, SOX17 and TFAP2C, Drive the Human Germ-Cell Specification Program', *Life Science Alliance*, 4/5 (2021), https://doi.org/10.26508/lsa.202000974.

Kojima, Yoji, Kaufman-Francis, Keren, Studdert, Joshua B., Steiner, Kirsten A., Power, Melinda D., Loebel, David A. F., et al., 'The Transcriptional and Functional Properties of Mouse Epiblast Stem Cells Resemble the Anterior Primitive Streak', *Cell Stem Cell*, 14/1 (2014), 107–120, https://doi.org/10.1016/j.stem.2013.09.014.

Koubova, Jana, Menke, Douglas B., Zhou, Qing, Capel, Blanche, Griswold, Michael D., and Page, David C., 'Retinoic Acid Regulates Sex-Specific Timing of Meiotic Initiation in Mice', *Proceedings of the National Academy of Sciences*, 103/8 (2006), 2474–2479, https://doi.org/10.1073/pnas.0510813103.

Kurimoto, Kazuki, Yabuta, Yukihiro, Hayashi, Katsuhiko, Ohta, Hiroshi, Kiyonari, Hiroshi, Mitani, Tadahiro, et al., 'Quantitative Dynamics of Chromatin Remodeling during Germ Cell Specification from Mouse Embryonic Stem Cells', *Cell Stem Cell*, 16/5 (2015), 517–532.

Kurlovich, Julia, Rodriguez Polo, Ignacio, Dovgusha, Oleksandr, Tereshchenko, Yuliia, Cruz, Carmela Rieline V., Behr, Rüdiger, et al., 'Generation of Marmoset Primordial Germ Cell-like Cells under Chemically Defined Conditions', *Life Science Alliance*, 7/6 (2024), https://doi.org/10.26508/lsa.202302371.

Lawson, Karen A., and Hage, Willem J., 'Clonal Analysis of the Origin of Primordial Germ Cells in the Mouse', in J. Marsh and J. Goode, eds., *Ciba Foundation Symposium 182: Germline Development* (London: John Wiley & Sons, 2007), 68–91, https://doi.org/10.1002/9780470514573.ch5.

Lawson, Kirstie A., Dunn, N. Ray, Roelen, Bernard A. J., Zeinstra, Laura M., Davis, Angela M., Wright, Christopher V. E., et al., 'Bmp4 Is Required for the Generation of Primordial Germ Cells in the Mouse Embryo', *Genes and Development*, 13/4 (1999), 424–436.

Lee, Heather J., A. Hore, Timothy, and Reik, Wolf, 'Reprogramming the Methylome: Erasing Memory and Creating Diversity', *Cell Stem Cell*, 14/6 (2014), 710–719, https://doi.org/10.1016/j.stem.2014.05.008.

Li, Ping, Tong, Chang, Mehran-Shai, Ruty, Jia, Li, Wu, Nancy, Yan, Youzhen, et al., 'Germline Competent Embryonic Stem Cells Derived from Rat Blastocysts', *Cell*, 135/7 (2008), 1299–1310, https://doi.org/10.1016/j.cell.2008.12.006.

Manku, Gurpreet, and Culty, Martine, 'Mammalian Gonocyte and Spermatogonia Differentiation: Recent Advances and Remaining Challenges', *Reproduction*, 149/3 (2015), 139–157, https://doi.org/10.1530/REP-14-0431.

Martinovitch, Petar. 'Development In Vitro of the Mammalian Gonad', *Nature*, 139 (1937), 413.

McGee, Elizabeth A., and Hsueh, Aaron J. W., 'Initial and Cyclic Recruitment of Ovarian Follicles*', *Endocrine Reviews*, 21/2 (2000), 200–214, https://doi.org/10.1210/edrv.21.2.0394.

Morohaku, Kanako, Tanimoto, Ren, Sasaki, Keisuke, Kawahara-Miki, Ryouka, Kono, Tomohiro, Hayashi, Katsuhiko, et al., 'Complete In Vitro Generation of Fertile Oocytes from Mouse Primordial Germ Cells', *Proceedings of the National Academy of Sciences*, 113/32 (2016), 9021–9026, https://doi.org/10.1073/pnas.1603817113.

Murase, Yusuke, Yokogawa, Ryuta, Yabuta, Yukihiro, Nagano, Masahiro, Katou, Yoshitaka, Mizuyama, Manami, et al., 'In Vitro Reconstitution of Epigenetic Reprogramming in the Human Germ Line', *Nature*, 631/8019 (2024), 170–178, https://doi.org/10.1038/s41586-024-07526-6.

Murase, Yusuke, Yabuta, Yukihiro, Ohta, Hiroshi, Yamashiro, Chika, Nakamura, Tomonori, Yamamoto, Takuya, et al., 'Long-Term Expansion with Germline Potential of Human Primordial Germ Cell-like Cells In Vitro', *The EMBO Journal*, 39/21 (2020), e104929, https://doi.org/10.15252/embj.2020104929.

Nagaoka, So I., Nakaki, Fumio, Miyauchi, Hidetaka, Nosaka, Yoshiaki, Ohta, Hiroshi, Yabuta, Yukihiro, et al., 'ZGLP1 Is a Determinant for the Oogenic Fate in Mice', *Science*, 367/6482 (2020), eaaw4115, https://doi.org/10.1126/science.aaw4115.

Nichols, Jennifer, and Smith, Austin, 'Naive and Primed Pluripotent States', *Cell Stem Cell*, 4/6 (2009), 487–492, https://doi.org/10.1016/j.stem.2009.05.015.

O'Brien, Marilyn J., Pendola, Janice K., and Eppig, John J., 'A Revised Protocol for In Vitro Development of Mouse Oocytes from Primordial Follicles Dramatically Improves Their Developmental Competence1', *Biology of Reproduction*, 68/5 (2003), 1682–1686, https://doi.org/10.1095/biolreprod.102.013029.

Ohinata, Yasuhide, Ohta, Hiroshi, Shigeta, Mayo, Yamanaka, Kaori, Wakayama, Teruhiko, Saitou, Mitinori, et al., 'A Signaling Principle for the Specification of the Germ Cell Lineage in Mice', *Cell*, 137/3 (2009), 571–584, https://doi.org/10.1016/j.cell.2009.03.014.

Ohta, Hiroshi, Yabuta, Yukihiro, Kurimoto, Kazuki, Nakamura, Tomonori, Murase, Yusuke, Yamamoto, Takuya, et al., 'Cyclosporin A and FGF Signaling Support the Proliferation/Survival of Mouse Primordial Germ Cell-like Cells In Vitro†', *Biology of Reproduction*, 104/2 (2021), 344–360, https://doi.org/10.1093/biolre/ioaa195.

Ohta, Hiroshi, Kurimoto, Kazuki, Okamoto, Ikuhiro, Nakamura, Tomonori, Yabuta, Yukihiro, Miyauchi, Hidetaka, et al., 'In Vitro Expansion of Mouse Primordial Germ Cell-like Cells Recapitulates an Epigenetic Blank Slate', *The EMBO Journal*, 36/13 (2017), 1888–1907, https://doi.org/10.15252/embj.201695862.

Oikawa, Mami, Kobayashi, Hisato, Sanbo, Makoto, Mizuno, Naoaki, Iwatsuki, Kenyu, Takashima, Tomoya, et al., 'Functional Primordial Germ Cell–like Cells from Pluripotent Stem Cells in Rats', *Science*, 376/6589 (2022), 176–179, https://doi.org/10.1126/science.abl4412.

Park, Tae Sub, Galic, Zoran, Conway, Anne E., Lindgren, Anne, van Handel, Benjamin J., Magnusson, Mattias, et al., 'Derivation of Primordial Germ Cells from Human Embryonic and Induced Pluripotent Stem Cells Is Significantly Improved by Coculture with Human Fetal Gonadal Cells', *Stem Cells*, 27/4 (2009), 783–795, https://doi.org/10.1002/stem.13.

Pfeffer, Peter L., 'Alternative Mammalian Strategies Leading towards Gastrulation: Losing Polar Trophoblast (Rauber's Layer) or Gaining an Epiblast Cavity', *Philosophical Transactions of the Royal Society B: Biological Sciences*, 377/1865 (2022), 20210254, https://doi.org/10.1098/rstb.2021.0254.

Piechota, Sabrina, Marchante, Maria, Giovannini, Alexa, Paulsen, Bruna, Potts, Kathryn S., Rockwell, Graham, et al., 'Human-Induced Pluripotent Stem Cell-Derived Ovarian Support Cell Co-culture Improves Oocyte Maturation In Vitro after Abbreviated Gonadotropin Stimulation', *Human Reproduction*, 38/12 (2023), 2456–2469, https://doi.org/10.1093/humrep/dead205.

Pierson Smela, Merrick, Sybirna, Anastasiya, Wong, Frederick C. K., and Azim Surani, M., 'Testing the Role of SOX15 in Human Primordial Germ Cell Fate', *Wellcome Open Research*, 4 (2019), 122, https://doi.org/10.12688/wellcomeopenres.15381.2.

Saitou, Mitinori, and Miyauchi, Hidetaka, 'Gametogenesis from Pluripotent Stem Cells', *Cell Stem Cell*, 18/6 (2016), 721–735, https://doi.org/10.1016/j.stem.2016.05.001.

Sakai, Yoshitake, Nakamura, Tomonori, Okamoto, Ikuhiro, Gyobu-Motani, Sayuri, Ohta, Hiroshi, Yabuta, Yukihiro, et al., 'Induction of the Germ Cell Fate from Pluripotent Stem Cells in Cynomolgus Monkeys†', *Biology of Reproduction*, 102/3 (2020), 620–638, https://doi.org/10.1093/biolre/ioz205.

Sanjo, Hiroyuki, Yao, Tatsuma, Katagiri, Kumiko, Sato, Takuya, Matsumura Takafumi, Komeya, Mitsuru, et al., 'Antioxidant Vitamins and Lysophospholipids Are Critical for Inducing Mouse Spermatogenesis under Organ Culture Conditions', *The FASEB Journal*, 34/7 (2020), 9480–9497, https://doi.org/10.1096/fj.202000245R.

Sasaki, Kotaro, Yokobayashi, Shihori, Nakamura, Tomonori, Okamoto, Ikuhiro, Yabuta, Yukihiro, Kurimoto, Kazuki, et al., 'Robust In Vitro Induction of Human Germ Cell Fate from Pluripotent Stem Cells', *Cell Stem Cell*, 17/2 (2015), 178–194, https://doi.org/10.1016/j.stem.2015.06.014.

Sato, Takuya, Katagiri, Kumiko, Yokonishi, Tetsuhiro, Kubota, Yoshinobu, Inoue, Kimiko, Ogonuki, Narumi, et al., 'In Vitro Production of Fertile Sperm from Murine Spermatogonial Stem Cell Lines', *Nature Communications*, 2/1 (2011), 472, https://doi.org/10.1038/ncomms1478.

Seita, Yasunari, Cheng, Keren, McCarrey, John R., Yadu, Nomesh, Cheeseman, Ian H., Bagwell, Alec, et al., 'Efficient Generation of Marmoset Primordial Germ Cell-like Cells Using Induced Pluripotent Stem Cells', *eLife*, 12 (2023), e82263, https://doi.org/10.7554/eLife.82263.

Seki, Yoshiyuki, Yamaji, Masashi, Yabuta, Yukihiro, Sano, Mitsue, Shigeta, Mayo, Matsui, Yasuhisa, et al., 'Cellular Dynamics Associated with the Genome-Wide Epigenetic Reprogramming in Migrating Primordial Germ Cells in Mice', *Development*, 134/14 (2007), 2627–2638, https://doi.org/10.1242/dev.005611.

Seki, Yoshiyuki, Hayashi, Katsuhiko, Itoh, Kunihiko, Mizugaki, Michinao, Saitou, Mitinori, and Matsui, Yasuhisa, 'Extensive and Orderly Reprogramming of Genome-Wide Chromatin Modifications Associated with Specification and Early Development of Germ Cells in Mice', *Developmental Biology*, 278/2 (2005), 440–458, https://doi.org/10.1016/j.ydbio.2004.11.025.

Seol, Dong-Won, Park, Seah, Young Shin, Eun, Chang, Jae Ho, and Lee, Dong Ryul, 'In Vitro Derivation of Functional Sertoli-Like Cells from Mouse Embryonic Stem Cells', *Cell Transplantation*, 27/10 (2018), 1523–1534, https://doi.org/10.1177/0963689718797053.

Shirane, Kenjiro, Kurimoto, Kazuki, Yabuta, Yukihiro, Yamaji, Masashi, Satoh, Junko, Ito, Shinji, et al., 'Global Landscape and Regulatory Principles of DNA Methylation Reprogramming for Germ Cell Specification by Mouse Pluripotent Stem Cells', *Developmental Cell*, 39/1 (2016), 87–103, https://doi.org/10.1016/j.devcel.2016.08.008.

Shirasawa, Atsushi, Hayashi, Masafumi, Shono, Mayumi, Ideta, Atsushi, Yoshino, Takashi, and Hayashi, Katsuhiko, 'Efficient Derivation of Embryonic Stem Cells and Primordial Germ Cell-like Cells in Cattle', *Journal of Reproduction and Development*, 70/2 (2024), 82–95, https://doi.org/10.1262/jrd.2023-087.

Shono, Mayumi, Kishimoto, Keiko, Hikabe, Orie, Hayashi, Masafumi, Semi Katsunori, Takashima, Yasuhiro, et al., 'Induction of Primordial Germ Cell-like Cells from Common Marmoset Embryonic Stem Cells by Inhibition of WNT and Retinoic Acid Signaling', *Scientific Reports*, 13/1 (2023), 3186, https://doi.org/10.1038/s41598-023-29850-z.

Smith, Austin, 'Formative Pluripotency: The Executive Phase in a Developmental Continuum', *Development*, 144/3 (2017), 365–373, https://doi.org/10.1242/dev.142679.

Sosa, Enrique, Chen, Di, Rojas, Ernesto J., Hennebold, Jon D., Peters, Karen A., Wu, Zhuang, et al., 'Differentiation of Primate Primordial Germ Cell-like Cells Following Transplantation into the Adult Gonadal Niche', *Nature Communications*, 9/1 (2018), 5339, https://doi.org/10.1038/s41467-018-07740-7.

Spiller, Cassy, Koopman, Peter, and Bowles, Josephine, 'Sex Determination in the Mammalian Germline', *Annual Review of Genetics*, 51 (2017), 265–285, https://doi.org/10.1146/annurev-genet-120215-035449.

Steinberger, Elisabeth, 'Maintenance of Adult Human Testicular Tissue in Culture', *The Anatomical Record*, 157 (1967), 327.

Steinberger, Alan, and Steinberger, Elisabeth, 'Differentiation of Rat Seminiferous Epithelium in Organ Culture', *Journal of Reproduction and Fertility*, 9 (1965), 243–248.

Steinberger, Alan, and Steinberger, Elisabeth, 'Factors Affecting Spermatogenesis in Organ Cultures of Mammalian Testes', *Journal of Reproduction and Fertility*, 2 (1967), 117–124.

Steinberger, Elisabeth, Steinberger, Alan, and Perloff, William H., 'Initiation of Spermatogenesis in Vitro1', *Endocrinology*, 74/5 (1964), 788–792, https://doi.org/10.1210/endo-74-5-788.

Steinberger, Alan, Steinberger, Elisabeth, and Perloff, William H., 'Mammalian Testes in Organ Culture', *Experimental Cell Research*, 36 (1964), 19–27.

Sybirna, Anastasiya, Tang, Walfred W. C., Smela, Merrick Pierson, Dietmann, Sabine, Gruhn, Wolfram H., Brosh, Ran, et al., 'A Critical Role of PRDM14 in Human Primordial Germ Cell Fate Revealed by Inducible Degrons', *Nature Communications*, 11/1 (2020), 1282.

Takahashi, Kazutoshi, Tanabe, Koji, Ohnuki, Mari, Narita, Megumi, Ichisaka, Tomoko, Tomoda, Kiichiro, et al., 'Induction of Pluripotent Stem Cells from Adult Human Fibroblasts by Defined Factors', *Cell*, 131/5 (2007), 861–872, https://doi.org/10.1016/j.cell.2007.11.019.

Tam, Patrick P. L., and Zhou, Shang-Xian, 'The Allocation of Epiblast Cells to Ectodermal and Germ-Line Lineages Is Influenced by the Position of the Cells in the Gastrulating Mouse Embryo', *Developmental Biology*, 178/1 (1996), 124–132, https://doi.org/10.1006/dbio.1996.0203.

Tang, Walfred W. C., Kobayashi, Toshihiro, Irie, Naoko, Dietmann, Sabine, and Azim Surani, M., 'Specification and Epigenetic Programming of the Human Germ Line', *Nature Reviews Genetics*, 17/10 (2016), 585–600, https://doi.org/10.1038/nrg.2016.88.

Tesar, Paul J., Chenoweth, Josh G., Brook, Frances A., Davies, Timothy J., Evans, Edward P., Mack, David L., et al., 'New Cell Lines from Mouse Epiblast Share Defining Features with Human Embryonic Stem Cells', *Nature*, 448/7150 (2007), 196–199, https://doi.org/10.1038/nature05972.

Toyooka, Yayoi, Tsunekawa, Naoki, Akasu, Ryuko, and Noce, Toshiaki, 'Embryonic Stem Cells Can Form Germ Cells In Vitro', *Proceedings of the National Academy of Sciences*, 100 (2003), 11457–11462, https://doi.org/10.1073/pnas.1932826100.

Wen, Lele, and Tang, Fuchou, 'Human Germline Cell Development: from the Perspective of Single-Cell Sequencing', *Molecular Cell*, 76/2 (2019), 320–328, https://doi.org/10.1016/j.molcel.2019.08.025.

Yamashiro, Chika, Sasaki, Kotaro, Yabuta, Yukihiro, Kojima, Yoji, Nakamura, Tomonori, Okamoto, Ikuhiro, et al., 'Generation of Human Oogonia from Induced Pluripotent Stem Cells In Vitro', *Science*, 362/6412 (2018), 356–360, https://doi.org/10.1126/science.aat1674.

Ying, Qi-Long, Wray, Jason, Nichols, Jennifer, Batlle-Morera, Laura, Doble, Bradley, Woodgett, James, et al., 'The Ground State of Embryonic Stem Cell Self-renewal', *Nature*, 453/7194 (2008), 519–523, https://doi.org/10.1038/nature06968.

Yokobayashi, Shihori, Okita, Keisuke, Nakagawa, Masato, Nakamura, Tomonori, Yabuta, Yukihiro, Yamamoto, Takuya, et al., 'Clonal Variation of Human Induced Pluripotent Stem Cells for Induction into the Germ Cell Fate†', *Biology of Reproduction*, 96/6 (2017), 1154–1166, https://doi.org/10.1093/biolre/iox038.

Yoshida, Michio, Kajikawa, Eriko, Kurokawa, Daisuke, Tokunaga, Tomoyuki, Onishi, Akira, Yonemura, Shigenobu, et al., 'Conserved and Divergent Expression Patterns of Markers of Axial Development in Eutherian Mammals', *Developmental Dynamics*, 245/1 (2016), 67–86.

Yoshimizu, Toru, Obinata, Msauo, and Matsui, Yoshitake, 'Stage-Specific Tissue and Cell Interactions Play Key Roles in Mouse Germ Cell Specification', *Development*, 128 (2001), 481–490.

Yoshino, Takashi, Suzuki, Takahiro, Nagamatsu, Go, Yabukami, Haruka, Ikegaya, Mika, Kishima, Mami, et al., 'Generation of Ovarian Follicles from Mouse Pluripotent Stem Cells', *Science*, 373/6552 (2021), eabe0237, https://doi.org/10.1126/science.abe0237.

Yu, Leqian, Wei, Yulei, Sun, Hai-Xi, Mahdi, Ahmed K., Pinzon Arteaga, Carlos A., Sakurai, Masahiro, et al., 'Derivation of Intermediate Pluripotent Stem Cells Amenable to Primordial Germ Cell Specification', *Cell Stem Cell*, 28/3 (2021), 550–567.e12, https://doi.org/10.1016/j.stem.2020.11.003.

2
Reproduction without Constraints
How IVG Will Reshape Reproduction

Christopher Gyngell, Julian Savulescu, and Tsutomu Sawai

The Emergence of Sex

Life appeared on Earth around 4.1 billion years ago.[1] At first, all life reproduced asexually. The only existing life forms—single-celled organisms—replicated themselves by duplicating their internal components, including genetic material, and then dividing into two new cells. This process allowed organisms to rapidly produce genetically identical offspring. Then, around 1.2 billion years ago, something remarkable happened: a radically new form of reproduction emerged—sexual reproduction.[2] Instead of simply splitting into two, some organisms evolved the ability to create specialized reproductive cells, called gametes, designed to fuse with other gametes to form novel new cells.

Sexual reproduction provides key evolutionary advantages over asexual reproduction. While asexual reproduction produces genetically identical offspring, sexual reproduction reshuffles genetic material through recombination, generating genetically diverse descendants. This heightened variability bolsters a population's resilience against a range of unpredictable challenges, including deleterious mutations, parasites, environmental fluctuations, and competition with other groups. Furthermore, sexual reproduction refines the evolutionary process by promoting the assembly of gene combinations that interact beneficially—that is, those exhibiting favourable epistatic interactions when present within a single individual.[3]

However, this mode of reproduction also imposed several constraints on organisms. For humans and other sexually reproducing species, procreation

[1] Elizabeth A. Bell et al., 'Potentially Biogenic Carbon Preserved in a 4.1 Billion-Year-Old Zircon', *Proceedings of the National Academy of Sciences*, 112/47 (2015), 14518–14521.

[2] Nicholas J. Butterfield, 'Bangiomorpha Pubescens n. Gen., n. Sp.: Implications for the Evolution of Sex, Multicellularity, and the Mesoproterozoic/Neoproterozoic Radiation of Eukaryotes', *Paleobiology*, 26/3 (2000), 386–404, https://doi.org/10.1666/0094-8373(2000)026<0386:BPNGNS>2.0.CO;2.

[3] Alexander Feigel, Avraham Englander, and Assaf Engel, 'Sex Is Always Well Worth Its Two-Fold Cost', *PLoS ONE*, 4/7 (2009), e6012.

Christopher Gyngell, Julian Savulescu, and Tsutomu Sawai, *Reproduction without Constraints* In: *Revolutionizing Reproduction*. Edited by: Christopher Gyngell, Julian Savulescu, and Tsutomu Sawai, Oxford University Press.
 DOI: 10.1093/9780197810415.003.0002

traditionally requires two (and only two) individuals of opposite sexes to mate, each with specialized reproductive organs. Finding and attracting a suitable mate can demand significant time and energy, and failure to do so means leaving no offspring—an evolutionary dead end. Moreover, reproduction through sex results in only half of an individual's DNA being transmitted to each descendant, whereas asexual reproduction produces offspring that are nearly identical copies of the parent's genome.

In sum, evolution's pathway for creating genetic variation in offspring came with trade-offs: the necessity of finding a mate, the development of separate male and female anatomies, and a finite supply of gametes (especially eggs). These biological constraints have shaped not only the animal kingdom's mating behaviours and life histories, but also human social structures and cultural norms around family and reproduction.

Emerging biotechnologies are redefining reproductive possibilities by bypassing traditional evolutionary constraints. *In vitro* gametogenesis (IVG)—a process where researchers reprogram adult somatic cells (such as skin or blood cells) into induced pluripotent stem cells (iPSCs) and then direct their transformation into gametes—offers new avenues in reproductive medicine. For example, in mice, scientists have successfully generated functional eggs from male-derived iPSCs, leading to the birth of healthy offspring. Although deriving sperm from female-derived iPSCs is more difficult, advances in genome editing and cellular reprogramming could eventually overcome these hurdles.[4] In human studies, while fully mature IVG gametes have not yet been achieved, early progress—including the creation of primordial germ cell-like cells and immature eggs—underscores the transformative potential of these approaches.[5,6]

IVG stands poised to become a new assisted reproductive technology (ART) that could supplement *in vitro* fertilization (IVF).[7] What makes IVG so transformative is its potential to decouple reproduction from the traditional requirements of sex and biology. In principle, gametes derived from any person's cells at any stage of life could be used to create embryos. This means that the fundamental limitations evolution built into sexual procreation might no longer apply.

This chapter explores how IVG could allow humanity to overcome four major evolutionary constraints of sexual reproduction: (1) the need to find a mate

[4] Mitinori Saitou and Katsuhiko Hayashi, 'Mammalian in Vitro Gametogenesis', *Science*, 374/6563 (2021), eaaz6830.

[5] Chika Yamashiro et al., 'Generation of Human Oogonia from Induced Pluripotent Stem Cells in Vitro', *Science*, 362/6412 (2018), 356–360.

[6] For a detailed account of the technical development of IVG, see Hayashi '*In Vitro* Gametogenesis' (Chapter 1, this volume).

[7] Henry T. Greely, *CRISPR People: The Science and Ethics of Editing Humans* (Cambridge, MA: MIT Press, 2021).

(of the opposite sex) in order to reproduce, (2) the reliance on specialized reproductive systems and time-bound fertility, (3) the finite supply of oocytes (eggs) in female individuals, and (4) the limitation of genetic parenthood to two individuals at a time. For each constraint, we discuss the evolutionary context and burdens it imposes, then examine how IVG can potentially eliminate or work around that limitation. We then discuss the ethics of crossing these boundaries and the ethics of IVG more broadly.

Constraint 1: Needing to Find a Mate (of the Opposite Sex)

One of the most obvious burdens of sexual reproduction is that it requires two compatible individuals to cooperate in creating offspring. Finding and attracting a suitable mate can demand significant time and energy. Competition for mates can drive the development of elaborate displays, behaviours, or disputes among animals—all of which carry energetic costs and impose risks. The need to find a willing and genetically complementary partner creates a significant barrier to reproduction. Humans are no exception to these dynamics—throughout history, factors like sex ratios, geography, social status, and cultural taboos have all determined who can find a mate and reproduce. Those unable to find a suitable partner (e.g., due to personal circumstances or sexual orientation) have traditionally been unable to have biologically related children. Recently, technologies like IVF and gamete donation have made it easier to have children with little input from others. IVG has the potential to remove the need for a partner in procreation altogether.

Solo Reproduction

One person's skin cell could be used to derive iPSCs that are then differentiated into both eggs and sperm. Fertilizing one's own IVG-derived egg with one's own IVG sperm would produce an embryo that has one genetic parent instead of two.[8] This scenario resembles cloning in its outcome (100% genetic relatedness between parent and child) but differs in that the child is not an identical copy of the parent. Due to meiotic recombination, a solo-IVG child's genome would be a shuffled combination of the single progenitor's two sets of chromosomes.

[8] Daniela Cutas and Anna Smajdor, '"I Am Your Mother and Your Father!" In Vitro Derived Gametes and the Ethics of Solo Reproduction', *Health Care Analysis*, 25/4 (2017), 354–369.

The prospect of solo reproduction is perhaps the most controversial aspect of IVG.[9] One immediate ethical concern are the health risks that could arise because of solo reproduction. Solo reproduction is essentially an extreme form of inbreeding, the equivalent of two identical twins having a child.[10] This raises obvious genetic risks: harmful recessive genes present in a parent might become homozygous in the child. Studies have shown that inbreeding can lead to higher rates of multiple genetic disorders, including blindness, hearing loss, neonatal diabetes, limb malformations, sex development disorders, schizophrenia, and several others.[11] The potential for increased genetic disease in offspring resulting from solo reproduction raises significant ethical concerns regarding its implementation. Robust safety data is therefore a prerequisite for any potential use of IVG for solo reproduction.[12,13] This will depend on whether we can use other technologies, like whole genome sequencing and gene editing, to negate its safety risks.

Absent safety risks, are there more fundamental reasons to object to solo reproduction?

One possibility is that there are harmful psychosocial implications for a child who does not have two genetic parents. The same concerns have been raised about donor-conceived children. Indeed, some have concluded that having an anonymous gamete donation as a parent, meaning one would not know the origin of half their genome, raises more serious identity concerns than solo parenting, as the children will always be deprived of knowing part of their ancestry.

It's important to highlight, though, that children born via anonymous gamete donation still have the same type of heritage as everyone else—they are the combination of two genetic lineages. Their family tree would look like everyone else's; the difference is that they don't know the identities of certain branches.

A child born through solo reproduction would have an ancestry unlike any other, which could create feelings of confusion and alienation. While we cannot know how this will affect someone psychologically, it seems plausible that the effect might be categorically different to gamete donation.

A further concern is that IVG could be harmful for children because it is preferable for children to be raised by two parents rather than by a single parent. Critics often point to empirical research suggesting that children raised

[9] Anne Le Goff et al., 'Anticipating *In Vitro* Gametogenesis: Hopes and Concerns for IVG among Diverse Stakeholders', *Stem Cell Reports*, 19/7 (2024), 933–945.

[10] Lauren Notini, Christopher Gyngell, and Julian Savulescu, 'Drawing the Line on In Vitro Gametogenesis', *Bioethics*, 34/1 (2020), 123–134.

[11] A. H. Bittles and M. L. Black, 'Consanguinity, Human Evolution, and Complex Diseases', *Proceedings of the National Academy of Sciences*, 107/suppl. 1 (2010), 1779–1786.

[12] Notini, Gyngell, and Savulescu, 'Drawing the Line on in Vitro Gametogenesis'.

[13] Dorian Accoe and Seppe Segers, '"False Hope" in Assisted Reproduction: The Normative Significance of the External Outlook and Moral Negotiation', *Journal of Medical Ethics*, 50/3 (2024), 181–184; Accoe and Segers, 'Critical Bioethics', (Chapter 13, this volume).

in single-parent homes, on average, experience worse social, educational, and economic outcomes compared to those raised in two-parent households.[14] Here, though, we can point to differences between solo reproduction and solo parenting. Endorsing solo reproduction as a means of bringing a child into the world need not amount to an endorsement that it is better for children to be raised by single caregivers. Solo reproduction is compatible with lots of different post-birth caring arrangements. If we are just focussed on the ethics of solo reproduction, concern about solo parenting should not be seen as relevant.

Ethicists Daniela Cutas and Anna Smajdor argue that while solo reproduction challenges our intuitions, it is not necessarily any more ethically problematic than other reproductive technologies.[15] They suggest that some individuals might be motivated to reproduce alone—for example, to preserve their lineage without involving another person, or to avoid passing on a heritable disease by using only their healthy gametes (a scenario analogous to self-cloning). IVG increases people's reproductive options and can therefore be seen to support forms of procreative liberty. Anantharaman Muralidharan explores the issues involved in solo reproduction in greater depth in Chapter 8.

Same-Sex Reproduction

If each person can, in the future, contribute both egg and sperm (either alone or in combination with someone of any gender), then reproduction is no longer locked to heterosexual coupling. A gay male couple could have a child genetically related to both of them by using IVG to create an egg from one partner's somatic cells and fertilizing it with the other partner's sperm; similarly, a lesbian couple could generate sperm from one female partner's cells to fertilize the other's egg. While these possibilities are still speculative, recent research has made their realization more plausible. As we have noted, researchers have already achieved 'sex reversal' gametogenesis in laboratory animals (e.g., converting XY cells to functional oocytes in mice).[16] Should these developments be made possible in humans, it would mean same-sex couples would no longer require a donor gamete to have biologically related children, enjoying the same genetic parenthood that has historically only be available to heterosexual couples.

[14] Matthew D. Bramlett and Stephen J. Blumberg, 'Family Structure and Children's Physical and Mental Health', *Health Affairs (Project Hope)*, 26/2 (2007), 549–558.

[15] Cutas and Smajdor, 'I Am Your Mother and Your Father!'.

[16] Kenta Murakami et al., 'Generation of Functional Oocytes from Male Mice In Vitro', *Nature*, 615/7954 (2023), 900–906, https://doi.org/10.1038/s41586-023-05834-x.

Some writers indicate that the use of IVG and other reproductive technologies for the purpose of same-sex reproduction should not be seen as a high priority use of the technology, especially when compared to their use for 'therapeutic' purposes such as restoring fertility.[17]

However, others point out that the use of IVG to allow same-sex reproduction can be seen as a therapeutic use in much the same way as it would be allowing heterosexual couples to have babies.[18] Under some medical definitions of infertility, same-sex couples qualify just as much as different-sex couples and can be seen as having the medical rights to access IVG as IVF.

Consideration of other ethical principles provide additional reasons to support the use of IVG to allow same-sex reproduction. As genetic parenthood is a socially valued good, the principle of equality suggests same-sex couples should have the same opportunity to pursue the genetic family bonds as opposite-sex couples.[19] Furthermore, same-sex couples have traditionally been discriminated against in most human societies. The principle of compensatory justice, which suggests we owe special consideration to groups who have the targets of past injustices, would indicate same-sex couples have additional claims on to use IVG for reproductive purposes.[20]

In sum, IVG directly targets the first evolutionary constraint of sex: the requirement of finding an opposite-sex mate. Being freed from this basic limitation of sexual reproduction will increase reproductive autonomy for individuals and could help promote equality for same-sex couples. At the same time, the potential of IVG to take us well beyond our current biological limits, as in solo reproduction, raises many difficult ethical challenges.

Constraint 2: The Need for Functional Reproductive Systems

Sexual reproduction requires not only two people, but also that each person has specialized biological equipment to produce gametes and, in the case of females, to carry a pregnancy. These reproductive systems are often complex, requiring time to develop and energy to maintain. Many diseases and illnesses interfere with workings of the reproductive systems and leave individuals unable to have

[17] Camisha Russell, 'Rights-Holders or Refugees? Do Gay Men Need Reproductive Justice?', *Reproductive Biomedicine & Society Online*, 7 (2018), 131–140, https://doi.org/10.1016/j.rbms.2018.07.001; Françoise Baylis, '"No" to Lesbian Motherhood Using Human Nuclear Genome Transfer', *Journal of Medical Ethics*, 44/12 (2018), 865–867.

[18] Notini, Gyngell, and Savulescu, 'Drawing the Line on In Vitro Gametogenesis'.

[19] Timothy F. Murphy, 'Pathways to Genetic Parenthood for Same-Sex Couples', *Journal of Medical Ethics*, 44/12 (2018), 823–824.

[20] Murphy, 'Sex and Gender Minorities and *In Vitro* Gametes' (Chapter 4, this volume).

genetically related children. In humans, reproductive systems tend to only function well during certain life periods, limiting fertility to defined life stages. IVG has the power to decouple gamete production from the body's reproductive system and timeline. Because IVG uses somatic cells (like skin cells) as starting material, it no longer matters whether the donor has functioning ovaries or testes—or even whether they are past their fertile years.

It's important to stress that IVG will not free humans from all biological limits on reproduction. Crucially all human children still require a uterus. While gestational surrogacy can mean some couples can have children without either having a functional uterus, it merely shifts the physical requirements from one individual to another. A long-predicted future technology, the artificial womb,[21] could potentially remove this constraint completely in the future and result in a more complete decoupling of human reproduction from human biology.

Restoring Fertility

One of the most immediate potential benefits of IVG is improving outcomes for individuals already using IVF. Instead of retrieving eggs directly from a woman's ovaries, clinicians could derive hundreds—or even thousands—of eggs from a small skin biopsy. This approach would offer multiple opportunities for successful fertilization and implantation, significantly boosting IVF success rates while also reducing risks associated with egg retrieval. Similarly, in cases where male infertility is due to low sperm count or reduced sperm motility, IVG could provide an effectively limitless supply of sperm cells.

By eliminating the need for fully functioning reproductive systems, IVG could also expand IVF eligibility to a broader range of people. Certain conditions—such as congenital diseases or a history of childhood cancer—may leave individuals unable to produce viable gametes, thereby ruling out standard IVF. IVG circumvents this challenge by generating gametes from somatic cells, making parenthood a possibility for those who would otherwise have no genetic pathway to reproduction.

From an ethical perspective, using IVG to address fertility problems caused by disease aligns closely with standard medical treatments. An individual's right to access technologies that enable them to reproduce is a natural extension of the right to reproductive healthcare.

[21] Seppe Segers, 'The Path toward Ectogenesis: Looking beyond the Technical Challenges', *BMC Medical Ethics*, 22/1 (2021), 59.

Reproduction in Old Age

IVG also promises to remove age-related constraints on human fertility. A notable example is the potential for post-menopausal motherhood. Menopause typically begins at fifty years of age and is marked by the cessation of ovarian function and the end of natural fertility.[22] Men's fertility also declines with age, though less abruptly. Older men frequently experience reductions in sperm quality—including motility and morphology—which can lower pregnancy rates and slightly increase the risk of genetic abnormalities.[23]

Although current assisted reproductive technologies, particularly IVF, already enable reproduction at advanced ages, IVG would allow fertility windows to be extended indefinitely for both men and women.

The use of IVG specifically for facilitating post-menopausal motherhood has been distinguished from its application in restoring fertility to individuals within typical reproductive ages.[24] Utilizing IVG to extend reproductive capability beyond natural age limits is often viewed not as a 'treatment' but as an 'enhancement'. Consequently, claims or entitlements to such interventions might be considered less compelling. Nonetheless, drawing such a rigid distinction between treatment and enhancement can be problematic in some scenarios. For example, is restoring the reproductive capacity of someone who has gone through menopause earlier than average a treatment or an enhancement? Furthermore, empirical data show that public acceptance of IVG varies significantly by cultural context,[25] and resistance to older or non-traditional parents often reflects cultural biases rather than principled ethical objections.[26]

Even if some intervention is an enhancement, this does not inherently make it unethical. In an era of increasing longevity, technologies that enable later-life parenthood can significantly enhance reproductive autonomy and broaden family-building opportunities.[27] Many individuals prefer to delay parenthood but feel compelled to start earlier due to biological time constraints. Technologies like IVG may enhance reproductive autonomy, by making reproduction compatible with many life plans. However, as Suter (Chapter 3) shows, increased reproductive options do not always equate to increased reproductive autonomy.

[22] Susan R. Davis et al., 'Menopause—Biology, Consequences, Supportive Care, and Therapeutic Options', *Cell*, 186/19 (2023), 4038–4058.

[23] Jason R Kovac et al., 'The Effects of Advanced Paternal Age on Fertility', *Asian Journal of Andrology*, 15/6 (2013), 723–728.

[24] Sonia M. Suter, 'In Vitro Gametogenesis: Just Another Way to Have a Baby?', *Journal of Law and the Biosciences*, 3/1 (2015), 87–119.

[25] Akatsuka et al., 'Japanese Perspectives on *In Vitro* Gametogenesis' (Chapter 12, this volume).

[26] Mertes, 'Gendered and Ageist Norms of Parenthood Manifested in Public Attitudes Towards IVG' (Chapter 11, this volume).

[27] Daniela Cutas and Anna Smajdor, 'Postmenopausal Motherhood Reloaded: Advanced Age and In Vitro Derived Gametes', *Hypatia*, 30/2 (2015), 386–402.

Accepting that parenthood beyond natural reproductive limits might be ethically permissible does not imply unlimited acceptance at any age or under all circumstances. Ethical concerns about older parents' ability to adequately care for and support their children remain important. Yet these concerns are not exclusive to IVG; they equally apply to other reproductive technologies already in use.

Reproduction in Youth

A more alarming potential use of IVG lies in expanding fertility windows at the other end of the age spectrum. Technically, IVG could enable the creation of functional gametes from the cells of a child at any stage of development, which then could be used in combination with gestational surrogacy to enable children to reproduce before sexual maturation.

While this may sound far-fetched, there are conceivable historical or cultural contexts—such as royalty, aristocracy, or other inheritance-based systems—where the ability to create a descendant from a child's cells would have been sought after. Imagine a scenario in which a child king occupies the throne as the sole heir. Being able to produce heirs from that child could be seen as vital to securing a dynasty's future.

Such actions may seem clearly morally objectionable. Children do not have the capacity to consent to their cells being used for reproduction. Unlike adult cases—where procreative liberty might justify using IVG—there is no justification in allowing prepubescent reproduction, as it infringes on the child's future autonomy and undermines their right to make informed reproductive decisions as adults. Ethical considerations strongly support preserving reproductive decisions until individuals are capable of meaningful consent.

Posthumous Reproduction

A more radical way in IVG could life biological contracts on production is removing the constraint of being alive. Posthumous reproduction refers to conceiving a child using the gametes (sperm or eggs) or other genetic material of an individual who has died. It is already possible without the use of IVF when individuals have sperm or eggs cells available after their death.[28]

[28] Belinda Bennett, 'Posthumous Reproduction and the Meanings of Autonomy', *Melbourne University Law Review* (1999), https://www.austlii.edu.au/cgi-bin/viewdoc/au/journals/MULR/1999/13.html.

There have been document cases of courts permitting posthumous reproduction for men (the removal of sperm and use of stored sperms for use in IVF), and posthumous reproduction for women (the use of frozen eggs for use in IVF).

Posthumous reproduction raises concerns as the deceased will not be able to rear the child, raising the question of whether someone can have an interest in reproducing when they cannot participate in parenting. Cases of posthumous reproduction where there has not been clear consent from the deceased individual are often agreed to be unethical. However, when there is evidence that a person wants to be a posthumous parent and has taken steps to freeze gametes for this purpose, this case is more tricky. Here the crucial question is whether one's procreative autonomy should be seen as extending to after one's death? If so, some cases of posthumous reproduction may be seen as ethically acceptable.[29]

IVG will make posthumous reproduction more technically feasible. With IVG it will be possible to take cells from recently deceased person and generate gametes that could be used in IVF. Once IVG is available, then we can expect there to be more cases of people wanting to use the cells from deceased individuals for posthumous reproduction. It is important that we carefully consider the conditions in which posthumous reproduction is ethically acceptable before IVG is clinically available.

Constraint 3: Limited Oocyte Supply

One of the most striking asymmetries in mammalian reproduction is the disparity between male and female gamete production. Evolution has endowed males with the ability to produce an essentially unlimited number of sperm throughout their lives. Females, however, are born with a finite stock of oocytes (egg cells) that diminishes over time and is never replenished. In humans, women ovulate around 300 times over their lifespan, on average. In contrast, males can produce over 500 million sperm in a single day.

A limited supply of oocytes is a common factor affecting the success of assisted reproductive technologies like IVF. Some couples fail to produce a viable embryo in a single IVF cycle, to a limited number of eggs harvested, particularly as women delay reproduction to pursue careers. IVG will free us from these limitations. Through a single biopsy, it would in principle be possible to generate thousands of eggs via IVG—far more than a woman naturally produces over her lifetime. This will allow couples (where there is no limit on

[29] Kelton Tremellen and Julian Savulescu, 'A Discussion Supporting Presumed Consent for Posthumous Sperm Procurement and Conception', *Reproductive Biomedicine Online*, 30/1 (2015), 6–13.

sperm number) to produce hundreds or thousands of embryos, therefore dramatically increasing the chance of couples producing viable embryos through a single IVF cycle.

In addition to boosting IVF success rates, IVG will increase our ability to select embryos for implantation on the basis of their resistance to disease, or their predisposition to develop other complex traits.

Improved Selection

Currently, most genetic screening programs focus on monogenic conditions, which involve selecting against a single gene variant. Under typical circumstances, an IVF cycle produces enough embryos to facilitate this kind of selection. However, in about 18% of IVF cycles, only one embryo is produced, and that single embryo may not be viable for implantation. By increasing the number of available oocytes (and sperm), IVG could generate more embryos in a single IVF cycle, thereby ensuring there was always enough embryos to select against monogenic conditions.

However, the potential benefits of IVG extend well beyond monogenic diseases. Polygenic conditions—such as heart disease, cancer, and diabetes—are influenced by multiple genetic variants, as well as by environmental factors. These conditions are among the leading causes of premature death and disability worldwide. Cardiovascular diseases alone account for a third of all deaths globally and consume trillions of dollars in health resources.

Selecting against such polygenic diseases is highly challenging under current technologies. For instance, there are at least 163 genomic loci implicated in cardiovascular disease.[30] Theoretically, a couple could produce billions of possible embryos with different combinations of 163 genes; the probability and a particular embryo having the absolute lowest risk is vanishingly small. Likewise, the difference in risk among the few embryos produced in a typical IVF cycle is usually minimal.[31]

For example, there may be individuals in the current population who are homozygous for the protective allele for all of the ten most influential genes for common diseases. But this would only be one per 2 billion for Alzheimer's disease, one per 3 billion of schizophrenia, and one per million for manic

[30] Khairul Anwar Zarkasi et al., 'Genetic Factors for Coronary Heart Disease and Their Mechanisms: A Meta-Analysis and Comprehensive Review of Common Variants from Genome-Wide Association Studies', *Diagnostics*, 12/10 (2022), 2561.

[31] Alex Polyakov et al., 'Polygenic Risk Score for Embryo Selection-Not Ready for Prime Time', *Human Reproduction (Oxford, England)*, 37/10 (2022), 2229–2236.

depression.[32] If IVG makes it possible to create thousands of embryos, the odds of finding one with the optimal combination of parental alleles rise dramatically.

Furthermore, the combination of IVG and genome editing technologies offers a powerful method for selecting against polygenic disease. iPSCs could be edited at multiple polygenic loci. A single edited cell could then be expanded into a clonal cell line. These cells could subsequently be sequenced to confirm the success of the gene edits. Once validated, the cells could be differentiated into gametes carrying a broad set of protective alleles against polygenic disease.

In the same way that IVG could help select against polygenic diseases, it will also increase our ability to select non-medical traits. Some IVF clinics already offer embryo selection based on projected IQ or height. However, like polygenic disease, these traits are affected by variants across hundreds of loci. The low number of embryos in standard IVF cycles severely limits the current power of screening for intelligence, height, and other non-medical trial. IVG's capacity to produce very large numbers of embryos would greatly increase this power and enable meaningful selection for these traits.

The ethics of applying reproductive technologies to enhance non-medical traits is complex and is discussed in a substantial body of literature. Some argue that parents have a moral responsibility to use genetic selection to help their children live better lives,[33] while others point to novel risks of enhancement[34] of and invoking the spectre of eugenics.[35] Like gene editing, IVG brings these once largely theoretical debates into sharper focus, giving them new practical urgency.

Iterative Embryo Selection

Beyond single-generation embryo selection, IVG may enable iterative embryo selection, sometimes referred to as '*in vitro* eugenics' (see Chapter 5). Here, lab-generated embryos would be used to derive new stem cell lines, which could, in turn, become gametes used to creature further embryos. This multi-generational process could accelerate human breeding by selecting the 'best' genetic traits over repeated lab-controlled generations, without involving a single pregnancy

[32] Peter M. Visscher et al., 'Heritable Polygenic Editing: The Next Frontier in Genomic Medicine?', *Nature*, 637/8046 (2025), 637–645.

[33] J. Savulescu, 'Procreative Beneficence: Why We Should Select the Best Children', *Bioethics*, 15/5–6 (2001), 413–426.

[34] Robert Sparrow, 'Yesterday's Child: How Gene Editing for Enhancement Will Produce Obsolescence-and Why It Matters', *The American Journal of Bioethics: AJOB*, 19/7 (2019), 6–15.

[35] Felipe E. Vizcarrondo, 'Human Enhancement: The New Eugenics', *The Linacre Quarterly*, 81/3 (2014), 239–243, https://doi.org/10.1179/2050854914Y.0000000021.

until a final embryo is created and chosen for implantation. This technique could be used for research (studying how certain genes are inherited or how multi-generational selection might combat a genetic disease). More controversially, the technique could be used to enhance humans by selecting desirable genotypes over several lab-controlled generations.

Why would people be interested in using iterative embryos selection, when they could just use polygenic selection discussed above? Consider again a couple who wants a child who is maximally protected against cardiovascular disease. Even with thousands of embryos from a single couple's cells via IVG, the odds of finding all 163 protective alleles in one embryo are negligible—simply because the couple themselves do not carry every beneficial variant. However, if many donors with additional protective alleles were available, labs could systematically combine desirable variants through successive embryo generations, and generate gametes with very specific genetic characteristics.

One hypothetical approach could involve gathering somatic cells from a large donor pool, for instance, around 1,000 individuals, each possessing various protective alleles across the 163 loci of interest. Gametes would then be generated from these donors' cells using IVG. Subsequently, embryos would be created through random pairings, and each embryo's polygenic risk score for cardiovascular disease would be assessed. Those embryos carrying the most favourable alleles would be identified, and new gametes would be derived from their cells to create a second generation of embryos. This process could be repeated until a population of embryos emerges that collectively carry the optimal combination of protective genetic variants across all 163 loci. Finally, gametes from these optimized lines could be combined with gametes derived from couples who want to have genetically related children who shared high protection against polygenic conditions.

This technique could theoretically offer parents great control over a future child's health profile and help reduce the disease. However, it also raises ethical challenges. Rob Sparrow argues that this iterative selection over many cellular generations is reminiscent of the motivations of twentieth-century eugenics programs, effectively turning humans into the object of radical genetic experimentation.[36]

In sum, by eliminating the constraint on oocyte availability, IVG not only boosts fertility treatment success but also opens the door to a degree of genetic selection—and even iterative enhancement—unimaginable under conventional IVF. Whether these capabilities are used responsibly or otherwise will depend on social values, regulatory frameworks, and ongoing ethical debate.

[36] Sparrow, 'Eugenics under the Microscope' (Chapter 5, this volume).

Constraint 4: Sex Is Limited to Two Organisms

Under natural conditions, sexual reproduction involves exactly two genetic contributors—one egg and one sperm fuse to form a zygote. This two-parent limit has been a defining feature of sexual species throughout their evolutionary history.

Recent biotechnological advances, however, have raised the possibility of creating children with multiple genetic progenitors.[37] For instance, mitochondrial replacement therapy substitutes the mitochondrial DNA a mother would pass on to their children with those of a donor, resulting in a child who inherits DNA from three individuals. Some scholars argue that this might complicate the child's sense of identity,[38] with others noting that mitochondrial DNA represents only a tiny fraction of the overall genome.

IVG makes the creation of children who inherit nuclear DNA from more three or more people possible. This has been given the moniker 'multiplex parenting'.[39] For example, imagine that four people in a relationship want to parent a child while all being genetically related to them. IVG would enable the following scenario: first, two embryos would be created from each couple through IVF using either naturally or *in vitro*-generated gametes. Cells from both embryos would then be used to generate new gametes, which would be combined in a second round of IVF. The resulting embryo would be genetically related to all four prospective parents, who would technically be the child's genetic grandparents.

Imagine that individuals A, B, and C want to have a child genetically related to all three of them. Using iterative embryo selection, embryos could be created that share DNA equally among the three individuals. Initially, genetic material from two people would be combined into an embryo, and gametes derived from this embryo would then be combined with genetic material from the third person. The resulting embryo would have approximately 25% DNA from individuals A and B, and 50% from individual C. Through additional rounds of this process, embryos could be produced that share genetic material more equally among all three individuals, eventually resulting in a child genetically related to each person.

While multiplex parenting may sound radical, César Palacios-González (Chapter 7) shows how it mimics naturally occurring processes like tetragametic

[37] For discussion of why to use the term 'progenitors' and not 'parent' here, see Bowman-Smart, '*In Vitro* Gametogenesis and the Concept of Genetic Parenthood' (Chapter 9, this volume).

[38] Rebecca Dimond, 'Social and Ethical Issues in Mitochondrial Donation', *British Medical Bulletin*, 115/1 (September 2015), 173–182, https://doi.org/10.1093/bmb/ldv037.

[39] César Palacios-González, John Harris, and Giuseppe Testa, 'Multiplex Parenting: IVG and the Generations to Come', *Journal of Medical Ethics*, 40/11 (November 2014), 752–758, https://doi.org/10.1136/medethics-2013-101810.

chimerism. He challenges the idea that there is something intrinsically problematic about multiplex parenting. There are many reasons why people might be motivated to pursue multiplex parenting. They could empower polyamorous families seeking to share equal genetic ties with their children or serve broader social and political aims to bind communities together genetically.

Nevertheless, multiplex parenting introduces novel ethical and practical questions. How should parental rights and responsibilities be legally recognized and allocated? How would family law frameworks adapt to these novel forms of parenthood? Additionally, ethical considerations arise regarding the status of intermediate embryos used in the process. Helen Watt (Chapter 6) argues that using embryos in this manner demonstrates an unethical 'manufacturer's mindset', treating embryos as mere objects rather than entities deserving respect. This view contrasts sharply with positions denying embryos any moral status.

These complex issues demand further philosophical, ethical, and social exploration to ensure that multiplex parenting aligns with both individual rights and broader societal values.

Conclusion

IVG represents a profound shift in reproductive possibilities, effectively circumventing evolutionary constraints that have historically defined human reproduction. By overcoming the limitations of needing an opposite-sex partner, functional reproductive systems, finite egg supplies, and restricting genetic parenthood to two individuals, IVG can significantly expand reproductive autonomy and reshape societal concepts of family and parenthood. However, with these new reproductive frontiers come substantial ethical challenges, including genetic health risks, psychological and identity implications for offspring, potential commodification of embryos, and societal consequences around parenting norms and family structures. Indeed, even the categories of 'gamete' and 'reproduction' may require re-examination in the IVG era.[40] While the benefits of IVG—such as increased reproductive equality, autonomy, and improved health outcomes—are compelling, careful ethical scrutiny is essential. A nuanced, case-by-case approach to regulating IVG is necessary, balancing innovation with responsibility to ensure technology serves human flourishing without undermining ethical standards.

[40] Smajdor, 'Artificial Gametes and the Deconstruction of Reproduction' (Chapter 10, this volume).

Bibliography

Accoe, D., and Segers, S., '"False Hope" in Assisted Reproduction: The Normative Significance of the External Outlook and Moral Negotiation', *Journal of Medical Ethics*, 50/3 (2024), 181–184, https://doi.org/10.1136/jme-2023-108916.

Baylis, F., '"No" to Lesbian Motherhood Using Human Nuclear Genome Transfer', *Journal of Medical Ethics*, 44/12 (2018), 865–867, https://doi.org/10.1136/medethics-2018-104860.

Bell, E. A., Boehnke, P., Harrison, T. M., and Mao, W. L., 'Potentially Biogenic Carbon Preserved in a 4.1 Billion-Year-Old Zircon', *Proceedings of the National Academy of Sciences*, 112/47 (2015), 14518–14521, https://doi.org/10.1073/pnas.1517557112.

Bennett, B., 'Posthumous Reproduction and the Meanings of Autonomy', *Melbourne University Law Review* (1999), https://www.austlii.edu.au/cgi-bin/viewdoc/au/journals/MULR/1999/13.html.

Bittles, A. H., and Black, M. L., 'Consanguinity, Human Evolution, and Complex Diseases', *Proceedings of the National Academy of Sciences*, 107/suppl. 1 (2010), 1779–1786, https://doi.org/10.1073/pnas.0906079106.

Bramlett, M. D., and Blumberg, S. J., 'Family Structure and Children's Physical and Mental Health', *Health Affairs*, 26/2 (2007), 549–558, https://doi.org/10.1377/hlthaff.26.2.549.

Butterfield, N. J., '*Bangiomorpha pubescens* n. gen., n. sp.: Implications for the Evolution of Sex, Multicellularity, and the Mesoproterozoic/Neoproterozoic Radiation of Eukaryotes', *Paleobiology*, 26/3 (2000), 386–404, https://doi.org/10.1666/0094-8373(2000)026<0386:BPNGNS>2.0.CO;2.

Cutas, D., and Smajdor, A., '"I Am Your Mother and Your Father!": In Vitro Derived Gametes and the Ethics of Solo Reproduction', *Health Care Analysis*, 25/4 (2017), 354–369, https://doi.org/10.1007/s10728-016-0321-7.

Cutas, D., and Smajdor, A., 'Postmenopausal Motherhood Reloaded: Advanced Age and In Vitro Derived Gametes', *Hypatia*, 30/2 (2015), 386–402, https://doi.org/10.1111/hypa.12151.

Davis, S. R., Pinkerton, J., Santoro, N., and Simoncini, T., 'Menopause—Biology, Consequences, Supportive Care, and Therapeutic Options', *Cell*, 186/19 (2023), 4038–4058, https://doi.org/10.1016/j.cell.2023.08.016.

Feigel, A., Englander, A., and Engel, A., 'Sex Is Always Well Worth Its Two-Fold Cost', *PLoS One*, 4/7 (2009), e6012, https://doi.org/10.1371/journal.pone.0006012.

Greely, H. T., *CRISPR People: The Science and Ethics of Editing Humans* (Cambridge, MA: MIT Press, 2021).

Kovac, J. R., Moustakli, E., Zikopoulos, A., Georgiou, I., Dimitriadis. F., Symeonidis, E., et al., 'The Effects of Advanced Paternal Age on Fertility', *Asian Journal of Andrology*, 15/6 (2013), 723–728, https://doi.org/10.1038/aja.2013.92.

Le Goff, A., Jeffries Hein, R., Hart, A.N., Roberson, I., and Landecker, H. L., 'Anticipating In Vitro Gametogenesis: Hopes and Concerns for IVG among Diverse Stakeholders', *Stem Cell Reports*, 19/7 (2024), 933–945, https://doi.org/10.1016/j.stemcr.2024.05.002.

Murakami, K., Hamazaki, N., Hamada, N., Nagamatsu, G., Okamoto, I., Ohta, H., et al., 'Generation of Functional Oocytes from Male Mice In Vitro', *Nature*, 615/7954 (2023), 900–906, https://doi.org/10.1038/s41586-023-05834-x.

Murphy, T. F., 'Pathways to Genetic Parenthood for Same-Sex Couples', *Journal of Medical Ethics*, 44/12 (2018), 823–824, https://doi.org/10.1136/medethics-2017-104291.

Notini, L., Gyngell, C., and Savulescu, J., 'Drawing the Line on In Vitro Gametogenesis', *Bioethics*, 34/1 (2020), 123–134, https://doi.org/10.1111/bioe.12679.

Polyakov, A., Amor, D. J., Savulescu, J., Gyngell, C., Georgiou, E. X., Ross, V., et al., 'Polygenic Risk Score for Embryo Selection—Not Ready for Prime Time', *Human Reproduction*, 37/10 (2022), 2229–2236, https://doi.org/10.1093/humrep/deac159.

Russell, C., 'Rights-Holders or Refugees? Do Gay Men Need Reproductive Justice?', *Reproductive Biomedicine & Society Online*, 7 (2018), 131–140, https://doi.org/10.1016/j.rbms.2018.07.001.

Saitou, M., and Hayashi, K., 'Mammalian In Vitro Gametogenesis', *Science*, 374/6563 (2021), eaaz6830, https://doi.org/10.1126/science.aaz6830.

Savulescu, J., 'Procreative Beneficence: Why We Should Select the Best Children', *Bioethics*, 15/5–6 (2001), 413–426.

Segers, S., 'The Path toward Ectogenesis: Looking beyond the Technical Challenges', *BMC Medical Ethics*, 22/1 (2021), 59, https://doi.org/10.1186/s12910-021-00630-6.

Sparrow, R., 'Yesterday's Child: How Gene Editing for Enhancement Will Produce Obsolescence—and Why It Matters', *The American Journal of Bioethics*, 19/7 (2019), 6–15, https://doi.org/10.1080/15265161.2019.1618943.

Suter, S. M., 'In vitro gametogenesis: just another way to have a baby?', *Journal of Law and the Biosciences*, 3/1, 2015, pp. 87–119, doi:10.1093/jlb/lsv057.

Tremellen, K., and Savulescu, J., 'A Discussion Supporting Presumed Consent for Posthumous Sperm Procurement and Conception', *Reproductive Biomedicine Online*, 30/1 (2015), 6–13, https://doi.org/10.1016/j.rbmo.2014.10.001.

Visscher, P. M., Gyngell, C., Yengo, L., and Savulescu, J., 'Heritable Polygenic Editing: The Next Frontier in Genomic Medicine?', *Nature*, 637/8046 (2025), 637–645, https://doi.org/10.1038/s41586-024-08300-4.

Vizcarrondo, F. E., 'Human Enhancement: The New Eugenics', *The Linacre Quarterly*, 81/3 (2014), 239–243, https://doi.org/10.1179/2050854914Y.0000000021.

Yamashiro, C., Sasaki, K., Yabuta, Y., Kojima, Y., Nakamura, T., Okamoto, I., et al., 'Generation of Human Oogonia from Induced Pluripotent Stem Cells In Vitro', *Science*, 362/6412 (2018), 356–360, https://doi.org/10.1126/science.aat1674.

Zarkasi, K. A., Abdullah, N., Abdul Murad, N. A., Ahmad, N., and Jamal, R., 'Genetic Factors for Coronary Heart Disease and Their Mechanisms: A Meta-analysis and Comprehensive Review of Common Variants from Genome-wide Association Studies', *Diagnostics*, 12/10 (2022), 2561, https://doi.org/10.3390/diagnostics12102561.

3

IVG and Relational Procreative Autonomy

Sonia M. Suter

Introduction

I have argued in previous work that IVG is 'just another way to have a baby'.[1] Even so, it has the potential to expand those possibilities in new ways if it could help same-sex couples and large groups of individuals reproduce together, and individuals reproduce alone. It might also refine reproductive selection. This chapter examines these uses of IVG through a lens of relational procreative autonomy. On first glance, these reproductive uses of IVG would seem only to expand procreative autonomy. I argue, however, that in some contexts they undermine and even restrict procreative autonomy as understood relationally. In some cases, context or motivations could thwart relational procreative autonomy. But if IVG emerges as a clinical reality in a legal, political, and cultural climate like the one we currently inhabit, we could see significant threats to procreative autonomy not just with respect to IVG, but more generally.

This chapter begins by explaining the various reproductive uses of IVG and then develops the relational procreative framework that shapes my analysis. It concludes by demonstrating the complex ways in which IVG can both promote and hinder relational procreative autonomy.

Potential Clinical Uses of IVG

The primary clinical aim of IVG is to overcome physical or social barriers to reproduction.[2] If research advances as some predict it might, IVG could derive gametes *in vitro* from the adult cells of individuals.[3] In some cases, it could

[1] Sonia M. Suter, 'In Vitro Gametogenesis: Just Another Way to Have a Baby?', *Journal of Law and the Biosciences*, 3/1 (2016), 87.

[2] National Academies of Sciences, Engineering, and Medicine, *In Vitro–Derived Human Gametes as a Reproductive Technology: Scientific, Ethical, and Regulatory Implications: Proceedings of a Workshop* (Washington: The National Academies Press, 2023).

[3] Suter, 'In Vitro Gametogenesis: Just Another Way to Have a Baby?'.

Sonia M. Suter, *IVG and Relational Procreative Autonomy* In: *Revolutionizing Reproduction*. Edited by: Christopher Gyngell, Julian Savulescu, and Tsutomu Sawai, Oxford University Press.
DOI: 10.1093/9780197810415.003.0003

overcome physical infertility.[4] IVG could also help people, whether or not they are physically infertile, to have genetically related children when their social circumstances would prevent that. Thus, it could allow opposite-sex couples (when one member of the couple is physically infertile) or same-sex couples to have children genetically related to both partners.[5]

In addition, IVG could overcome more unusual types of social infertility. Individuals who desire to reproduce alone, for example, could use solo IVG to create gametes derived only from themselves, obviating the need for donor gametes.[6] And if more than two people want to reproduce as a group, multiplex IVG could make that possible.[7]

Finally, IVG could theoretically enhance reproductive selection intended to reduce disease risk or to select for (or against) certain traits by making preimplantation genetic testing (PGT) more potent. First, it would make *in vitro* fertilization (IVF) less burdensome by eliminating the need for egg retrieval, which is invasive, physically onerous, and not without risks.[8] Second, IVG could theoretically create an unlimited number of eggs, overcoming the biological limitations as to how many eggs can be retrieved and fertilized and potentially leading to vast supplies of embryos for PGT. The more embryos available for screening, the higher the odds of identifying those with 'desirable' genetic predispositions.[9] If polygenic risk screening develops to the point of providing meaningful genomic profiles of embryos,[10] future parents could obtain

[4] This could include diseases of the reproductive system, 'injury, surgery . . . cancer treatment', menopause, or some forms of sterilization. Suter, 'In Vitro Gametogenesis: Just Another Way to Have a Baby?', 92.

[5] Suter, 'In Vitro Gametogenesis: Just Another Way to Have a Baby?'

[6] A female could, for example, use her cells to produce sperm that would fertilize her naturally or IVG-created ova. And a male could use his cells to produce ova to be fertilized with his naturally or IVG-created sperm. In the latter instance, however, the male would depend on others to gestate the pregnancy, unless ectogenesis was a viable option.

[7] An example of multiplex IVG might involve eight individuals (of any sex). The process would begin with the creation of embryos from four pairs of these individuals, using IVG when necessary. Using IVG again, eggs would be derived from two of the four embryos, and sperm from the other two embryos, which would be used to create two more embryos. Finally, sperm would be derived from one of the two embryos to fertilize the egg that was derived from the other embryo, leading to the final embryo, which would be transferred to a woman's womb and, if all goes well, lead to the birth of child. Each individual would share one-eighth of their genes with the child, making them the genetic equivalent of great grandparents; the four embryos would share one-fourth of their genes with the child, making them the genetic equivalent of grandparents; and the last two embryos would share half of their genes with the child, making them the genetic equivalent of parents. Multiplex IVG effectively compresses generations.

[8] Suter, 'In Vitro Gametogenesis: Just Another Way to Have a Baby?'.

[9] Suter, 'In Vitro Gametogenesis: Just Another Way to Have a Baby?'.

[10] Currently, there is reason to be sceptical about the value of polygenic risk scoring in embryos. Dov Fox et al., 'Choosing Your "Healthiest" Embryo After *Dobbs*: Polygenic Screening and Distinctive Challenges for Truth in Advertising and Informed Consent', *Harvard Journal of Law & Technology*, 38/2 (2024), 471–472. But in the future, our understanding of the relationship between genotype and phenotype may become more sophisticated and nuanced to the point that information about certain genotypes identified through PGT could provide some meaningful information about predispositions to certain traits.

comprehensive probabilistic data for each embryo regarding certain physical and cognitive medical and non-medical traits.[11]

As gene editing improves, it could theoretically be used in combination with IVG to edit precursor cells to gametes.[12] Rather than genetically edit an embryo or an egg or sperm, scientists could edit spermatogonial stem cells or patient-derived pluripotent stem cells from which gametes with 'desirable' genetic profiles could be derived.[13] In theory, those gametes could be used to create embryos with a greater or lower chance of developing particular traits.

An important caveat, of course, is that all of these uses of IVG depend on scientific developments that may or may not ultimately be possible. Despite the hopes of some, IVG may never become clinically viable for reproduction. And we may never be able to discern a child's ultimate phenotype from an embryo's genomic profile in a meaningful way for many traits. Nevertheless, this chapter considers the implications for procreative autonomy *if* the science eventually advances to a point where these uses of IVG become feasible. To make sense of that question, the next section describes my understanding of procreative autonomy, which I define relationally.

A Relational Conception of Procreative Autonomy

Procreative autonomy, of course, refers to autonomy in the reproductive context. However, because I use the term relationally, I begin by explaining the value and richness of a relational conception of autonomy in general. I then explore how it applies to procreative decisions.

Relational Autonomy

The etymological roots of the term 'autonomy' connote self ('autos') rule ('nomos').[14] As a result, autonomy often describes a liberty or 'independence from something or other'.[15] In bioethics, autonomy has become a central

[11] Sonia M. Suter, 'The Tyranny of Choice: Reproductive Selection in the Future,' *Journal of Law and the Biosciences* 5/2 (2018), 262.

[12] Judith Daar, I. Glenn Cohen, Seema Mohapatra, and Sonia M. Suter, *Reproductive Technologies and the Law*, 3rd ed. (Durham: Carolina Academic Press, 2022), 1057.

[13] Of course, this would be limited to traits strongly influenced by identifiable genetic variants.

[14] Gerald Dworkin, *The Theory and Practice of Autonomy* (New York: Cambridge University Press, 1988), 12.

[15] Dworkin, *The Theory and Practice of Autonomy*, 12, 28; Isaiah Berlin, *Four Essays on Liberty* (London: Oxford University Press, 1969), 131 (describing it as being 'a subject, not an object . . . deciding, not being decided for, self-directed and not acted upon by external nature or by other men').

principle to combat abuses in human research and medical paternalism.[16] It underlies the right to refuse unwanted medical treatment and the need for consent before administering treatment.[17] In liberal political theory, autonomy is often conceptualized as a negative right that protects us from interference with the exercise of moral, political, personal, or religious choices.[18]

Although autonomy includes independence from restraints in decision-making, it is 'not a matter of mere, sheer independence', nor is it just the right to make a choice in all instances.[19] As Gerald Dworkin suggests, it 'cannot be identical to liberty'.[20] While '[l]iberty, power, control over important aspects of one's life' are necessary for autonomy, they are not equivalent to autonomy.[21]

For many scholars, autonomy involves and protects the capacity to exercise choice in certain instances as a form of 'self-legislation'.[22] Most suggest that the kinds of choices that manifest true expressions of autonomy derive from self-reflection and an understanding of one's motivations.[23] Onora O'Neill distinguishes '[a]utonomous choices . . . from mere choices', describing the former as choices that 'follow from and reflect a great degree of self-knowledge . . . self-control, or . . . capacities to review, revise and endorse other desires'.[24] Similarly, for many, autonomy involves acting according to higher-order preferences, rather than first-order preferences.[25]

Exercising the competencies necessary for autonomy, like self-reflection and deliberation, is tied to self-definition. In using these capacities to make choices based on our higher-order preferences, one expresses 'one's "true self"', which is defined by our 'genuine desires' or higher-order preferences.[26]

But how do we understand the self at the heart of autonomy? Autonomy in Western modern bioethics tends to be dominated by the individualism fundamental to 'the social imaginary of Western societies'.[27] This view of autonomy sees the deciding agents as 'atomistic or separate, or radically individualistic',[28]

[16] Sonia M. Suter, 'A Brave New World of Designer Babies?', *Berkeley Technology Law Journal*, 22 (2007), 952.

[17] Dworkin, *The Theory and Practice of Autonomy*, 4–5.

[18] Catriona MacKenzie and Natalie Stoljar, 'Introduction: Autonomy Refigured', in *Relational Autonomy: Feminist Perspectives on Autonomy, Agency, and the Social Self*, ed. Catriona MacKenzie and Natalie Stoljar (New York: Oxford University Press, 2000), 5.

[19] Onora O'Neill, *Autonomy and Trust in Bioethics* (Cambridge: Cambridge University Press, 2002), 25–26.

[20] Dworkin, *The Theory and Practice of Autonomy*, 14

[21] Dworkin, *The Theory and Practice of Autonomy*, 18.

[22] MacKenzie and Stoljar, 'Introduction: Autonomy Refigured', 5, 20.

[23] MacKenzie and Stoljar, 'Introduction: Autonomy Refigured', 13.

[24] O'Neill, *Autonomy and Trust in Bioethics*, 33.

[25] Dworkin, *The Theory and Practice of Autonomy*, 15.

[26] Diana Meyers, 'Personal Autonomy and the Paradox of Feminine Socialization', *Journal of Philosophy*, 84/11 (1987), 619–628; Dworkin, *The Theory and Practice of Autonomy*, 15.

[27] MacKenzie and Stoljar, 'Introduction: Autonomy Refigured', 25.

[28] MacKenzie and Stoljar, 'Introduction: Autonomy Refigured', 7.

and it 'underestimate[s] the importance of intimate and personal relationships' by relying 'on a misleading image of the independent self'.[29] It tends to 'suggest that values, social practices, relationships and communities that are based on co-operation and interdependence threaten, or at least compromise, autonomy'.[30]

The Western conception of autonomy ignores the fact that self-definition depends largely on the communities and relationships of which we are a part. MacIntyre writes: 'the story of my life is always embedded in the story of those communities from which I derive my identity'.[31] And Taylor argues that '[o]ne is a self among other selves. A self can never be described without reference to those who surround it'.[32] Donchin asserts that self-definition 'exists fundamentally in relation to others' and is constructed and reconstructed throughout our lives by these 'interconnections'.[33]

A richer, less individualistic account of autonomy therefore considers not just the role of social relationships, but also the role of 'social norms and institutions', 'cultural practices', and 'historical contexts in which agents are embedded' in our definition of ourselves.[34] It conceives of self-determination and self-definition in terms of our 'moral identity in and through its membership in communities such as those of the family, the neighborhood, the city and the tribe'.[35]

A relational understanding of autonomy is consistent with the reality of our lived experiences and suggests that more is at stake than protecting people's right to make decisions in all cases. Instead, it focusses on the ability to make the kinds of choices that are self-defining as understood relationally and that allow the relationships that define us to flourish. It also has implications for the ways in which autonomy can be restricted.

Obviously, overt barriers to one's ability to make choices central to our relational self-definition impede autonomy. But relational autonomy can be impeded by more than just explicit prohibitions. Autonomy can be thwarted when 'illegitimate external influences'[36] or oppressive societal structures institutions impede individuals' ability 'to imagine themselves otherwise'.[37]

[29] Anne Donchin, 'Autonomy and Interdependence: Quandaries in Genetic Decision Making', in *Relational Autonomy: Feminist Perspectives on Autonomy, Agency, and the Social Self*, 238.

[30] MacKenzie and Stoljar, 'Introduction: Autonomy Refigured', 6.

[31] Alasdair MacIntyre, *After Virtue: A Study in Moral Theory*, 3rd ed. (Notre Dame: University of Notre Dame Press, 2007), 221.

[32] Charles Taylor, *Sources of the Self: The Making of the Modern Identity* (Cambridge: Harvard University Press, 1989), 35.

[33] Donchin, 'Autonomy and Interdependence: Quandaries in Genetic Decision Making', 239–240; see also Evelyn Fox Keller, *Reflections on Gender and Science* (New Haven: Yale University Press, 1985), 99.

[34] MacKenzie and Stoljar, 'Introduction: Autonomy Refigured', 22; see also Donchin, 'Autonomy and Interdependence: Quandaries in Genetic Decision Making', 239.

[35] MacIntyre, *After Virtue: A Study in Moral Theory*, 205.

[36] John Christman, 'Autonomy: A Defense of the Split-Level Self', *Southern Journal of Philosophy* 25/3 (1987), 281–293.

[37] MacKenzie and Stoljar, 'Introduction: Autonomy Refigured', 24.

Social norms and institutions can also undermine self-reflection and thereby autonomy.[38] But even when there are no barriers to self-reflection, meaningful autonomy requires external circumstances that actually *allow* for a 'range of *significant* options'.[39]

Relational Procreative Autonomy

Just as self-definition is central to autonomy in general, it is also integral to procreative autonomy, as many scholars[40] and even some Supreme Court opinions have noted. Ronald Dworkin writes that at the heart of the 'right of procreative autonomy' is a Western 'belief . . . that people have the moral right—and moral responsibility—to confront the most fundamental questions about the meaning and value of their own lives for themselves: answering their own conscience and conviction'.[41] John Harris states that procreative freedom is a 'freedom to choose one's own way of life, according to one's own deeply held beliefs'.[42] Similarly, when affirming *Roe v. Wade*'s essential holding that there is a constitutional right to abortion before viability,[43] the Supreme Court's plurality opinion in *Planned Parenthood v. Casey* stated the following: '[A]t the heart of liberty is the right to define one's own concept of existence, of meaning, of the universe, and of the mystery of human life. Beliefs about these matters should not define the attributes of personhood were they formed under compulsion of the State'.[44] These conceptualizations of reproductive autonomy as central to self-expression and self-definition have been understood to protect not just a right to prevent 'unwanted fertility' but also a right to try to remedy 'unwanted infertility'.[45] John Harris argues, for example, that 'it seems invidious to require that people who need assistance with reproduction meet tests to which those who need no such assistance are not subjected'.[46] His view of procreative autonomy would protect a broad range of assisted reproductive technologies (ART), even cloning oneself.[47] John Robertson also espouses a capacious view of procreative autonomy that protects almost all forms of procreation, whether 'achieved coitally or

[38] MacKenzie and Stoljar, 'Introduction: Autonomy Refigured' 22.
[39] MacKenzie and Stoljar, 'Introduction: Autonomy Refigured' 26.
[40] O'Neill, *Autonomy and Trust in Bioethics*, 22, 55.
[41] Ronald Dworkin, *Life's Dominion: An Argument About Abortion, Euthanasia, and Individual Freedom* (New York: Alfred A. Knopf, 1993), 166–167.
[42] John Harris, 'Rights and Reproductive Choices', in *The Future of Human Reproduction: Ethics, Choice and Regulation*, ed. John Harris and Soren Holm (Oxford: Clarendon Press, 1998), 35.
[43] *Roe v. Wade* 410 US. 113 (1973).
[44] *Planned Parenthood of Southeast Pennsylvania v. Casey* 505 U.S. 833, 851 (1992).
[45] O'Neill, *Autonomy and Trust in Bioethics*, 57.
[46] Harris, 'Rights and Reproductive Choices', 7.
[47] Harris, 'Rights and Reproductive Choices', 34–35.

noncoitally'.[48] Robertson's view of procreative autonomy, however, does not include enhancement or cloning because 'they deviate too far from the experiences that make reproduction a valued activity'.[49]

Some have argued that procreative autonomy as self-expression cannot encompass the use of assisted reproductive technologies. Onora O'Neill, for example, distinguishes the *prevention* of fertility from *assistance* in treating infertility because the latter implicates third parties—the children people wish to bring into the world.[50] She argues that procreative autonomy as a form of self-expression fails to consider whether the intended parents seeking ART can fulfil the needs of the future child.[51] Although she does not believe that 'coercive intervention' is appropriate for those using ART, she believes they have 'no extra claim to reproductive assistance, such as others' co-operation in or subsidy for any reproductive plan that may harm a fetus, an infant, or a child'.[52] She also rejects the idea that procreative autonomy includes the use of ART because the demands of parenthood itself will thwart autonomy.[53]

A relational view of procreative autonomy and self-definition, however, addresses both concerns. First, it considers the impact of reproductive decisions on the relationships that define the person, which requires consideration of the broader impacts of one's choices, including whether they will allow those self-defining relationships to flourish. Thus, to describe procreative autonomy as self-defining does not suggest that the implications for the future child are irrelevant.

Second, the notion that parenting itself thwarts autonomy reflects a thin, individualistic understanding of autonomy. In contrast, a relational view of autonomy sees parenthood as self-defining and a form of self-expression, even with its demands. Indeed, those demands are tied to the second-order desires of connecting deeply with others, which is at the heart of relational autonomy, even if (or perhaps because) it involves commitment, expectations, and sometimes even burdens.

Finally, a narrow view of procreative autonomy that is limited to 'freedom from coercion in reproduction' and excludes efforts *to* reproduce[54] is fundamentally inconsistent with reproductive justice, a concept related to relational autonomy.

[48] John A. Robertson, *Children of Choice: Freedom and the New Reproductive Technologies* (Princeton: Princeton University Press, 1994), 39.

[49] Robertson, *Children of Choice: Freedom and the New Reproductive Technologies*, 167.

[50] O'Neill, *Autonomy and Trust in Bioethics*, 61.

[51] O'Neill, *Autonomy and Trust in Bioethics*, 62 ('Reproduction aims to create a dependent being, and reproductive decisions are irresponsible unless those who make them can reasonably offer adequate and lasting care and support to the hoped-for child'. They must be able to 'ensure that each child is born not just to an individual who seeks to express himself or herself, but to person who can reasonably intend and expect to be present and active for the child across many years'.).

[52] O'Neill, *Autonomy and Trust in Bioethics*, 64.

[53] O'Neill, *Autonomy and Trust in Bioethics*, 65–66.

[54] O'Neill, *Autonomy and Trust in Bioethics*, 50–51.

Reproductive justice is 'the human right to maintain personal bodily autonomy, have children, not have children, and parent the children we have in safe and sustainable communities'.[55] Reproductive justice requires more than just access to abortion and contraception; it also requires access to care that promotes fertility, healthy pregnancies, and healthy families.[56] The reproductive justice movement also addresses the ways that social structures can limit access to the full range of reproductive care, especially for women of colour and other marginalized groups of women, which undermines their procreative autonomy.[57] Reproductive justice, in other words, connotes a fulsome and *relational* understanding of procreative autonomy by recognizing that meaningful autonomy requires consideration of how societal structures and relationships can enhance or impede that autonomy.

***

To summarize, my conception of relational procreative autonomy incorporates a reproductive justice framework that rejects the view that procreative autonomy only supports the decisions to prevent, but not also assist, procreation. But it does not sweep as broadly as some more individualistic conceptions of procreative autonomy, which view most uses of ART as central to an individualistic notion of self-definition. While my view of procreative autonomy is linked to self-definition, the relational focus of self-definition means that some procreative decisions will be consistent with and promote relational procreative autonomy, while others may hinder it. I turn now to a consideration of whether the potential uses of IVG enhance or thwart this relational procreative autonomy.

Reproductive Uses of IVG and Relational Procreative Autonomy

I begin with the caveat that my analysis in this section presumes that IVG has been shown to be safe enough for clinical use, which is a large assumption indeed. Establishing IVG's clinical safety is a baseline requirement under virtually any ethical theory, including relational procreative autonomy.[58] In the interests

[55] 'Reproductive Justice', SisterSong, accessed 31 January 2025, https://www.sistersong.net/reproductive-justice [https://perma.cc/WZ7A-QMDB].

[56] That care includes 'contraception, comprehensive sex education, sexually transmitted illness prevention and care, alternative birth options, adequate prenatal and pregnancy care, domestic violence assistance, adequate wages to support [their] families, safe homes, and so much more'. 'Reproductive Justice', SisterSong, accessed 31 January 2025, https://www.sistersong.net/reproductive-justice [https://perma.cc/WZ7A-QMDB].

[57] 'Reproductive Justice', SisterSong, accessed 31 January 2025, https://www.sistersong.net/reproductive-justice [https://perma.cc/WZ7A-QMDB].

[58] Suter, 'In Vitro Gametogenesis: Just Another Way to Have a Baby?'. For respondents in a recent survey about their 'hopes and concerns for IVG', participants 'viewed safety for children born through IVG as the most fundamental requirement for its use'. Anne Le Goff et al., 'Anticipating In

of space, however, I do not grapple with the thorny question of how to establish IVG's safety.

I will also only briefly explore the use of IVG to address physical infertility because it is so similar in principle to other forms of ART, like IVF. By making biological reproduction possible for individuals who are physically unable to produce gametes, IVG promotes self-expression and self-definition with respect to the deeply held value of having genetically related children.[59] This desire is so great that many devote vast resources, energy, and time to pursue this goal,[60] and many suffer great distress when those efforts fail.[61] In a world where parents are usually genetically related to their children, it is not surprising that, for many, their sense of self is partially defined by those relationships. After all, our self-definition is shaped by society and culture.

The focus of this section is on the ways IVG may advance or hinder relational procreative autonomy when used to address social infertility for same-sex couples who want to reproduce together, for individuals who want to reproduce by themselves, and for individuals who want to reproduce as a group. The last section explores the autonomy interests related to using IVG to enhance reproductive selection.

Same-Sex IVG

I begin with same-sex couples, who may or may not have the physical capacity to reproduce with someone of the opposite sex,[62] but who cannot do so with their chosen partner. IVG promotes relational procreative autonomy here just as ART does for opposite-sex couples by allowing them to create genetically related families.[63] Although it is currently biologically impossible for same-sex couples

Vitro Gametogenesis: Hopes and Concerns for IVG Among Diverse Stakeholders', *Stem Cell Reports*, 19 (2024), 937.

[59] Suter, 'In Vitro Gametogenesis: Just Another Way to Have a Baby?'; Saskia Hendricks et al., 'The Importance of Genetic Parenthood for Infertile Men and Women', *Human Reproduction* 39/10 (2017), 2076–2087. However, some studies have shown that although '[i]nfertile patients prefer genetic over non-genetic parenthood', they may prefer treatments that do 'not enable genetic parenthood in return' for other benefits such as 'child health risk reduction of 3.6%, a cost reduction of €3500, an ovarian hyperstimulation risk reduction of 4.6%, a maternal cancer risk reduction of 2.7% or a pregnancy rate increase of 18%'. Saskia Hendricks et al., 'The Relative Importance of Genetic Parenthood', *Reproductive BioMedicine Online*, 103/1 (2019), 103.

[60] See generally National Academy of Sciences, Engineering, and Medicine, *In Vitro Derived Human Gametes as a Reproductive Technology: Scientific, Ethical, and Regulatory Implications* (2023).

[61] Laura Notini, Christopher Gyngell, and Julian Savulescu, 'Drawing the Line on In Vitro Gametogenesis', *Bioethics*, 34 (2019), 127.

[62] Notini, Gyngell, and Savulescu, 'Drawing the Line on In Vitro Gametogenesis'.

[63] See above text accompanying notes 59–61.

to procreate together, this does not mean that many of them do not share the same deep desires to have children genetically related to their partners. There is no reason, therefore, to think this ability would be any less self-defining for same-sex couples than for opposite-sex couples. Indeed, a recent study found that many LGBTQ+ participants saw IVG as providing the same reproductive options heterosexual couples 'naturally have access to', and an expansion of their reproductive options.[64] The capacity to procreate genetically is somewhat analogous to the ability to be legally married in that much of our culture is centred around these rites of passage. As a result, the desire to have the option to participate in these activities may be strong given these cultural norms, in spite of the current biological inability to do so. Just as the legalization of gay marriage promoted relational autonomy by allowing same-sex couples to engage in this relationally self-defining ritual, so too would IVG promote relational autonomy.

IVG would also eliminate the reliance on gamete donors that is often part of family formation for same-sex couples today. As a result, IVG would avoid potential conflicts over the status and role of gamete donors that can threaten procreative and parental autonomy, and potentially even the integrity of the family. IVG also avoids the deeply thorny relational autonomy issues concerning the disclosure of donor identity that looms large in this area.[65]

But in some ways IVG raises relational autonomy concerns in this context. One would expect many same-sex couples to pursue IVG instead of using donor gametes to form families, which would seem to privilege genetic relatedness. Norms may start to shift in ways that undermine social support for and acceptance of socially constructed same-sex families, which is problematic from a relational autonomy perspective.[66]

These shifting norms might also indirectly undermine legal recognition of same-sex families. While bringing a child into existence and creating familial relationships through ART promotes relational procreative autonomy, those relationships are strongest when they are legally recognized and have community support. But today, same-sex families are socially constructed in part (at

[64] Le Goff et al., 'Anticipating In Vitro Gametogenesis: Hopes and Concerns for IVG Among Diverse Stakeholders', 937.

[65] Suter, 'In Vitro Gametogenesis: Just Another Way to Have a Baby?'; Rebecca W. O'Connor, 'Examining the Politics, Ethics, and Legality of Anonymity in Gamete Donation', in *Seminal*, ed. Rene Almeling, Lisa Campo-Engelstein, and Brian T. Nguyen (New York: New York University Press, 2025), 199–202; Naomi Cahn and Sonia M. Suter, 'Developing a Reproductive Justice Approach to Regulating Formal and Informal Sperm Donation', in *Seminal*, ed. Rene Almeling, Lisa Campo-Engelstein, and Brian T. Nguyen (2025), 228–242.

[66] If the parents' main connection with children brought into the world through IVG were based primarily on genetic links as opposed to social bonds, these relational autonomy concerns are enhanced. However, because social bonds usually play an enormous role in relational connections between parents and child, it is unlikely that genetic relatedness will dominate the parent-child connection in IVG. Suter, 'In Vitro Gametogenesis: Just Another Way to Have a Baby?'.

most, only one partner can be genetically related to each child). This makes these families—and especially the relationship between the child and intended parent who is not the child's genetic progenitor—vulnerable to interference and disruption.[67] In the United States, advocacy on behalf of same-sex families has led to model legislation that recognizes parentage based on intent.[68] A growing number of jurisdictions have passed legislation based on or consistent with this model law.[69] Would the adoption of IVG by same-sex families undermine these efforts, legally and socially? And would advocacy on behalf of such families diminish if many same-sex couples turn to IVG to create families? A recent survey found that many of its LGBTQ+ participants worried that IVG could potentially undermine queer families, including 'indirectly undermin[ing] the legal status of a queer parent who does not have a biological link to their child'.[70]

IVF raises similar issues for opposite-sex couples who rely on gamete donors or adoption to create families, but because those families reflect heteronormative norms, the law has been more much more willing to legally recognize those parent-child relationships than in the same-sex context.[71] Thus, if same-sex IVG is widely used, efforts should be made to ensure that socially constructed same-sex families receive the same legal protection as genetically connected families.[72]

Adding to the procreative autonomy concerns in this context are recent legal, political, and cultural shifts. First, *Dobbs v. Jackson Women's Health Organization* overturned a constitutionally recognized right to abortion in the United States.[73] While the Supreme Court claimed to address just abortion,[74] the decision effectively precluded constitutional protections of procreative autonomy writ large through the legal methodology it employed. In short, it relied on the fact that the abortion right is neither expressly referenced in the Constitution nor 'deeply rooted in this Nation's history and tradition' because it was not legally protected in 1868 when the Fourteenth Amendment was ratified.[75] Under this analysis, there is also no constitutional right to technologies like IVF or IVG: neither is

[67] See Douglas NeJaime, 'The Nature of Parenthood,' *Yale Law Journal*, 126/8 (2017), 2260–2249.

[68] Uniform Parentage Act 2017 (U.S.).

[69] See 'Bill List: Parentage Act', Uniform Law Commission, accessed 31 January 2025, https://www.uniformlaws.org/committees/community-home?communitykey=c4f37d2d-4d20-4be0-8256-22dd73af068f.

[70] Le Goff et al., 'Anticipating In Vitro Gametogenesis: Hopes and Concerns for IVG Among Diverse Stakeholders'.

[71] It was not until 2017 that the Uniform Parentage Act recognized the possibility of same-sex parentage. Prior versions revealed the norms of opposite-sex parentage. See Courtney G. Joslin, 'Nurturing Parenthood Through the UPA', *Yale Law Journal Forum*, 127 (2017), 2017–2018.

[72] Suter, 'In Vitro Gametogenesis: Just Another Way to Have a Baby?'.

[73] *Dobbs v. Jackson Women's Health Organization*, 597 U.S. 215 (2022).

[74] *Dobbs v. Jackson Women's Health Organization*, 597 U.S. 215, 290 (2022) (The 'decision concerns the constitutional right to abortion and no other right').

[75] *Dobbs v. Jackson Women's Health Organization*, 597 U.S. 215, 231 (2022) (citing *Washington v. Glucksberg*, 521 U.S. 702, 721 [1997]).

mentioned in the Constitution, and neither was in existence, let alone protected, in 1868.[76] Thus, under this form of constitutional analysis, US legislatures are free to prohibit anyone's access to IVG.

Dobbs' methodology also threatens the Supreme Court's earlier recognition of same-sex couples' constitutional right to marry in *Obergefell v. Hodges.*[77] Notably, that decision focussed in part on the strong interests same-sex couples have in creating families.[78] It is no secret that some of the conservative justices harbour strong animosity towards that decision.[79] *Dobbs* gives them the tools to overturn it, which itself threatens the parental and procreational interests of same-sex families.

Additionally, the success of Donald Trump's 2024 presidential candidacy, with its propaganda about the 'threats' of gender-affirming surgery,[80] reflects a mindset that would likely view IVG for same-sex procreation as threatening or 'unnatural'. Indeed, one of Trump's first executive orders in his second term explicitly rejects the existence of intersex individuals and those with variations in sex characteristics. The order declares that 'It is the policy of the United States to recognize two sexes, male and female. These sexes are not changeable and are grounded in fundamental and incontrovertible reality'.[81] The full reach of this executive order is not yet clear, but it is manifestly part of a larger strategy to combat legal protections for the LGBTQ+ communities.

With the cultural and political winds in many parts of the world blowing against policies that deviate from heteronormative norms, one might expect legislatures in many US states (and other countries) to ban the use of IVG for same-sex couples to procreate. In fact, this use of IVG might be so threatening to these prejudiced segments of society that it inspires a backlash against legal recognition of *any* type of family formation by same-sex couples, including through adoption or use of donor gametes. In fact, an Oklahoma court recently ruled that a sperm donor, and not the intended second mother of a lesbian married couple, was the legal parent of the child conceived using the donor's sperm and the other wife's egg.[82] This is notable given a Supreme Court ruling that found

[76] I. Glenn Cohen, Judith Daar, and Eli Y. Adashi, 'What Overturning Roe v Wade May Mean for Assisted Reproductive Technologies in the US', *JAMA Network*, 328 (2022), 15–16, https://doi.org/10.1001/jama.2022.10163.

[77] *Obergefell v. Hodges*, 576 U.S. 644 (2015).

[78] *Obergefell v. Hodges*, 576 U.S. 644, 667 (2015) (basing the holding in part on the value of marriage in safeguarding the children and families of same-sex couples and the view that marriage 'draws meaning from related rights of childrearing [and] procreation').

[79] Judith Daar et al., *Reproductive Technologies and the Law*, 1101.

[80] Bill Barrow, 'Trump and Vance Make Anti-Transgender Attacks Central to Their Campaign's Closing Argument', *AP*, 1 November 2024, https://apnews.com/article/trump-harris-transgender-politics-61cff97a64fac581ffc5f762be4c57d3.

[81] 'Defending Women from Gender Ideology Extremism and Restoring Biological Truth to the Federal Government', Executive Order 14168, *Federal Register* 90 (20 January 2025) 8615.

[82] Kate Sosin and Sara Luterman, 'An Oklahoma Judge just Transferred a Lesbian Mom's Parental Rights to Her Son's Sperm Donor', *The 19th*, 14 February 2023, https://19thnews.org/2023/02/oklah

it unconstitutional to treat married same-sex couples differently from married opposite-sex couples with respect to presumptions of parentage.[83] If applied more broadly, the Oklahoma court's reasoning could significantly hinder same-sex couples' procreative autonomy despite technological advances that might otherwise seem to expand their procreative autonomy.

Solo IVG

The analysis with respect to solo IVG and multiplex IVG is more complex, with solo IVG being the most problematic. One concern with solo IVG is that it could be rooted in egoism, selfishness, and a belief that no one else can contribute anything of value to the child. If so, this exercise of 'autonomy' is at odds with relational autonomy. However, if the goal were to preserve the integrity of family unit without the potential interference of donors,[84] it would be more consistent with relational autonomy.[85]

Solo IVG, more than any other type of IVG (or ART), is unique in the absence of 'true' genetic parents, that is, people who each share 50% of the child's genes.[86] The future child would have only one plausible genetic parent; there would be no second genetic parent, a relationship that defines many of us.[87] In addition, the relationship with the one putative genetic parent would be novel. All of the child's genes would come from a single genome, although the child would not be a clone because she would not have the same combination of genes as the solo parent: at some loci, she would have the same pair of genes; at others, she would have two copies of only one of the genes. But the child would share more than 50% of this individual's genes. Unlike other forms of procreation, solo IVG would not involve the reshuffling of two genomes that promotes genetic diversity.[88]

Because of these features, solo IVG heightens the genetic risks associated with consanguineous reproduction. The more shared genes between two individuals who procreate together, the greater the chance they carry the same recessive genetic mutation, which presents a 25% risk of having a child homozygous for that

oma-judge-parental-rights-lgbtq-same-sex-marriage/?utm_campaign=19th-social&utm_content=an_oklahoma_judge_has_rev&utm_medium=social&utm_source=twitter.

[83] *Pavan v. Smith*, 582 U.S. 563 (2017).

[84] See above text accompanying note 64.

[85] Suter, 'In Vitro Gametogenesis: Just Another Way to Have a Baby?'.

[86] Robert Sparrow, 'Cloning, Parenthood, and Genetic Relatedness', *Bioethics*, 20 (2006), 308–318.

[87] Suter, 'In Vitro Gametogenesis: Just Another Way to Have a Baby?', 107. Although this raises the question of whether it is worse if a second genetic parent existed whom the child never knew, or worse if a second genetic parent never existed.

[88] Suter, 'In Vitro Gametogenesis: Just Another Way to Have a Baby?'.

mutation. Reproduction between first cousins, for example, increases the risk of congenital anomalies—most of which are autosomal recessive disorders—by 1.7 to 2.8%.[89] Because first cousins share only roughly one-eighth (12.5%) of the same DNA by descent,[90] this means solo IVG multiplies that risk approximately eight-fold given that the genetic source of both gametes that produced the child is the same person. While PGT or other forms of prenatal testing might identify homozygosity for known recessive mutations, that might not be possible if the individual carries unknown recessive disease variants.[91]

The riskiness of this unique form of reproduction therefore raises significant relational autonomy concerns. Indeed, some have argued that solo IVG is never ethically acceptable, 'no matter how laudable the motivation'.[92] I have previously claimed it is 'the least justifiable and the most problematic reproductive use of IVG'.[93] However, not all motivations for solo IVG are contrary to procreative relational autonomy, and it is theoretically possible that we could one day have the capacity to identify all genetic risks in IVG-created embryos. Thus, my ultimate assessment of whether solo IVG is consistent with relational autonomy is contingent on both motivation and the ability to ameliorate the currently significant risk to future children.

Multiplex IVG

The second unusual form of IVG is multiplex IVG—reproduction among three or more people that results in a child genetically related to the group. Although reproduction typically involves only two people,[94] multiplex IVG is more like natural reproduction than solo IVG because it combines unique genomes. But it differs from typical reproduction by effectively compressing generations. Depending on the number of progenitors (which affects the degree of genetic relatedness), the intended 'parents' would be the genetic equivalent of grandparents, great-grandparents, great-great grandparents, and so on.[95]

[89] Hanan Hamamy et al., 'Consanguineous Marriages, Pearls And Perils: Geneva International Consanguinity Workshop Report', *Genetics in Medicine*, 13/9 (2011), 841, 845 https://doi.org/10.1097/GIM.0b013e318217477f ('An increased 2% risk that first cousin couples will bear a child with an autosomal recessive disorder indicates that approximately 8% of these couples have an increased risk of 25% or more, whereas 92% of first cousin couples will not be at increased risk of the birth of an affected child').

[90] Hamamy, 'Consanguineous Marriages: Preconception Consultation in Primary Health Care Settings', 185, 187.

[91] Suter, 'In Vitro Gametogenesis: Just Another Way to Have a Baby?'.

[92] Notini, Gyngell, and Savulescu, 'Drawing the Line on In Vitro Gametogenesis'.

[93] Suter, 'In Vitro Gametogenesis: Just Another Way to Have a Baby?'.

[94] Of course, ART can involve many individuals, gamete donors and gestational carriers.

[95] The degree of genetic connection is inversely related to the number of individuals reproducing together: four progenitors would be the genetic equivalent of biological grandparents, sharing 25% of their DNA with the child; 32 people would be the genetic equivalent of great-great-great grandparents, sharing 3.125% of their DNA with the child. If odd numbers of progenitors were

Reproduction involving many individuals raises questions about their legal relationship to the future child, which implicates relational procreative autonomy. As noted, the relational autonomy interests of ART are not merely to bring a child into existence, but to create a legally recognized family. Yet because the law typically recognizes only two legal parents,[96] multiplex IVG may lead to uncertainty and potential conflict over parentage. While this is not a new issue with ART,[97] it would be heightened here. In addition, the large number of 'parents' could potentially minimize the relational connections between parents and child, which might become inversely thinner as the size of the parenting groups increase. With so many parents, conflict over parental roles is also a risk.[98]

On the other hand, with a 'village' to raise the child, particularly if the goal is to do so collaboratively, the relational connections could be enhanced.[99] As with all of the IVG procreative scenarios, intentions matter. If thirty-two individuals were to use IVG to create an uber-race, multiplex reproduction is problematic.[100] But multiplex reproduction is more likely to be motivated by a genuine joint commitment to parent together and create a family, which is consistent with relational autonomy.[101] While the 'notion of the couple (and coupling) currently dominates sentimental relationships worldwide', as some scholars note, 'several societies still practice or tolerate other bonding models through various forms of polygamy'.[102] It is not clear, then, why the dominant cultural norm should define the kind of bonding that supports relational autonomy. Furthermore, multiplex parenting might also provide cultural support for a diversity of family forms by helping to normalize deviations from traditional, nuclear families. Courtney Cahill argues, for example, that IVG has the potential to challenge norms about what kind of procreation is granted supremacy,[103] and therefore what kinds of families will be recognized and valued.

Additional Threats to Procreative Autonomy

Finally, reproductive uses of IVG that differ from 'typical' reproduction, like solo and multiplex IVG, pose threats to procreative autonomy in the backlash

involved, the math is more complex. With three progenitors, for example, two would be 25% related and the third, 50% related to the resulting child.

96 Suter, 'In Vitro Gametogenesis: Just Another Way to Have a Baby?'.
97 Suter, 'In Vitro Gametogenesis: Just Another Way to Have a Baby?'.
98 Suter, 'In Vitro Gametogenesis: Just Another Way to Have a Baby?'.
99 Suter, 'In Vitro Gametogenesis: Just Another Way to Have a Baby?'.
100 Suter, 'In Vitro Gametogenesis: Just Another Way to Have a Baby?'.
101 Suter, 'In Vitro Gametogenesis: Just Another Way to Have a Baby?'.
102 César Palacios-González, John Harris, Giuseppe Testa, 'Multiplex Parenting: IVG and the Generations to Come', *Journal of Medical Ethics* 40/11 (2014), 752–757.
103 Courtney Cahill, 'After Sex', *Nebraska Law Review* 97/1, 1–64, 16.

they could instigate. Just as unease about 'unnatural' procreation could motivate legislatures to prohibit same-sex uses of IVG, so too could it inspire laws prohibiting solo and multiplex IVG, and possibly *any* use of IVG. In fact, in pushing the limits of ART, IVG could lead to restrictions of most forms of ART, even IVF.

There are many signals that the appetite to limit ART is growing in the United States. The Alabama Supreme Court ruled two years ago that IVF-created embryos are persons for purposes of wrongful death claims.[104] As a result of the ruling, some fertility clinics paused their activities for fear of legal liability.[105] The Alabama legislature quickly enacted an IVF immunity statute, but it failed to address the personhood status of embryos. As a result, the law does not necessarily immunize providers or even future parents from liability for deliberate destruction of excess embryos.[106] In addition, repeated efforts to protect a nationwide right to IVF have failed.[107] At the same time, the *Dobbs* decision provides a green light to legal personhood for embryos and foetuses,[108] which may motivate growing restrictions on ART. The impetus to regulate in this space may also be driven by cultural pressures to return to 'traditional' family values. We see those views among many voters who helped Trump earn the popular vote in the presidential election.[109] Regulation rooted in these perspectives might well preclude solo IVG, multiplex IVG and, same-sex IVG.

Even without outright bans on reproductive uses of IVG, procreative autonomy is an empty concept if people cannot afford the technology. IVG will likely be expensive, just as IVF is. As commercial entities develop and market IVG,[110] profit motives may enhance that possibility. While legislation mandating insurance coverage for IVF is slowly expanding in the United States, even now,

[104] LePage v. Center for Reproductive Medicine, 2024 WL 656591 (2024).

[105] Aria Bendix, 'Three Alabama Clinics Pause IVF Services after Court Rules that Embryos Are Children', NBC News, 21 February 2024, https://www.nbcnews.com/health/health-news/university-alabama-pauses-ivf-services-court-rules-embryos-are-childre-rcna139846.

[106] Sonia Suter and Naomi Cahn, 'Lawmakers Rushed to Protect IVF in Alabama-Why Wasn't Abortion Afforded the Same Urgency?', *BMJ* (2024), 384.

[107] Sarah Owermohle, 'Senate IVF bill Fails Again as Democrats Pressure GOP on Reproductive Health', STAT, 17 September 2024, https://www.statnews.com/2024/09/17/democrats-senate-ivf-treatment-bill-fails-republicans/.

[108] Not only did the court repeatedly reference the 'unborn child', but it also declared that the state has a legitimate interest in 'respect for and preservation of prenatal life at *all* stages of development', *Dobbs v. Jackson Women's Health Organization*, 597 U.S. 215, 301 (2022) (emphasis added), which certainly would include embryos, in and outside the womb.

[109] Pew Research Center, 'Cultural Issues and the 2024 Election', 6 June 2024, https://www.pewresearch.org/politics/2024/06/06/cultural-issues-and-the-2024-election/ (finding wide gaps between republicans and democrats regarding gender identity, same-sex marriage, the importance of family and children, the value of women taking their husband's name, and whether women's gains in society have come at men's expense).

[110] Emily Witt, 'The Future of Fertility: A New Crop of Biotech Startups Want to Revolutionize Human Reproduction', *New Yorker*, 17 April 2023, https://www.newyorker.com/magazine/2023/04/24/the-future-of-fertility.

only a minority of jurisdictions have such mandates.[111] And some of those that do limit the protection to married, heterosexual couples.[112] These restrictions undermine relational autonomy by limiting the ability to exercise procreative autonomy to those with the greatest means, creating inequities for marginalized groups who already experience the kind of social oppression that can undermine autonomy more generally.

Enhancing Reproductive Selection

Finally, I turn to IVG to enhance reproductive selection. On first glance, it might seem to promote procreative autonomy by maximizing reproductive options, whether to prevent disease in a future child, or, if technologically possible, to select for or against traits that would help the future child be 'better than' average in some respect or enhance the future child beyond what is normal or 'natural'.

An individualistic notion of procreative autonomy might allow for all of these forms of selection, although even John Robertson concluded that enhancement à la *Bladerunner* is not protected by procreative autonomy.[113] John Harris' concern about drawing 'invidious' distinctions between people who reproduce using fertility assistance and those who do without it, however, does not directly address reproductive selection. After all, reproductive selection is not limited to those who need assistance with fertility. IVG in this context is more about controlling the results of procreation than truly *assisting* procreation.

As I have argued, the motivations can be more or less consistent with relational autonomy.[114] If IVG is used here to protect the best interest of the child and one's family, it aligns with relational autonomy because it promotes self-definition understood in relation to one's family. It is also consistent with higher-order preferences of promoting the flourishing of self-defining relationships.

But if the motivations are rooted in prejudice or a commodifying view of the child as something to 'perfect', rather than a gift to be accepted, they undermine relational autonomy. Prejudice divides the community by which one is defined. Moreover, these perspectives promote a fragmented conception of the child that weakens the familial relationship; failing to see the child in the fullness of her humanity undercuts the flourishing of all involved.[115]

[111] Resolve, 'Insurance Coverage by State', 30 September 2024, https://resolve.org/learn/financial-resources-for-family-building/insurance-coverage/insurance-coverage-by-state/.

[112] For example, Hawaii's coverage only applies to heterosexual married couples. Haw. Rev. Stat. § 431:10A-116.5 (2013).

[113] Robertson, *Children of Choice: Freedom and the New Reproductive Technologies*, 167.

[114] Suter, 'A Brave New World of Designer Babies?', *Berkeley Technology Law Journal* 22, (2007), 897–969.

[115] Suter, 'A Brave New World of Designer Babies?'.

Context also matters. If only the privileged few can access this technology to select for beneficial traits, their children, who presumably already benefit from extensive familial resources, will be further advantaged. IVG, like many other benefit-accruing technologies that are not equally accessible, could create a vicious cycle that perpetuates and enhances inequity.[116]

Additionally, if this use of IVG became routinized, it could potentially shift the motivations underlying reproductive selection. That some use the technology to enhance their children's advantages might pressure others to use it similarly, so their children are not left behind. As more people are driven by these goals, the social pressures could intensify and influence first-order and perhaps even higher-order desires, thereby undermining procreative autonomy. This spiralling may also reinforce discrimination against those with 'undesirable' traits or who lack 'desirable' traits, further disrupting relationships within the community and weakening relational autonomy.[117]

The implementation of such reproductive selection poses additional threats to procreative autonomy. The technology may be so expensive that some people have no true capacity to exercise their procreative autonomy for the uses of reproductive selection that are consistent with relational autonomy. Because there is limited insurance coverage of IVF, and even less for PGT,[118] it seems unlikely that insurance would overcome this barrier, leaving, autonomy illusory for many.

Even those with the financial ability to access IVG for embryo selection may find it difficult to exercise choice consistent with their values because of the 'tyranny of choice'.[119] If it becomes possible to create large numbers of embryos and test them for genetic variants that influence health and non-medical traits, autonomy might be undermined in several ways. First, the dizzying amount of information could be too much for many people to comprehend and absorb.[120] Second, even if they could process the information, the actual decision-making could be overwhelming or paralyzing.[121] Third, social pressures of the sort I described above might impede people's capacity to focus on choices consistent with their higher-order preferences.[122]

Previously, I examined possible remedies to those challenges, including algorithms that could ostensibly assist future parents in making choices. Algorithms based on people's expressed preferences might be most helpful

[116] Suter, 'In Vitro Gametogenesis: Just Another Way to Have a Baby?'.

[117] Suter, 'In Vitro Gametogenesis: Just Another Way to Have a Baby?'.

[118] Ruben Alvero and Jamie Grifo, 'Costs of PGT-A', *Fertility IQ*, https://www.fertilityiq.com/fertilityiq/pgs-embryo-genetic-screening/costs-of-pgs#what-does-pgs-cost.

[119] Sonia M. Suter, 'The Tyranny of Choice: Reproductive Selection in the Future', *Journal of Law and the Biosciences* 5/2 (2018), 266.

[120] Suter, 'The Tyranny of Choice: Reproductive Selection in the Future'.

[121] Suter, 'The Tyranny of Choice: Reproductive Selection in the Future'.

[122] Suter, 'The Tyranny of Choice: Reproductive Selection in the Future'.

in trying to assist them in making choices consistent with their higher-order preferences. But this refined approach to selection could heighten expectations regarding the future child and lead to the 'choice overload effect', which often results in dashed expectations.[123] Marketing might intensify this risk by further heightening parental expectations.[124] All of this exacerbates the relational autonomy concerns of commodifying reproduction, failing to see the child as a whole, and reinforcing underlying inequities.

Algorithms based on generic preferences raise even greater threats to autonomy. In one sense, the decision to let someone else decide (like Odysseus' request to be tied to the mast) is an exercise of autonomy. But in another sense, the biases that shape the development of the algorithms could lead to choices that conflict with people's higher-order preferences. And because biases in algorithms are often hidden,[125] people may not understand the ways in which the choices do not reflect their true preferences, further thwarting autonomy.[126]

Finally, any concerns about embryo destruction associated with IVF and PGT would be magnified multi-fold with IVG-reproductive selection because its success depends on the creation of large numbers of embryos. In today's climate, one could easily imagine many jurisdictions banning the procedure based on concerns about enormous amounts of embryo destruction. Even if the procedure itself weren't banned, legislatures might ban the destruction of embryos, which would effectively make this form of reproductive selection impossible, highly inefficient, or exceedingly expensive if people had no control over disposition of the extra embryos.

Conclusion

As this chapter argues, the potential emergence of IVG as a new reproductive technology could expand procreative options to allow for new kinds of family formations. It could also offer greater control over the procreative decisions we make. The mere expansion of choice, however, is insufficient to conclude that IVG will necessarily promote procreative autonomy, at least as understood relationally. The determination of whether IVG will, in fact, enhance such autonomy depends on numerous factors that are shaped not only by the motivations of those want to use this technology, but also by societal attitudes and norms as well as legal principles. If IVG becomes possible in a cultural climate like the one

123 Suter, 'The Tyranny of Choice: Reproductive Selection in the Future'.
124 Fox et al., 'Choosing Your "Healthiest" Embryo After *Dobbs*'.
125 Suter, 'The Tyranny of Choice: Reproductive Selection in the Future'.
126 Suter, 'The Tyranny of Choice: Reproductive Selection in the Future'.

we are currently experiencing, there is reason to worry that IVG could undercut procreative autonomy with respect to not only IVG but also other forms of ARTs.

Bibliography

Alvero, R., and Grifo, J., 'Costs of PGT-A', *Fertility IQ*, https://www.fertilityiq.com/fertilityiq/pgs-embryo-genetic-screening/costs-of-pgs#what-does-pgs-cost.

Bendix, A., 'Three Alabama Clinics Pause IVF Services after Court Rules that Embryos Are Children', NBC News, 21 February 2024, https://www.nbcnews.com/health/health-news/university-alabama-pauses-ivf-services-court-rules-embryos-are-childre-rcna139846.

Berlin, I., *Four Essays on Liberty* (London: Oxford University Press, 1969).

Barrow, B., 'Trump and Vance Make Anti-transgender Attacks Central to Their Campaign's Closing Argument', *Associated Press*, 1 November 2024, https://apnews.com/article/trump-harris-transgender-politics-61cff97a64fac581ffc5f762be4c57d3.

Cahill, C., 'After Sex', *Nebraska Law Review*, 97/1 (2018), 1–64.

Cahn, N., and Suter, S. M., 'Developing a Reproductive Justice Approach to Regulating Formal and Informal Sperm Donation', in R. Almeling, L. Campo-Engelstein and B. T. Nguyen, eds., *Seminal* (New York: New York University Press, 2025), 228–242.

Christman, J., 'Autonomy: A Defence of the Split-Level Self', *Southern Journal of Philosophy*, 25/3 (1987), 281–293.

Cohen, I. G., Daar, J., and Adashi, E. Y., 'What Overturning *Roe v Wade* May Mean for Assisted Reproductive Technologies in the US', *JAMA*, 328 (2022), 15–16, https://doi.org/10.1001/jama.2022.10163.

Cohen, I. G., Daley, G. Q., and Adashi, E. Y., 'Disruptive Reproductive Technologies', *Science Translational Medicine*, 9 (2017), eaag2959.

Daar, J., Cohen, I. G., Mohapatra, S., and Suter, S. M., *Reproductive Technologies and the Law*, 3rd ed. (Durham: Carolina Academic Press, 2022).

'Defending Women from Gender Ideology Extremism and Restoring Biological Truth to the Federal Government', Executive Order 14168, *Federal Register*, 90 (20 January 2025), 8615.

Dobbs v. Jackson Women's Health Organization, 597 U.S. 215 (2022).

Dworkin, R., *Life's Dominion: An Argument about Abortion, Euthanasia, and Individual Freedom* (New York: Knopf, 1993).

Dworkin, G., *The Theory and Practice of Autonomy* (New York: Cambridge University Press, 1988).

Fox, D., Suter, S. M., Mukherjee, M., Pereira, S., and Lázaro-Muñoz, G., 'Choosing Your "Healthiest" Embryo After *Dobbs*: Polygenic Screening and Distinctive Challenges for Truth in Advertising and Informed Consent', *Harvard Journal of Law & Technology*, 38 (forthcoming), 1–21.

Hamamy, H., 'Consanguineous Marriages: Preconception Consultation in Primary Health Care Settings', *Journal of Community Genetics*, 3 (2012), 185, https://doi.org/10.1007/s12687-011-0072-y.

Hamamy, H., Antonarakis, S. E., Cavalli-Sforza, L. L., Temtamy, S., Romeo, G., Kate, L. P., et al., 'Consanguineous Marriages, Pearls and Perils: Geneva International Consanguinity Workshop Report', *Genetics in Medicine*, 13/9 (2011), 841, https://doi.org/10.1097/GIM.0b013e318217477f.

Harris, J., 'Rights and Reproductive Choices', in J. Harris and S. Holm, eds., *The Future of Human Reproduction: Ethics, Choice and Regulation* (Oxford: Clarendon Press, 1998), 5–37.

Haw. Rev. Stat. § 431:10A-116.5 (2013)A.

Hendriks, S., Peeraer, K., Bos, H., Repping, S., and Dancet, E. A. F., 'The Importance of Genetic Parenthood for Infertile Men and Women', *Human Reproduction*, 39/10 (2017), 2076–2087.

Hendriks, S., van Wely, M., D'Hooghe, T. M., Meissner, A., Mol, F., Peeraer, K., et al., 'The Relative Importance of Genetic Parenthood', *Reproductive Biomedicine Online*, 103/1 (2019), 103–110.

Joslin, C. G., 'Nurturing Parenthood through the UPA', *Yale Law Journal Forum*, 127 (2017), 2017–2018.

Horer, S., Feichtinger, M., Rosner, M., and Hengstschläger, M., 'Pluripotent Stem Cell-Derived In Vitro Gametogenesis and Synthetic Embryos—It Is Never Too Early for an Ethical Debate', *Stem Cells Translational Medicine*, 12/9 (2023), 569–575, https://doi.org/10.1093/stcltm/szad042.

Keller, E. F., *Reflections on Gender and Science* (New Haven: Yale University Press, 1985).

LePage v. Center for Reproductive Medicine, 2024 WL 656591 (2024).

Le Goff, A., Jeffries Hein, R., Hart, A. N., Roberson, I., and Landecker, H. L., 'Anticipating In Vitro Gametogenesis: Hopes and Concerns for IVG among Diverse Stakeholders', *Stem Cell Reports*, 19 (2024), 933–945.

MacIntyre, A., *After Virtue: A Study in Moral Theory*, 3rd ed. (Notre Dame: University of Notre Dame Press, 2007).

MacKenzie, C., and Stoljar, N., eds., *Relational Autonomy: Feminist Perspectives on Autonomy, Agency, and the Social Self* (New York: Oxford University Press, 2000).

Meyers, D., 'Personal Autonomy and the Paradox of Feminine Socialisation', *Journal of Philosophy*, 84/11 (1987), 619–628.

National Academy of Sciences, Engineering, and Medicine, *In Vitro Derived Human Gametes as a Reproductive Technology: Scientific, Ethical, and Regulatory Implications* (Washington, DC: National Academies Press, 2023).

NeJaime, D., 'The Nature of Parenthood', *Yale Law Journal*, 126/8 (2017), 2249–2260.

Notini, L., Gyngell, C., and Savulescu, J., 'Drawing the Line on In Vitro Gametogenesis', *Bioethics*, 34 (2019), 127–134.

Obergefell v. Hodges, 576 U.S. 644 (2015).

O'Connor, R. W., 'Examining the Politics, Ethics, and Legality of Anonymity in Gamete Donation', in R. Almeling, L. Campo-Engelstein, and B. T. Nguyen, eds., *Seminal* (New York: New York University Press, 2025), 199–204.

O'Neill, O., *Autonomy and Trust in Bioethics* (Cambridge: Cambridge University Press, 2002).

Owermohle, S., 'Senate IVF Bill Fails Again as Democrats Pressure GOP on Reproductive Health', *STAT*, 17 September 2024, https://www.statnews.com/2024/09/17/democrats-senate-ivf-treatment-bill-fails-republicans/.

Palacios-González, C., Harris, J., and Testa, G., 'Multiplex Parenting: IVG and the Generations to Come', *Journal of Medical Ethics*, 40/11 (2014), 752–758.

Pavan v. Smith, 582 U.S. 563 (2017).

Pew Research Center, 'Cultural Issues and the 2024 Election', 6 June 2024, https://www.pewresearch.org/politics/2024/06/06/cultural-issues-and-the-2024-election/.

Planned Parenthood of Southeast Pennsylvania v. Casey, 505 U.S. 833 (1992).

Resolve, 'Insurance Coverage by State', 30 September 2024, https://resolve.org/learn/financial-resources-for-family-building/insurance-coverage/insurance-coverage-by-state/.

Romualdez Tan, M. V., 'Modelling In Vitro Gametogenesis Using Induced Pluripotent Stem Cells: A Review', *Cell Regeneration*, 12 (2023), 33, https://doi.org/10.1186/s13619-023-00176-5.

Robertson, J. A., *Children of Choice: Freedom and the New Reproductive Technologies* (Princeton: Princeton University Press, 1994).

Roe v. Wade, 410 U.S. 113 (1973).

Saitou, M., and Hayashi, K., 'Mammalian In Vitro Gametogenesis', *Science*, 374/6563 (2021), eaaz6830, https://doi.org/10.1126/science.aaz6830.

Sosin, K., and Luterman, S., 'An Oklahoma Judge Just Transferred a Lesbian Mom's Parental Rights to Her Son's Sperm Donor', *The 19th*, 14 February 2023, https://19thnews.org/2023/02/oklahoma-judge-parental-rights-lgbtq-same-sex-marriage/.

Sparrow, R., 'Cloning, Parenthood, and Genetic Relatedness', *Bioethics*, 20 (2006), 308–318, https://doi.org/10.1111/j.1467-8519.2006.00509.x.

Suter, S. M., 'A Brave New World of Designer Babies?', *Berkeley Technology Law Journal*, 22 (2007), 897–969.

Suter, S. M., 'In Vitro Gametogenesis: Just Another Way to Have a Baby?', *Journal of Law and the Biosciences*, 3/1 (2016), 87–119.

Suter, S. M., 'The Tyranny of Choice: Reproductive Selection in the Future', *Journal of Law and the Biosciences*, 5/2 (2018), 262–300.

Suter, S. M., and Cahn, N., 'Lawmakers Rushed to Protect IVF in Alabama—Why Wasn't Abortion Afforded the Same Urgency?', *BMJ*, 384 (2024), https://doi.org/10.1136/bmj.q581.

Taylor, C., *Sources of the Self: The Making of the Modern Identity* (Cambridge: Harvard University Press, 1989).

Uniform Parentage Act 2017 (U.S.).

Witt, E., 'The Future of Fertility: A New Crop of Biotech Startups Want to Revolutionise Human Reproduction', *The New Yorker*, 17 April 2023, https://www.newyorker.com/magazine/2023/04/24/the-future-of-fertility.

4

Sex and Gender Minorities and *In Vitro* Gametes

Ethics, Access, and Status Equality

Timothy F. Murphy

Who Should Benefit from *In Vitro* Gametes?

The imagined beneficiaries of synthetic gametes or *in vitro* gametes (IVG) are most often characterized as cisgender, different-sex couples with impediments to fertility rooted in anatomy, physiology, disease, or disease treatment.[1] For example, some commentators portray synthetic gametes as a potential benefit for women with cancer whose circumstances preclude treatment with gonadotropic-releasing hormone agonists and cryopreservation of oocytes, embryos, and ovarian tissue.[2] For women in this situation, IVG would be a pathway to genetically related children that would not be possible otherwise. In general, securing genetic relatedness to children is the theory and practice of fertility medicine as far as possible. A standard moral presumption at work in having children is that different-sex couples experiencing infertility have a *prima facie* claim on access to ARTs, a claim that would ostensibly extend to IVG should it become safe and effective in humans.[3] Despite this general focus, some researchers and commentators have also suggested that IVG might enable same-sex couples to have genetically related children (GRC) together,[4] and some commentators have defended the right of these couples to use synthetic gametes that way.[5] Part of

[1] Immaculada Moreno, José Manuel Míguez-Forjan, and Carlos Simón, 'Artificial Gametes from Stem Cells', *Clinical and Experimental Reproductive Medicine*, 42 (2015), 33–44.

[2] Victoria G. Wesevich, Christopher Arkfeld, and David B. Seifer, 'In Vitro Gametogenesis in Oncofertility: A Review of Its Potential Use and Present-Day Challenges in Moving toward Fertility Preservation and Restoration', *Journal of Clinical Medicine*, 12/9 (2023), 3305.

[3] Lauren Notini, Christpher Gyngell, and Julian Savulescu, 'Drawing the Line on In Vitro Gametogenesis', *Bioethics*, 34/1 (2018), 123–134.

[4] Toshihiro Kobayashi et al., 'Blastocyst Complementation Using Prdm14-Deficient Rats Enables Efficient Germline Transmission and Generation of Functional Mouse Spermatids in Rats, *Nature Communications*, 12/1 (2021), 1328.

[5] Giuseppe Testa and John Harris, 'Ethics and Synthetic Gametes', *Bioethics*, 19/2 (2005), 146–166; Timothy F. Murphy, 'Pathways to Genetic Parenthood for Same-Sex Couples', *Journal of Medical Ethics*, 44/12 (2017), 823–824.

Timothy F. Murphy, *Sex and Gender Minorities and* In Vitro *Gametes* In: *Revolutionizing Reproduction*.
Edited by: Christopher Gyngell, Julian Savulescu, and Tsutomu Sawai, Oxford University Press.
 DOI: 10.1093/9780197810415.003.0004

what makes this separate defence necessary is that the infertility of different-sex couples is not what obstructs conception for two men, two women, or parties in a cis-and-trans relationship in which the trans partner has not taken steps to preserve gametic fertility. IVG stands to overcome these obstructions in having GRC. For example, one kind of IVG might enable two same-sex partners to rely on a naturally produced sperm (or ovum) from one member while relying on an ovum (or sperm) produced synthetically from the somatic cells of the other. In that case, the parents would share a 50-50% genetic relatedness to their children. In another kind of IVG, one partner would donate a naturally produced gamete while the other produced a gamete derived from embryonic stem cells. Those parents would share a 50-25% genetic relatedness to their children because the gametes derived from embryonic stem cells would involve a donor gamete in the production of the embryo, contributing 25% of the *in vitro* gamete's DNA.[6]

Several commentators have raised objections to the extension of ARTs to same-sex/queer couples (hereafter: SSQ couples) hoping to have children genetically related to both parents. These commentators object to the idea that SSQ couples have any strong claim on access to assisted reproductive interventions (ARTs) because their 'infertility' is only metaphorical, their infertility distress not compelling, and any entitlement they have rests on only a thin notion of justice. More searchingly, some commentators object to the very motive for wanting IVG, namely the interest in having children of shared genetic parentage, characterizing it as objectionable in itself and in its effects.[7] (I will refer to this motive as geneticism, although other terms are used, such as biologism.) One commentator characterizes the interest of two women having a biogenetic relation to their children as 'a mere wish'.[8] and another warns against 'pandering' to two women's desire to share genetic filiation with their children.[9] These analyses reject either in whole or in part any strong moral entitlement to ARTs for SSQ couples, IVG included. Some of them therefore recommend against extending government or insurance subsidies to SSQ couples interested in shared biogenetic relatedness to their children.[10]

Rather than conceptualizing the integrity of IVG according to whether it qualifies as treatment for infertility in a strict sense, I suggest that we should imagine it as an instrument of status equality in general and for SSQ couples

[6] Seppe Segers et al., 'Balancing Ethical Pros and Cons of Stem Cell Derived Gametes', *Annals of Biomedical Engineering*, 54/7 (2017), 1620–1632.

[7] Mianna Lotz, 'Public Funding of Uterus Transplantation: Deepening the Socio-moral Critique', *Bioethics*, 35/7 (2021), 664–671.

[8] Ezio Di Nucci, 'IVG, Same-Sex Couples, and the Value of Biological Ties', *Journal of Medical Ethics*, 42 (2016), 784–787.

[9] Françoise Baylis, '"No" to Lesbian Motherhood Using Human Nuclear Genome Transfer', *Journal of Medical Ethics*, 44/12 (2018), 865–867.

[10] Christina Richie, 'What Would an Environmentally Stable Reproductive Technology Industry Look Like?', *Journal of Medical Ethics*, 41/5 (2015), 383–387.

in particular. To make this case, I first consider whether a strictly naturalistic notion of 'infertility' and its 'treatment' can be relied on to ground entitlement to IVG. I also consider the significance of 'infertility distress' for SSQ couples as a possible way to enfold IVG within the definition of 'therapeutic.' More expansively, I consider the objection that the values that underlie interest in GRC are objectionable in a way that effectively disqualifies SSQ couples from access to ART practices in general and IVG in particular. I make the case that these considerations do not raise a morally significant bar to the use of IVG by same-sex couples. On the contrary, as a matter of respect for reproductive autonomy and status equality, SSQ couples have a presumptive entitlement to access IVG should it become safe and effective. That entitlement at least opens the door to possible subsidy by third-party payers.

Infertility and Infertility Distress as Justification for Entitlement to IVG

One rationale for access to IVG is the view that disease and disorder justify biomedical fertility interventions for those who cannot produce gametes. According to several commentators, SSQ couples would not qualify this way because their infertility is not an effect of inability to produce gametes. On Christina Richie's account of IVF, people who suffer from anatomic or physiological disorders that obstruct gamete production or conception are entitled to infertility interventions, whereas same-sex couples exhibit no equivalent condition.[11] Along similar lines, Camisha Russell argues that 'infertility is an individual physical state' and that gay men are not ordinarily infertile on this definition as they ordinarily produce gametes. It is as coupled that they are infertile, which she describes as being 'dysfertile.'[12] That dysfertility is not the result of disease and disorder and therefore grounds no prima facie entitlement to ARTs. Françoise Baylis adopts a therapeutic justification to argue against the 'backdoor' use of novel therapeutic techniques, namely 'slippage' from treatment proper to broader social uses.[13] Giuseppe Testa and John Harris have characterized the production of gametes from embryonic stem cells for purposes of fertilization—joining them with naturally produced gametes—as 'therapeutic' since the technique 'treats infertility' in a way analogous 'to the host of medical technologies that restore other deficient functions of the human body.'[14] They describe the

[11] Richie, 'What Would an Environmentally Sustainable Reproductive Technology Industry Look Like?'.

[12] Camisha Russell, 'Rights-Holders or Refugees? Do Gay Men Need Reproductive Justice?', *Reproductive Biomedicine and Society* Online, 16/7 (2018), 131–140.

[13] Bayliss, '"No" to Lesbian Motherhood', 865.

[14] Testa and Harris, 'Ethics and Synthetic Gametes', 155.

intervention as an assisted reprogramming effort. That characterization would exclude a man or woman relying on IVG to produce an ovum or sperm, respectively, since that technique does not restore functions characteristic of their sex. Testa and Harris defend access to IVG by SSQ couples for reasons distinct from the question of therapeutics, but I will suggest below that the idea of 'therapeutic' work in these accounts does not survive close analysis, for reasons that undercut reliance on notions of disorder, disease, and dysfunction as criteria of entitlement to ARTs.

Testa and Harris describe IVG as therapeutic, but that is misleading because the intervention does not 'treat' or 'cure' any underlying anatomical or physiological deficits any more than IVF does. IVF does not 'cure' or 'treat' low sperm count, for example, and neither does intracytoplasmic sperm injection. Neither would IVG cure or treat anatomical and physiological dysfunctions that obstruct gamete production. The obstacles that anatomy and physiology put in the way of gamete production and/or fertilization will remain in place after IVG, just as they remain in place after IVF. It would be fairer to call IVF or IVG 'prosthetic' rather than therapeutic. IVF no more cures infertility than a prosthetic arm 'cures' an amputation. In this sense, SSQ couples will stand in the same relation to IVG as anyone else looking for help with fertility: trying to bypass limitations of anatomy and physiology, not to cure diseases and disorders. To underscore this point, note that clinicians might eventually be able to rely on IVG to produce gametes (sperm and ova or both) from anyone, even the braindead, so long as the relevant embryonic and/or somatic cells are available. In that case, in the context of safe and effective IVG, infertility would be situational for everyone and not just SSQ couples even if people are in some other way fertile.

Another possible way to argue that IVG is justified as therapeutic benefit would be to show that it remedies infertility distress, in which case it treats psychological dissonance, not anatomic or physiological disorder. We might even understand this dissonance as a disability, as some commentators do.[15] If treatment of infertility distress justifies fertility treatment for cisgender, opposite-sex couples, it is unclear why—as a matter of moral consistency—it would not also justify treatment for SSQ couples. However, Notini et al. do not assume that all couples are equivalently situated with respect to fertility distress. On their account, SSQ couples would historically have understood that their bodies did not offer the possibility of having children together, and by extension they would not have the experienced the distress characteristic of different-sex couples. Different-sex couples, they say, would by contrast have borne the weight of social expectation for GRC. That being the case, Notini et al. say, 'same-sex couples may experience

[15] David Orentlicher, 'Discrimination Out of Dismissiveness: The Example of Infertility', *Indiana Law Journal*, 85/143 (2009), http/www.repository.law.indiana.edu/ilj/vol85/iss1/4.

much less infertility-related distress than is experienced by opposite-sex couples, who may feel that they need to live up to social expectations to have genetically related children with their partners'.[16] If there is less infertility distress, there is comparatively less justification for understanding IVG as therapeutic.

But that was then—what about now? Notini et al. tread cautiously when trying to estimate the extent to which fertility distress might be present in SSQ couples now in the context of changed social circumstances. Even so, they suggest there may still be less fertility distress and therefore less reason to justify IVG as therapeutic.[17] Of course, the available evidence shows that LGBTQ people in the United States have wanted to have children, even well before the broad social and legal changes that normalized SSQ relationships.[18] Against such a background, Notini et al. allow that because of the availability of ARTs at present, LGBT people might now have a stronger expectation of having children, no matter the past. At the very least, SSQ couples will have to consider whether to have children by relying on certain ARTs, when no such consideration was necessary before the innovation of those ARTs. Even if infertility distress did not exist in the past, it will make its presence felt as the normalization of SSQ couples having children becomes entrenched. That desire for children might be interpreted as 'buying into' social normalcy, as Russell characterizes the desire of gay men to have children,[19] but even so one can expect that infertility distress among SSQ couples will increase in the wake of novel mechanisms for having children and the expectations they engender.

If fertility distress follows in the wake of IVG and other ARTs, one can make a plausible case that it qualifies for available therapeutic remedies. This conclusion would not by itself mean that IVG is the only appropriate response, since there are a variety of ways by which SSQ couples can have children, and in the future, there may be several options for having children genetically related to both parents in some degree. Neither would it mean that we necessarily have to establish the presence of fertility distress in a strict sense. As Notini et al. point out, even if we reject the idea that only 'disease' or 'disorder' justify biomedical interventions, considerations of beneficence would still offer prima facie support to SSQ couples wanting to have GRC by way of IVG. IVG may be meaningful on moral grounds, that is, for reasons other than its strictly 'therapeutic' value. But

[16] Notini, Gyngell, and Savulescu, 'Drawing the Line', 127.

[17] Notini, Gyngell, and Savulescu, 'Drawing the Line', 129.

[18] From work done in the 1960s and 1970s, Bell and Weinberg reported that of the white homosexual men who had ever been married 50% of the white homosexual men and 71% of the black homosexual men reported having one or more children from that different-sex marriage. Of those who had ever been married, half the white homosexual female and significantly more of black homosexual females reported having one or more children from their first marriage. Alan P. Bell and Martin S. Weinberg, *Homosexualities: A Study of Diversity among Men and Women* (New York: Simon and Schuster, 1978), 164, 168.

[19] Russell, 'Rights-Holders or Refugees?', 138.

if one demands a therapeutic rationale as justification for opening IVG to SSQ couples, infertility distress will serve that purpose for at least some.

For her part, Lotz calls into question the very idea that infertility harms people in a morally significant sense.[20] She interprets the possible harms of infertility as artefacts rooted in objectionable background social values; it is not that infertility distress is not real, it is that it is socially manufactured in the context of ideological valuations of fertility. She argues that if society shed its sexist and pronatalist expectations regarding children, infertility would lose its significance as people would be able to value their lives without relying on GRC, adopted children, or any children at all. Fertility distress would wither as the desire for children and specifically for GRC lost its power in the face of changing expectations that adult human lives and relationships are valuable only by and through children. I will consider this line of analysis next, that the desire for GRC lacks moral significance, which in certain ways undercuts the claim of entitlement to IVG.

Presupposing and Reinforcing the Objectionable Value of Geneticism

Notini et al. briefly consider the ethics of people's desire to have GRC. They acknowledge that the value of this kind of desire is 'hard to cash out'. That said, they make short shrift of objections to the desire to have GRC, holding that 'many people value' that kind of relatedness 'highly,' with GRC being a 'dominant part of many autonomous life plans'.[21] On that view, they endorse a prima facie justification for respecting adults' interests in having GRC, but according to other commentators the damages of geneticism—the very desires for GRC—are harmful in a morally significant way.

Damien Riggs has analysed three studies reporting the interest Australian gay men have in GRC.[22] He interprets his findings as evidence that many gay men who have donated sperm or relied on egg donors and/or surrogates to have children exhibit objectionable presumptions about the value of genetic relatedness to children. He identifies presumptions that genetic relatedness matters as the most authentic kind of parenthood, that this kind of relatedness both asserts and secures the men's standing as full citizens of the reproductive state, and that the interest in GRC overrides—or at least eclipses—women's interests and

[20] Mianna Lotz, 'Commentary on Stephen Wilkinson and Nicola Williams: "Should Uterus Transplantation be Publicly Funded"', *Journal of Medical Ethics*, 42/9 (2016), 570–571.

[21] Notini, Gyngell, and Savulescu, 'Drawing the Line', 129–130.

[22] Damien W. Riggs, 'Making Matter Matter: Meanings Accorded to Genetic Material among Australian Gay Men', *Reproductive Biomedicine and Society Online*, 7 (2018), 150–157.

contributions as ova donors and gestational surrogates. Riggs therefore disputes the integrity of the interest in GRC, challenging the men's interests as discursive strategies that function as claims to status and value. The interest in genetic relatedness represents status-seeking in a world that disvalues 'lesser' forms of parent-child relationships. Riggs proposes to deconstruct this male privilege and explore 'a multitude of routes to parenthood'. Otherwise, the use of ARTs to confer genetic relatedness as far as possible reproduces 'the very inequalities that structure reproductive citizenship itself'.[23] Riggs draws no conclusions that the goal of GRC for gay men should be obstructed by law or policy, but his analysis effectively undercuts the moral value of desire for GRC as assisted by ARTs. Reliance on IVG by SSQ couples would not be exempt from this analysis.

Russell draws an equally unforgiving view of gay men's interest in GRC. She says that in their desire for GRC, gay and lesbian couples have been captured by the 'ideology'—the 'social imaginary'—of reproductive capacity: producing and rearing GRC. Russell says further: 'That gay men . . . cannot have biological children without assistance is not a punishment that society has issued against them, whether intentionally (via discrimination) or unintentionally (via neglect). It is a lived reality that may be experienced as a hardship, but not a result of injustice. . . . neither gay nor straight people (nor people who defy that binary) are owed access to reproductive technology by society as a matter of justice'.[24] She sees some good coming from the reliance of gay men on ARTs because those interventions afford gay men ways to create 'families on their own terms, without judgment or interference,' but the interest in GRC is problematic in its conceptual and historical origins: 'the privileging of biological relationships . . . in the USA has deep ties with the establishment of racial hierarchy and attempted maintenance of racial purity'.[25] Russell says further: 'Securing access to the norm of the nuclear family for middle-class gay men and calling it "justice" runs this very risk of stunting the potential for a more universal and authentic vision of freedom and equality in the realm of procreation'.[26] And that matters because, as Russell quotes Peter Gabel, 'every social movement by a marginalized group possesses an authentic self-understanding that seeks not merely to reform the existing system, but to transform it. . . . When a movement is seduced into simply demanding, and eventually receiving, particular rights from the state for its particular class of members, this transformative potential is obscured and goes unrealized'.[27] Russell does say that 'if society believes that biological children are an important social good, and if that society decides to do more to help people

[23] Riggs, 'Making Matter Matter', 156.
[24] Russell, 'Rights-Holders or Refugees?', 136.
[25] Russell, 'Rights-Holders or Refugees?', 137.
[26] Russell, 'Rights-Holders or Refugees?', 138.
[27] Russell, 'Rights-Holders or Refugees?', 138.

achieve biological parenthood, then I believe justice would demand that it do so for people of all races, ethnicities, classes and sexual orientations'.[28] Even so, merely putting equal opportunity in place fails, she holds, compared to the transformative justice she envisions, namely a justice under which any child could be the child of any parent without having to 'qualify' as genetically related or not or having any particular genome, among other potential specifications of dubious merit.

Mianna Lotz makes part of her case against geneticism by noting that reproductive interventions are 'profoundly social phenomena, deeply embedded in complex social norms and aspirations'.[29] I draw from the analysis she focussed on the meaning of those norms and aspirations for uterus transplantation and, specifically, for public funding of uterus transplantation, but I will examine the general dimensions of that analysis in relation to IVG because the moral scepticism she develops carries over to IVG without much alteration. As Lotz sees things, the desire for uterus transplantation and the experience of gestation is largely a social artefact: produced and sustained by background cultural and social values that are morally objectionable. They are sexist, pronatalist, and essentialist in their underpinnings. With Neil Levy, Lotz finds the value of GRC unproved. Together they argue that, among other things, GRC are not obviously necessary for a fulfilling personal history or self-knowledge, securing a greater assurance of parental bonding and care, expressing love and affirmation between people, being more 'natural,' or offering a greater assurance of shared outlook.[30] This is not to say that people do not want to have GRC; it is to say that the moral significance of that desire is largely unproven.[31] On the contrary, Levy and Lotz argue that the goal of GRC has harmful effects: it stigmatizes families without certain genetic ties, stigmatizes and marginalizes orphaned children, induces exposure to the risk of ARTs, and wrongly distinguishes 'real' from 'not real' parents. So long as the valuation of children depends on their being genetically relatedness, they argue, infertility distress will ensue while also leaving some children without parents.

When it comes to SSQ couples looking for help to have genetically related children, Heidi Mertes says that holding out the idea that same-sex couples should be related to their children in a 50-50% way is 'not only a wrong message to send but also creates a situation in which this new innovation not only meets demand but creates demand'.[32] According to Mertes: 'The optimal scenario would be

[28] Russell, 'Rights-Holders or Refugees?', 136.

[29] Mianna Lotz, 'Commentary on Stephen Wilkinson and Nicola Williams: "Should Uterus Transplantation be Publicly Funded"', *Journal of Medical Ethics*, 42/9 (2016), 570–571.

[30] Neil Levy and Mianna Lotz, 'Reproductive Cloning and a (Kind of) Genetical Fallacy', *Bioethics*, 19/3 (2005), 232–250.

[31] Lotz, 'Commentary on Stephen Wilkinson and Nicola Williams', 570.

[32] Heidi Mertes, 'Gamete Derivation from Stem Cells: Revisiting the Concept of Genetic Parenthood', *Journal of Medical Ethics*, 40 (2014), 744–747.

that people who are infertile can be convinced that they overestimate the importance of genetic parenthood and that there are other equally valuable means of constructing a family—or building a meaningful life without children—thus changing their desires'.[33]

The question at stake here is whether these kinds of criticism succeed in making a persuasive case that access by SSQ couples to GRC is a morally negligible or even objectionable benefit of IVF. I think they do not for the following reasons.

Respect for autonomy. Notini et al. have defended the interest in using IVG as a matter of respect for the autonomous life plans of SSQ couples. This is a forceful defence because respect for autonomy undergirds choices about, first, having children or not and, second, the people with whom to have children, not to mention the circumstances under which to have children and how many. In the United States, the Supreme Court has never ruled on whether gay and lesbian or trans people have a legal right to use ARTs because no legislature has denied them that right and the matter has never been litigated to clarify the merits of such a denial.[34] On the contrary, the law at present presumes strong rights of adults of all sexual orientations and genders to ARTs that assist them in having children. This right prevails over historical animus towards same-sex couples as parents and even over continuing moral criticism of ARTs from religious quarters, for example.[35]

As a moral matter, a strong presumption of reproductive autonomy is reflected in the ethics guidance of the American Society of Reproductive Medicine, which advises practitioners 'to treat all requests for assisted reproduction equally without regard to marital status, sexual orientation, or gender identity'.[36] It can also be argued that SSQ couples looking to IVG for help in having GRC together are just doing what opposite-sex couples already do in regard to the genes of their children. A kind of gene selection occurs when Woman A agrees to have children with Man A instead of Man B, or when unmarried Woman A makes a choice among sperm donors based on characteristics of the donor that resonate with her. Or again, genetic selection occurs when Couple A and B choose from among several embryos for implantation. Choosing the individual with whom to have GRC is already a morally tolerated practice whether it is done directly

[33] Mertes, 'Gamete Derivation from Stem Cells', 746.

[34] John A. Robertson, 'Reproductive Rights and Reproductive Technology in 2030', Governance Studies at Brookings, 21 January 2011, https://www.brookings.edu/articles/reproductive-rights-and-reproductive-technology-in-2030/.

[35] Congregation for the Doctrine of the Faith, 'Instruction *Dignitas Personae* on Certain Bioethical Questions', Vatican, 20 June 2008, https://www.vatican.va/roman_curia/congregations/cfaith/documents/rc_con_cfaith_doc_20081208_dignitas-personae_en.html.

[36] Ethics Committee of the American Society for Reproductive Medicine, 'Access to Fertility Treatment Irrespective of Marital Status, Sexual Orientation, or Gender Identity: An Ethics Committee Report', *Fertility and Sterility*, 16/2 (2021), 326–329.

or indirectly. Oddly enough, the attempt to secure GRC for same-sex couples even found acceptance by Mertes, who has otherwise criticized the impetus to value children by reason of their genetic relatedness to parents. While generally doubtful about the importance of shared genetic relatedness of parents to children, she nevertheless recommends one pathway to genetic relatedness over another for SSQ couples.[37] Holding out for 50-50% relatedness is likely to prove a difficult path for practical reasons, so she endorses a method that would confer 50-25% genetic relatedness, again, for practical reasons. Against the permissive background of gamete selection already tolerated, it is not clear that IVG is morally distinctive in a way that would justify an override of reproductive autonomy for SSQ couples on the basis that genetic relatedness of a certain kind is wanted.

To make a case that respect for reproductive autonomy should be constrained, ordinarily one must show that morally significant harms occur in consequence of the choice or that rights are violated. As to the question of harm, it is not clear that SSQ couples relying on IVG would cause morally significant harm. First, there is no reason to think that children born via safe and effective IVG would be harmed simply by being related to two same-sex or two queer parents. Neither is there reason to think that the parents themselves are harmed by relying on IVG to have children. It is arguable that women face harms through any gestational surrogacy that would be required, but the benefits that women gain for reasons important to them offset these risks. IVG for male same-sex couples would in fact reduce some risk for women, as no ova donation would be necessary. Is it possible to maintain that reliance on IVG by SSQ couples would harm orphaned children? They may remain unadopted because people who might have adopted them find other ways to have children. This outcome ensues, of course, in the wake of all other successful ARTs, just as it does for couples who have children without clinical assistance, who might have adopted rather than bring another human being into the world. Respect for reproductive freedom does seem to contribute to that outcome, but it is hardly the sole cause of unadopted children, and it is hard to see why any single ART should be prohibited in order to avoid this outcome. It is also morally presumptuous to assume that people who face certain kinds of infertility are themselves more responsible for the welfare of orphaned children than others. Research from 2014–2016 shows that in the United States lesbian and gay couples are seven times more likely to adopt children or raise foster children than the rest of the population.[38] In light of this research, it is hard to argue that these couples are not already shouldering a fair share of adoption and foster care responsibilities.

[37] Seppe Segers et al., 'Using Stem Cell Derived Gametes for Same-Sex Reproduction: An Alternative Scenario', *Journal of Medical Ethics*, 43 (2017), 688–691, 689.

[38] Williams Institute, 'How Many Same-Sex Couples in the US Are Raising Children?', July 2018, https://williamsinstitute.law.ucla.edu/publications/same-sex-parents-us/.

Another line of challenge to reproductive autonomy comes from commentators who see certain ARTs for SSQ couples as cutting against social interests. Baylis has objected to the extension of mitochondrial nuclear transfer as a way of securing a child genetically related to two women. Among other things, she says: 'The genetic connectedness is negligible: only 1% of the total amount of genetic material'.[39] The outlay of effort to achieve that degree of relatedness is not, she says, a 'judicious use of limited resources'—the effort does not map onto genuine 'social priorities' having a 'benefit for all'. From her perspective, 'society is better served by an understanding of parenting not as primarily a biological relationship but as primarily a social and legal responsibility, hopefully one that is willfully undertaken'. This objection turns on harm to society, and while that kind of harm is important to moral analysis, this approach perhaps requires too much of people choosing to have children. Especially worrisome is the idea that decisions about having children must be tested for their 'benefit for all'. Many people's choices about having children, without relying on ARTs, can hardly be said to meet that expectation. It does not happen either, that resources not used for IVG—money, clinical skill, research efforts chief among them—would necessarily and of their own accord go to social purposes having a benefit for all.

I also want to draw attention to Baylis' claim that there is 'no right to biological parenthood,' which she understands as meaning at least that 'there is no compelling need for human nuclear genome transfer for genetically related children'.[40] Let me assume for the sake of the argument that there is no right to biological parenthood. All that would mean in this context is that the parties in question may not call on others to provide mitochondrial nuclear transfer as if it were their duty. Yet clinicians and researchers would still be free to offer their support for other reasons. There does not need to be a 'right' to biologically related children to take the interest in GRC seriously. Even if there were no right to GRC, it does not follow that others may obstruct the pursuit of that goal. Respect for reproductive autonomy secures the integrity that interest in children requires to be taken seriously. That said, on the basis of evidence so far, I suggest there is no good reason to oppose the use of IVG by SSQ couples, either on the grounds of harm or as involving violations of rights.

Focus on same-sex/queer couples amounts to selective enforcement. I have so far argued that there is nothing particularly distinctive about SSQ couples when it comes to respect for their reproductive autonomy. It is unclear to me, moreover, that SSQ couples should be expected to bear the responsibility for the social reforms recommended by various commentators, or even to bear it first. In

[39] Baylis, '"No" to Lesbian Motherhood', 865.
[40] Baylis, '"No" to Lesbian Motherhood', 865.

other words, it is not clear to me why commentators criticize the extension of novel ARTs to SSQ couples while all others may pursue GRC despite the moral criticisms in play. One possible answer is that a certain amount of gatekeeping is possible in the access of SSQ couples to ARTs where it is not possible for others. Most children are conceived without any oversight from clinicians who, for example, rely on professional standards giving them some degree of freedom to turn away some would-be parents.[41] Many of the commentators I have mentioned above are aware of this and try to explain why they focus instead on SSQ couples. This is a pressing concern because when it comes to complicity, the population of people needing no clinical assistance seems to be most responsible for the presence of pronatalist, essentialist, and geneticist interests in children.

Of this problem Richie says, 'a moratorium on all fertility clinics would be the most ecologically sound decision . . . [but] it is unlikely that established fertility procedures or treatments would be effectively "banned"'.[42] Lotz says that when it comes to objectionable values, the state 'is simply unable to impose modes of procreation that challenge rather than entrench bionormativity on those procreators who do not require assistance'.[43] Mertes says that 'the preference for a genetically related child seems too entrenched as a matter of both evolutionary biology and culture' to be set aside by rational debate'.[44] In these acknowledgments, moral criticism is characterized as powerless in face of the vast majority of morally problematic decisions. Indeed, Mertes even seems to say that the interest in GRC is intractable.

It is against this background that commentators have criticized the reliance of SSQ couples on techniques that would offer them GRC. It is to be wondered, however, whether this focus amounts to selective enforcement, especially when the number of SSQ couples turning to ARTs will always be dwarfed by the number of people who conceive children on their own. It is also unclear whether the imagined problems of ARTs and IVG would be brought under control—problems such as ideological pronatalism, geneticism, even the environmental effects of overpopulation—if SSQ couples had any fewer children than they do. A small percent of total births is due to LGBTQ parents turning to ARTs, and the number of those who might rely on IVG would be smaller still.

LGBTQ people have faced socially imposed obstacles without parallel for others in having children, particularly social impediments to forming

[41] Ethics Committee of the American Society for Reproductive Medicine, 'Child-Rearing Ability and the Provision of Fertility Services: An Ethics Committee Opinion', *Fertility and Sterility*, 108/6 (2017), 944–947.

[42] Richie, 'What Would an Environmentally Sustainable Reproductive Technology Industry Look Like', 386.

[43] Mianna Lotz, 'Public Funding of Uterus Transplantation: Deepening the Socio-moral Critique', *Bioethics*, 35/7 (2021), 664–671, 671.

[44] Seppe Segers et al., 'Using Stem Cell-Derived Gametes for Same-Sex Reproduction', 689.

relationships. Until 2017, some US states explicitly banned lesbian and gay people from adoption, but the US Supreme Court indirectly overruled those statutes by conferring the right of two same-sex partners to be listed as parents on a child's birth certificate.[45] Even now, though, some private adoption agencies decline to place children with LGBTQ parents. Why should people who—as a class—have faced socially imposed and morally unjustified obstacles to having children be expected to bear the cost of reform efforts that are unlikely to dent overall presuppositions about and effects of having children? If anything, perhaps because of past treatment of LGBTQ people, some degree of compensatory justice would put access to IVG by SSQ couples ahead of access by others.

ARTs in General Confer Status Equality on SSQ Couples

Access to social goods, opportunities, and certain social roles is a key component of status equality, which I will stipulate as an important moral goal that is, at minimum, free from invidious distinctions. By status equality in this analysis, I mean access to and equity in social goods, services, and certain social roles. I put the totality of ARTs in the former category and parental identities somewhere between the second and third categories. As I suggested above, there is not a clear moral or legal ground for closing off ARTs to people based on their sexual and gender identities. Yet some commentators would limit the extension of certain ARTs to SSQ couples relative to the identities they hope to secure through them. The motives and imagined effects of interest in GRC feature prominently in these arguments.

Mertes suggests that the origins of people's interest in GRC belongs to evolutionary biology (it is built in) and cultural values (it is artefactually produced and sustained). By themselves, neither of those explanations offer a moral defence of the interest. For moral reasons we frequently override objectionable interests that have evolutionary roots (e.g., impulses to kill), and for moral reasons we override interests shaped by social forces (e.g., slavery). I will suggest, however, that the interest in having GRC has a certain moral value in terms of its meaning for parental identity. Over and above being just biological, genetic relatedness offers a certain narrative relatedness between parents and children as well. Having GRC is one way for two people to author a relationship with each other and their offspring. Critics may say that parents overvalue this narrative, but as a matter of respect for reproductive autonomy, we are ordinarily obliged to respect the way adults author the nature and meaning of their relationships. This same respect is also owed, not incidentally, to parents who adopt children for

[45] *Pavan v. Smith*, 582 U.S. (2017).

reasons important to them or who mix biologically related and adopted children together in their families for the authorial value in that.

Conclusions

ARTs increase the number of people able to secure GRC, and IVG will be no different that way. Yet at the very time of increasing social equality for LGBTQ people, some commentators challenge the extension of ARTs to SSQ couples hoping to have biogenetic relationships to their children, or at least they challenge public subsidy for that purpose. These commentators do not argue that same-sex and queer parents are unsuitable as parents or that children face harm from having such parents. Instead, the objections turn on claims that the interventions are not justified as therapeutic, that a particular sense of justice does not oblige this kind of assistance, and that the interventions presuppose and reinforce objectionable values, among other things. These arguments offer little support for government or insurance subsidies for fertility assistance for SSQs, and IVG will be no exception.

By one line of objection, ARTs should be reserved for people whose diseases and disorders obstruct gamete production, fertilization, or gestation. By this standard, SSQ couples do not qualify because their 'infertility' is not pathological in origin but only situational. Healthcare does not, however, confine itself to corrective interventions in anatomy and physiology.[46] Both in practice and in theory, healthcare focuses on human well-being and opportunity in ways that justify certain public subsidies on those grounds.[47] Even though some theorists of ARTs characterize interventions such as IVF in therapeutic terms, that characterization is doubtful since IVF does not ordinarily treat any underlying diseases or disorders that obstruct gamete production and/or inhibit fertilization. The interventions are effectively prosthetic in the way that an artificial limb does not treat or cure an amputated arm. If that is the case, then SSQ couples stand in the same relation to options that open the prospect of GRC as do different-sex/cisgender couples, with those options not 'treating' or 'curing' in a strict sense. IVG stands to render moot any distinction between therapeutic and non-therapeutic interventions because—so long as the right resources are available—everyone's infertility will be effectively situational.

That said, one might make the case that IVG would be therapeutic if we understood it as treatment for infertility distress. As a matter of consistency, if we

[46] Carl Elliott, *Better than Well: Medicine Meets the American Dream* (New York: Norton, 2004).
[47] Stephen Wilkinson and Nicola J. Williams, 'Should Uterus Transplants Be Publicly Funded?', *Journal of Medical Ethics*, 42/9 (2016), 559–565.

imagine ARTs as techniques for the treatment of that kind of psychological dissonance, it is not clear why the infertility distress of SSQ couples should have any less standing—any less entitlement—than the infertility distress of different-sex couples. One challenge to this claim of equivalence is that because of historical circumstances, SSQ couples might not have ever expected to have children at all let alone children genetically related to them both. In that case, they might not experience infertility distress at levels of clinical significance and be disqualified as candidates for therapeutic IVG. Of course, SSQ couples have increasing options for children, as their relationships as couples and as their relationships with children become socially normalized. Under these circumstances one may reasonably expect an increase in the degree of fertility distress of SSQ couples, and, with that, a 'therapeutic' rationale for IVG gains traction. That 'fertility distress' among SSQ couples may be only a recent social phenomenon would make it no less compelling than other instances of infertility distress. As I have suggested, however, we do not need to rely on a criterion of access to IVG framed as a matter of therapeutic expectations. As Notini et al. have pointed out, moral responsibilities rooted in beneficence can be justification enough for offering to SSQ couples what is available to others in matters of their well-being.

That very idea of well-being as a moral criterion for access to certain medical interventions attracts challenges, however, from commentators who characterize the interest in GRC as morally problematic in itself and in its effects. These commentators hold, among other things, that fertility distress is an artefact of wrongful valuation of children in the first place and wrongful valuation of specific kinds of children in the second. More generally, they say the very interest in GRC is morally groundless, its benefits unproved, and its effects objectionable (such as children remaining orphaned). Women face disproportionate costs of GRC as they bear the risks of gamete donation, IVF, and gestational surrogacy. Also, the practice of securing GRC is problematic, it is argued, in that nothing of intrinsic moral value is secured by the practice of ARTs, IVG included.

It is not that these considerations lack relevance, rather the question is whether they rise to a level of significance that overrides the prospect of SSQ couples having moral entitlement to IVG. I have argued that they do not. First, it does not follow that an interest in GRC must be a right for it to be respected as a matter of reproductive autonomy. I have further argued that any limits to reproductive autonomy ought to demonstrate harm of moral significance in degree or kind, and I have argued that IVG does not foreseeably pose those kinds of harm. It is not after all harmful in itself to a child to be genetically related to both parents. Neither are parents inherently harmed by having a shared genetic relationship with a child. Social harms are more difficult to evaluate, but the effects of SSQ couples relying on IVG must be set against the background of the vast majority of people who have GRC without reliance on ARTs. It is idiosyncratic—if not also selective enforcement—to focus objections against SSQ couples using ARTs to have GRC when most people do exactly that without any gatekeepers making decisions about the merits of their choices. I have also tried to show that it is not clear why any one group should be obstructed in access to novel ARTs to resolve larger, shared social problems, including, for example, the welfare of orphaned children. My analysis does not overlook social effects in evaluating questions of access for IVG for SSQ couples. It does not, however, assign more responsibility for changing deeply seated, objectionable social values to SSQ couples than to the parties who in their values and practices are far more responsible for indulging in disputed values and their downstream damaging effects.

Demonstrating that SSQ couples are entitled to access IVG should it become safe and effective is one thing. Demonstrating that those couples are entitled to subsidy for IVG requires separate consideration. Richie has argued that governments and insurers should decline to subsidize ARTs for SSQ couples, on the grounds that the interventions do not serve health properly speaking, and have environmentally fraught consequences, among other reasons. Baylis has argued that subsidizing mitochondrial transplantation represents misplaced priorities. Di Nucci has set the default against state subsidy for ARTs whose goal is to secure biogenetically shared parentage because of the moral insignificance of that kind of relationship. Lotz has argued against subsidy for uterus transplantation because of the morally compromised values the practice presupposes and would reinforce. I expect that subsidies for IVG will face equivalent challenges.

In the United States, there is limited subsidy from private insurers or government agencies for ARTs, and IVG is unlikely to change this arrangement. Employers provide most private healthcare insurance, or at least a good portion of it, and some corporations do cover ova freezing, IVF, surrogacy costs, and even adoption costs, but most employers do not. Some states do require that insurance companies doing business in their jurisdictions subsidize a certain number of IVF attempts, but most do not.[48] To the extent health insurers look to disease and disorder as reasons to subsidize ARTs, SSQ couples are unlikely to find support on that rationale. That said, if third parties do make some subsidy available on broader grounds, it is not clear why SSQ couples should be excluded especially if, as does happen, legislatures and employers paying insurance costs put well-being at the centre of their decision-making. It would be morally consistent for states that oblige private insurers doing business in their jurisdiction to cover some IVF costs ought to make similar requirements regarding at least some IVG. (It would also be morally consistent if more states require certain

[48] Karen N. Bown, 'The State of IVF Insurance Coverage: Global Highlights and Where We Are Now', *GE Empowered Women's Health*, 26 October 2021. https://www.volusonclub.net/empowered-womens-health/the-state-of-ivf-insurance-coverage-global-highlights-and-where-we-are-now.

ART coverage of the health insurers doing business in their jurisdictions). Of course, employers offering health insurance as work benefits are free to offer whatever they wish regarding ARTs, and some of them do subsidize gamete retrieval, gamete storage, and IVF as a way of attracting employees. But those benefits may have to do more with business self-interest—attracting and retaining employees—than with moral considerations. Given that no good argument obstructs access by SSQ couples to IVG, the relevant moral starting point will be to ask whether the rationales used by private third-party payers do not also oblige them to offer some degree of subsidy IVG to SSQ couples in the name of status equality.

Bibliography

Baylis, F., '"No" to Lesbian Motherhood Using Human Nuclear Genome Transfer', *Journal of Medical Ethics*, 44/12 (2018), 865–867.

Bell, A. P., and Weinberg, M. S., *Homosexualities: A Study of Diversity among Men and Women* (New York: Simon and Schuster, 1978).

Bown, K. N., 'The State of IVF Insurance Coverage: Global Highlights and Where We Are Now', *GE Empowered Women's Health*, 26 October 2021, https://www.volusonclub.net/empowered-womens-health/the-state-of-ivf-insurance-coverage-global-highlights-and-where-we-are-now.

Congregation for the Doctrine of the Faith, 'Instruction *Dignitas Personae* on Certain Bioethical Questions', *Vatican*, 20 June 2008, https://www.vatican.va/roman_curia/congregations/cfaith/documents/rc_con_cfaith_doc_20081208_dignitas-personae_en.html.

Di Nucci, E., 'IVG, Same-Sex Couples, and the Value of Biological Ties', *Journal of Medical Ethics*, 42/12 (2016), 784–787.

Elliot, C., *Better than Well: Medicine Meets the American Dream* (New York: Norton, 2004).

Ethics Committee of the American Society for Reproductive Medicine, 'Access to Fertility Treatment Irrespective of Marital Status, Sexual Orientation, or Gender Identity: An Ethics Committee Report', *Fertility and Sterility*, 16/2 (2021), 326–329.

Ethics Committee of the American Society for Reproductive Medicine, 'Child-Rearing Ability and the Provision of Fertility Services: An Ethics Committee Opinion', *Fertility and Sterility*, 108/6 (2017), 944–947.

Goldberg, S. K., and Conron, K. J., 'How Many Same-Sex Couples in the US Are Raising Children?', *UCLA Williams Institute*, 1 July 2018, https://williamsinstitute.law.ucla.edu/publications/same-sex-parents-us/.

Kobayashi, T., Goto, T., Oikawa, M., Sanbo, M., Yoshida, F., Terade, R., et al., 'Blastocyst Complementation Using *Prdm14*-Deficient Rats Enables Efficient Germline Transmission and Generation of Functional Mouse Spermatids in Rats', *Nature Communications*, 12/1 (2021), 1328.

Levy, N., and Lotz, M., 'Reproductive Cloning and a (Kind of) Genetical Fallacy', *Bioethics*, 19/3 (2005), 232–250.

Lotz, M., 'Commentary on Stephen Wilkinson and Nicola Williams: "Should Uterus Transplantation Be Publicly Funded"', *Journal of Medical Ethics*, 42/9 (2016), 570–571.

Lotz, M., 'Public Funding of Uterus Transplantation: Deepening the Socio-moral Critique', *Bioethics*, 35/7 (2021), 664–671.

Mertes, H., 'Gamete Derivation from Stem Cells: Revisiting the Concept of Genetic Parenthood', *Journal of Medical Ethics*, 40/11 (2014), 744–747.

Moreno, I., Míguez-Forjan, J. M., and Simón, C., 'Artificial Gametes from Stem Cells', *Clinical and Experimental Reproductive Medicine*, 42/2 (2015), 33–44.

Murphy, T. F., 'Pathways to Genetic Parenthood for Same-Sex Couples', *Journal of Medical Ethics*, 44/12 (2017), 823–824.

Notini, L., Gyngell, C., and Savulescu, J., 'Drawing the Line on In Vitro Gametogenesis', *Bioethics*, 34/1 (2018), 123–134.

Orentlicher, D., 'Discrimination out of Dismissiveness: The Example of Infertility', *Indiana Law Journal*, 85/1 (2010), 143–185, http://www.repository.law.indiana.edu/ilj/vol85/iss1/4.

Pavan v. Smith, 582 U.S. 563 (2017).

Richie, C., 'What Would an Environmentally Stable Reproductive Technology Industry Look Like?', *Journal of Medical Ethics*, 41/5 (2015), 383–387.

Riggs, D. W., 'Making Matter Matter: Meanings Accorded to Genetic Material among Australian Gay Men', *Reproductive Biomedicine & Society Online*, 7 (2018), 150–157.

Robertson, J. A., 'Reproductive Rights and Reproductive Technology in 2030', *Governance Studies at Brookings*, 21 January 2011, https://www.brookings.edu/articles/reproductive-rights-and-reproductive-technology-in-2030/.

Russell, C., 'Rights-Holders or Refugees? Do Gay Men Need Reproductive Justice?', *Reproductive Biomedicine & Society Online*, 16/7 (2018), 131–140.

Segers, S, Mertes, H., de Wert, G., Dondorp, W., and Pennings, G., 'Balancing Ethical Pros and Cons of Stem Cell-Derived Gametes', *Annals of Biomedical Engineering*, 54/7 (2017), 1620–1632.

Segers, S., Mertes, H., Pennings, G., de Wert, G., and Dondorp, W., 'Using Stem Cell-Derived Gametes for Same-Sex Reproduction: An Alternative Scenario', *Journal of Medical Ethics*, 43/10 (2017), 688–691.

Testa, G., and Harris, J., 'Ethics and Synthetic Gametes', *Bioethics*, 19/2 (2005), 146–166.

Wesevich, V. G., Arkfeld, C., and Seifer, D. B., 'In Vitro Gametogenesis in Oncofertility: A Review of Its Potential Use and Present-Day Challenges in Moving toward Fertility Preservation and Restoration', *Journal of Clinical Medicine*, 12/9 (2023), 3305.

Wilkinson, S., and Williams, N. J., 'Should Uterus Transplants Be Publicly Funded?', *Journal of Medical Ethics*, 42/9 (2016), 559–565.

Williams Institute, 'How Many Same-Sex Couples in the US Are Raising Children?' (July 2018). https://williamsinstitute.law.ucla.edu/publications/same-sex-parents.us/.

5

Eugenics under the Microscope

Robert Sparrow

In Vitro Eugenics

Once it becomes possible to derive both sperm and egg from embryonic stem cells, an intriguing possibility will open up. By combining ES-cell-derived gametes in IVF, it would be possible to create embryos and derive new gametes from them. Combining these gametes would produce new embryos and allow the derivation of new gametes, which in turn could be used to create new embryos, new gametes—and so on for multiple 'generations'. If scientists performed genetic testing of the embryos created at each stage before deciding whether to derive gametes from them and which other embryos to 'mate' them with, they could use this technology to selectively 'breed' human beings *in vitro*. At the end of the process, having crossed as many cell lines as needed to bring about some combination of desirable genes, scientists could implant one or more embryos into the womb of a willing woman in order to bring a child into existence.

In previous work, I called this process of iterated embryo selection (IES) '*in vitro* eugenics' (IVE).[1] In this chapter, I examine what is at stake in identifying this technology as eugenic. While the idea of eugenics has a bad name owing to its history, philosophers have struggled to agree on what, if anything, is wrong with eugenics. The programs of racist murder and coercive sterilization in the service of bad science that were justified by eugenic ideals in the previous century were obviously evil and wrong. However, the idea that parents should care about the genetics of their children is widely endorsed today and plays a key role in motivating contemporary programs of prenatal and genetic screening. The idea that the state should care about the distribution of genes within the community is more controversial but arguably still plays a role in such programs. Moreover, insofar as the selective breeding enabled by artificial gametogenesis would take place entirely *in vitro*, the direct harms to individuals that were a product of historical enthusiasm for eugenics could not occur.

[1] Robert Sparrow, 'In Vitro Eugenics', *Journal of Medical Ethics*, 40/11 (2014), 725–731.

Robert Sparrow, *Eugenics under the Microscope* In: *Revolutionizing Reproduction.* Edited by: Christopher Gyngell, Julian Savulescu, and Tsutomu Sawai, Oxford University Press. © Oxford University Press 2026.
DOI: 10.1093/9780197810415.003.0005

Nevertheless, I will argue that, except—possibly—in one special case, it is accurate to describe this project as eugenic. Where it is used to bring children into existence, its goal would be to improve the genetics of the future child on the basis of controversial ideas about the relative merits of different genomes. Any program of IES would involve repeated instances of negative selection intended to improve the distribution of those genes judged to be desirable in a population. The first observation implies that iterated embryo selection for reproductive purposes would be eugenic according to the weakest and broadest definition thereof: the second implies that it would be eugenic according to—what I will argue is—a more precise definition. Moreover, IES would have many of the features that make eugenics more generally an object of moral suspicion. The idea that scientists know which genes are the 'best' appears remarkably arrogant and offends against egalitarian sentiments that are widespread today. Any children born as a result of IES would be the product of an extended process of 'design', with a designer who would stand in a relationship to them that they could never have in relation to their designer. The power that the designer would have over the designed has implications for the extent to which the latter is free. Moreover, this process would undercut the distinction between the 'begotten' and the 'made'. Finally, while *in vitro* eugenics is unlikely to involve any direct harms to individuals, there are also good reasons to fear that it will reinvigorate scientifically dubious ideas about the relationship between genes and life outcomes in the community more broadly, which might in turn increase support for policies that would threaten the reproductive liberty of existing individuals. Placing eugenics 'under the microscope', both literally and figuratively, may productively illuminate discussions of the ethics of other reproductive technologies, including germline gene editing, by helping to clarify how concerns about eugenics might apply, or not, in each instance. I will argue that an important subset of the concerns that arise regarding these technologies also arise regarding philosophical discussions of their ethics.

In focussing on whether—and whether it matters whether—IES is a form of eugenics, I am deliberately bracketing a wider range of ethical questions raised by this technology. In particular, I will not attempt to provide a full account of the ethics of human enhancement.[2] I also ignore questions about the ethics of this technology arising out of a concern for the moral status of human embryos, about the permissibility of creating embryos without intending to implant them in a woman's womb, about the implications of rapid advances in the techniques of genetic

[2] I have gone some way towards doing so elsewhere. See: Robert Sparrow, 'Liberalism and Eugenics', *Australasian Journal of Philosophy*, 89/3 (2011), 499–517; and Robert Sparrow, 'A Child's Right to a Decent Future? Regulating Human Genetic Enhancement in Multicultural Societies', *Asian Bioethics Review*, 4/4 (2012), 355–373.

modification,[3] and about whether children brought into existence at the end of this process would have genetic parents[4] or whether the technology would violate the rights of embryos not to become parents.[5] Nor will I discuss the ethics of bringing people into existence using a technology that is necessary experimental and therefore potentially risky when first used in humans, the normative significance of the expectations that this technology might inculcate in potential parents, or the ethics of soliciting women to gestate the embryos produced via this process.[6]

Given that I chose to name this process '*in vitro* eugenics' when I first described this possibility, it should come as little surprise that I will conclude that this name is accurate and appropriate. However, in order to hold the question of whether it *is* appropriate open during the course of my discussion, for most of the remainder of this chapter I shall refer to it as iterated embryo selection. I would ask the reader to keep in mind that this usage is a temporary and reluctant one. As we shall see, one of the reasons why I think it is important to recognize IVE/IES as eugenic is precisely because the former name tells us something significant about the technology and its ethics, which is obscured by the use of other nomenclature.

Selection and Conception of Human Beings *In Vitro*

Like artificial gametogenesis, at the time of writing, IES is only a theoretical possibility. It is, further, contingent on the development of the capacity to derive gametes from embryonic stem (ES) cell lines and not just from induced pluripotent stem (IPS) cells. However, it is unlikely that it will prove possible to create gametes from IPS cells but not ES cells.

[3] Robert Sparrow, 'Yesterday's Child: How Gene Editing for Enhancement Will Produce Obsolescence—and Why It Matters', *American Journal of Bioethics*, 19/7 (2019), 6–15.

[4] Robert Sparrow, 'Orphaned at Conception: The Uncanny Offspring of Embryos', *Bioethics*, 26/4 (2012), 173–181; Thomas Douglas, 'Stem Cell-Derived Gametes, Iterated In Vitro Reproduction, and Genetic Parenthood', *Journal of Medical Ethics*, 40/11 (2014), 723–724; Heidi Mertes, 'A Moratorium on Breeding Better Babies', *Journal of Medical Ethics*, 40/11 (2014), 734–735.

[5] Helen Watt, 'Ancestor Embryos: Embryonic Gametes and Genetic Parenthood', *Journal of Medical Ethics*, 40/11 (2014), 759–761.

[6] A discussion of the implications of the fact that the first uses of reproductive technologies in humans are necessarily experimental—and that all uses remain such for at least a generation—may be found in Robert Sparrow, Julian Koplin, and Catherine Mills, 'Mitochondrial Replacement Techniques: A Critical Review of the Ethical Issues', in Diana Bowman, Karinne Ludlow, and Walter Johnson, eds., *Reproduction Reborn: How Science, Ethics and Law Shape Mitochondrial Replacement Therapies* (Oxford: Oxford University Press, 2023), 32–61. For discussion of the implication of concerns about risk in IES, see Cesar Palacios-González, John Harris, and Giuseppe Testa, 'Multiplex Parenting: IVG and the Generations to Come', *Journal of Medical Ethics*, 40/11 (2014), 752–758.

As I acknowledged in my original discussion of this possibility, there are also several further technical/scientific barriers that would need to be overcome before it might be possible to use IES to create children with desired combinations of genetic traits.[7]

First, it must not be the case that repeated iterations of this process would lead to an accumulation of genetic damage, transmitted epigenetically from cell line to gametes over multiple generations, that would make it irresponsible to bring a child into existence using IES. The fact that researchers will need to be able to tell whether gametes are safe to use in order for IVG to be used for reproduction at all is some comfort here.

Second, the amount of time required to move forward a 'generation' *in vitro* needs to be short enough to make it plausible to carry out an adequate number of genetic crosses to create the desired genotype in a realistically available time frame. That is, scientists would need to be able to derive sperm and eggs from stem cells, create a new embryo and allow it to develop to the blastocyst stage, take cells from the blastocyst and transform them into stem cells, and characterize the genetics of the stem cells in order to decide whether to proceed to derive gametes from them and/or how the cell line might contribute to achieving the desired phenotype. Each of these steps will take time, on the order of days or weeks, although it would be possible for the genetic testing of cell lines to go on simultaneously with the generation of stem cells. In my previous discussion, I suggested that 'four-to-six months' was a reasonable estimate of the time required to complete all of these steps, which would also allow scientists to create two to three generations of embryos each year and move forward as many as thirty generations in a decade—which would be enough to generate significant changes to the genomes of the embryos in the final generation.[8] The technology will be more powerful, the less time required to go through all these steps.

It must also be admitted that the use case for IES is much weaker than it appeared when I first started considering the technology, shortly before the invention of the CRISPR-Cas9 genome editing system.[9] To produce genetically modified human beings today it would be simpler to directly modify embryos using CRISPR or one of the other more recent gene-editing tools.[10] Unlike IES, gene editing would also allow the introduction of genes from other species, or synthetic genes, to create traits that would never arise naturally in humans.

[7] Sparrow, 'In Vitro Eugenics'; Flávio Guimarães da Fonseca et al., 'Human In Vitro Eugenics: Close, Yet Far Away', *Journal of Medical Ethics*, 40/11 (2014), 738–739.

[8] Sparrow, 'In Vitro Eugenics'.

[9] Debra J. H. Mathews, 'Language Matters', *Journal of Medical Ethics*, 40/11 (2014), 733–734.

[10] National Academies of Sciences, Engineering, and Medicine, *Human Genome Editing: Science, Ethics, and Governance* (Washington, DC: The National Academies Press, 2017); Antonio Regalado, 'Engineering the Perfect Baby', *MIT Technology Review*, 118/3 (2015), 26–33.

Nevertheless, there are at least still three reasons why scientists—or those seeking to raise a genetically enhanced child—might wish to pursue IES.

First, it is possible that the best way to achieve certain genetic modifications or combinations of traits would be to combine genome-editing with selective breeding via IES. Embryos, or embryonic stem cells, might first be genetically modified and then cross-bred and selected to combine traits associated with genes from different cell lines or to eliminate other unwanted genes that had been introduced in the course of the modification. A similar process of genetic modification and cross-breeding is often used to create genetically modified plants and animals.

Second, the argument that IES is limited in its capacity to generate novel or 'artificial' traits might be flipped on its head to suggest that IES is a more natural process than GGE and that the products of this process are also thereby more natural. Where GGE involves intervention into the genome of the editing child, IES only involves genetic selection. To the extent that one thinks of the human genome as possessing some natural—historical or aesthetic—integrity, this would be a reason to prefer IES. Moreover, any child born at the end of a process of IES would be one that might have arisen naturally, as the result of a historical series of matings between human beings.

Arguments about the moral significance of nature or naturalness are properly controversial. Clearly not everything that occurs in nature—cancer, for instance—is good.[11] Those interested in the pursuit of human enhancement are particularly unlikely to think that something being unnatural makes it bad, given that the most plausible definition of human enhancement holds that it is the provision of capacities that are beyond species-typical (i.e., natural) to human beings.[12] Nevertheless, IES might allow those using it to very significantly enhance the genomes of the people they bring into existence at the end of the process, while still insisting to critics that they are not doing anything that could not occur by chance in nature. Again, a similar argument is often made in defence of the creation of novel plants and animals by breeding programs rather than through genetic modification.[13]

Third, and finally, IES might also have a valuable role to play in fundamental research into reproductive biology and human genetics, allowing researchers to investigate how genes are related and their impact on the early stages of embryo development. Researchers could, for instance, use IES to deliberately bring about certain combinations of genes in order to determine their impacts on

[11] John Harris, *Enhancing Evolution: The Ethical Case for Making Better People* (Princeton: Princeton University Press, 2007); John Stuart Mill, 'Nature', in John Stuart Mill, *Three Essays on Religion* (Amherst: Prometheus Books, 1998), 3–68.

[12] Norman Daniels, 'Normal Functioning and the Treatment-Enhancement Distinction', *Cambridge Quarterly of Healthcare Ethics*, 9/3 (2000), 309–322.

[13] Wendy Russell and Robert Sparrow, 'The Case for Regulating Intragenic GMOs', *Journal of Agricultural and Environmental Ethics*, 21/2 (2008), 153–181.

conception and the early stages of embryonic development.[14] Importantly, this project would most likely be pursued entirely *in vitro*, without bringing children into existence at any point.

Eugenics: What's in a Name?

When I first drew attention to the possibility of IES, I called the technology '*in vitro* eugenics'. In doing so, I drew attention to features it shared with historical programs of eugenics, in particular the extent to which it would involve shaping future human beings in accordance with a conception of what counts as desirable traits in human beings and, consequently, what counts as 'good genes'. Several respondents to that paper were highly critical of my choice of nomenclature, variously suggesting that it was overly dramatic, or alarming, and/or that it would undermine public support for a technology that might have valuable applications.[15] Moreover, subsequent discussions in the literature have tended to use other names to refer to the same technology.[16]

Having described the process and suggested that it may have some utility even in the age of CRISPR, for the remainder of this chapter, I shall consider whether it is indeed, as my original discussion held, a form of eugenics.

One of the reasons why enthusiasts for IES and other novel ways of shaping future children have tended to resist the idea that these technologies are eugenic is because of the historical associations of the word, which have led it to be used as a term of criticism in most contexts.[17] In liberal public discourse, eugenics is a 'boo word': it implies that what it describes is 'bad'. As I discuss further below, though, since the development of technologies of prenatal testing and diagnosis, there has also been a movement to try to reclaim the word and argue for a more value-neutral—or even positively valenced—understanding of eugenics.

A plausible definition of eugenics should make it clear why the project is often thought to be a morally suspect one. However, insofar as much of the evidence 'for' the wrongness of eugenics consists in the historical evils done in its name, it remains open to defenders of eugenics to insist that this connection is

[14] Mathews, 'Language Matters'.

[15] Mathews, 'Language Matters'; Misao Fujita, Yoshimi Yashiro, and Mika Suzuki, 'Throwing the Baby Out with the Bathwater: A Critique of Sparrow's Inclusive Definition of the Term "In Vitro Eugenics"', *Journal of Medical Ethics*, 40/11 (2014), 735–736.

[16] See, e.g., Mathews, 'Language Matters'; Douglas, 'Stem Cell-Derived Gametes'; Ignacio Macpherson, María Victoria Roqué, and Ignacio Segarra, 'Ethical Challenges of Germline Genetic Enhancement', *Frontiers in Genetics*, 10 (2019), article 767; Carl Shulman and Nick Bostrom, 'Embryo Selection for Cognitive Enhancement: Curiosity or Game-Changer?', *Global Policy*, 5/1 (2014), 85–92.

[17] Daniel Wikler, 'Can We Learn from Eugenics?', *Journal of Medical Ethics*, 25/2 (1999), 183–194. It is striking that authors such as Douglas and Bostrom, who in other contexts have defended the moral permissibility of eugenics, nonetheless resist the idea that we should openly acknowledge when it is taking place—e.g., by referring to a technology as '*in vitro* eugenics'.

a contingent one—a product of other circumstances and historical dynamics—and that it is possible to affirm eugenic ideas without thereby endorsing human rights violations.[18] Although I am inclined to think that, as I will argue below, the connection between eugenics and a certain politics is not entirely contingent, I also believe that it is a mistake to foreclose the question of the ethics of technologies of genetic selection and modification by building a normative judgment into the definition of eugenics. In any case, in practice such attempts simply transform the substantive moral question around a particular technology from a debate about whether it is ethical to a debate about whether it is really eugenic. In what follows, then, I will first offer an initial definition of eugenics that does leave it open whether it is morally wrong. I will then briefly describe the history of how this idea has been taken up in practice, which has brought many to conclude that the project is deeply morally flawed. Finally, I will discuss what, if anything, might be wrong with eugenics as I have defined it.

Eugenics and Its History

As almost every discussion of eugenics informs us, the word itself derives from the Greek words *eu* meaning 'good' and *gene* meaning birth.[19] A 'thin'—but, as we shall see, controversial because of that—definition of eugenics, then, is that it is a concern with children being 'well born'. Historically, at various times, people have cared whether children were born of noble 'blood', whether they realized particular ideas of human health and/or excellence, or how they fitted into local schemas of racial hierarchy and character. Ideas about human breeding were undoubtedly informed by people's experience with plants and animals. Nevertheless, concerns about the 'breeding' or 'bloodlines' of other people flourished long before the biology of human reproduction was understood.[20]

[18] Jonathan Anomaly, 'Defending Eugenics', *Monash Bioethics Review*, 35/1 (2018), 24–35; Nicholas Agar, 'Liberal Eugenics', *Public Affairs Quarterly*, 12/2 (1998), 137–155; Nicholas Agar, *Liberal Eugenics: In Defence of Human Enhancement* (Oxford: Blackwell, 2004); Harris, *Enhancing Evolution*; Julian Savulescu and Guy Kahane, 'The Moral Obligation to Create Children with the Best Chance of the Best Life', *Bioethics*, 23/5 (2009), 274–290.

[19] Henk ten Have and Maria do Céu Patrão Neves, 'Eugenics', in Henk ten Have and Maria do Céu Patrão Neves, *Dictionary of Global Bioethics* (Cham: Springer Nature, 2021), 487–488; Rosemarie Garland-Thomson, 'Velvet Eugenics: In the Best Interests of Our Future Children?', in Neal Baer, ed., *The Promise and Peril of CRISPR* (Baltimore: Johns Hopkins University Press, 2024), 151–171.

[20] There are any number of excellent accounts of the history of eugenics, which I would encourage the interested reader to consult: this history is also different in different settings and national contexts. Useful starting points include: Alison Bashford and Philippa Levine, eds., *The Oxford Handbook of the History of Eugenics* (Oxford: Oxford University Press, 2010); Daniel J. Kevles, *In the Name of Eugenics: Genetics and the Uses of Human Heredity* (Cambridge, MA: Harvard University Press, 1995); Nancy Leys Stepan, *The Hour of Eugenics: Race, Gender, and Nation in Latin America* (Ithaca: Cornell University Press, 1996); Alexandra Minna Stern, *Eugenic Nation: Faults and Frontiers of Better Breeding in Modern America*, 2nd ed. (Oakland: University of California Press, 2016); and Stefan Kuhl, *The Nazi Connection: Eugenics, American Racism, and German National Socialism* (Oxford: Oxford University Press, 2002). In the current context, for reasons of space, I can

Precisely for that reason, though, choosing to treat any concern with the bloodlines—or, in more modern parlance, the genes—of children as eugenic is to indulge in anachronism. The person who introduced the term, and in doing so 'invented' eugenics, Francis Galton, defined it, in 1883, as the 'science of improving stock—not only by judicious mating, but whatever tends to give the more suitable races or strains of blood a better chance of prevailing over the less suitable than they otherwise would have had'.[21] Two features of this definition, as well as an observation about the history of eugenics, are essential to a proper understanding of its nature. First, eugenics is explicitly 'scientific': that is, it draws on the scientific method, as well as the knowledge produced by science, to pursue the goal of 'improving stock'. Second, as the references to 'improving stock', 'races or strains of blood', and the idea that the most suitable races or strains should prevail over the less suitable imply, eugenics is oriented towards the distribution of genes in a *population* rather than the welfare of a particular individual. In historical practice, the goal of eugenicists was to improve the 'race' or the 'blood' of the 'nation'. Third, relatedly, although enthusiasts for eugenics advocated for those with 'superior genes' to have more children and in favour of programs wherein, those judged to be, the 'fittest' were encouraged to marry and reproduce—that is, for so-called 'positive eugenics'—in practice eugenicists overwhelmingly focussed on 'negative eugenics'—that is, on discouraging or preventing those judged to have unwanted or 'inferior' traits from reproducing. As Robert Wilson has argued, this obsession with those judged to be 'unfit' was so prevalent in the real-world history of eugenic writing and policy that it should be understood to be a defining feature of eugenics.[22]

Given the bad name it has today, it is important to understand that eugenics had broad public support towards the end of the nineteenth—and in the early years of the twentieth—century. The idea that the scientific method should be used to improve the human stock was widely promulgated by both scientists and social reformers.[23] Indeed, eugenics was a project with significant support from the 'left'.[24] That is, many socialists were strongly in favour of the idea that the state should try to influence, or perhaps even direct, reproductive choice to improve human welfare in accordance with the latest scientific understanding of inheritance. Just as the state should try to ameliorate the social conditions that

only sketch a broad-brush picture of this history, focussing on the role played by the idea of eugenics in United States and Europe.

[21] Francis Galton, *Inquiries into Human Faculty and Its Development* (London: J M Dent and Sons, 1883), cited in Wikler, 'Can We Learn from Eugenics', 184.

[22] Robert A. Wilson, 'Eugenics Undefended', *Monash Bioethics Review*, 37/1 (2019), 68–75.

[23] Nathaniel Comfort, *The Science of Human Perfection: How Genes Became the Heart of American Medicine* (New Haven: Yale University Press, 2014); Kevles, *In the Name of Eugenics*; Wikler, 'Can We Learn from Eugenics?'.

[24] Diane Paul, 'Eugenics and the Left', *Journal of the History of Ideas*, 45/4 (1984), 567–590.

perpetuate poverty and ignorance, so too should it try to address the biological component of intergenerational transmission of these evils. So-called 'Left Eugenics' was an important strand of eugenic thinking.[25]

However, when eugenic ideas entered into public policy, they were used to grant a smoke-screen of scientific respectability to social programs that were both the products and the expression of social bigotry and that involved profound violations of human rights. Claims about the role of heredity in determining human traits dovetailed all too nicely with Social Darwinist claims that where people were located within existing social and economic arrangements was a product of individual merit.[26] Whole populations were stigmatized and denied access to migration and/or citizenship. Individuals deemed at risk of having 'deficient' children were institutionalized and/or sterilized against their will. Notoriously, the German government led by the Nazi party legislated programs of compulsory sterilization of 'undesirables', including people with mental illness and homosexuals. Shortly thereafter Hitler authorized the murder of those deemed 'unworthy of life'. Eugenic ideas also played a role in motivating and mobilizing public support for the mass murder of Jews, who were held to be poisoning the blood of the German people, in the Holocaust.[27]

A key question when it comes to our evaluation of eugenics is whether the monstrous wrongness of fascist eugenics is fully accounted for by the murder and human rights abuses it motivated or whether the arrogance of the 'scientists' guiding the programs and the racism in their application also plays a role. The horrors associated with the Nazi eugenics program were so grotesque that it is tempting to understand the eugenic vision of the Nazis as being exceptional. However, again, it is important to emphasize that 'scientific racism' had widespread currency outside of Germany and played a key role in motivating restrictions on immigration into many countries as well as in still more sinister and repugnant policies, including infanticide and involuntary sterilization, directed against those marked as socially deviant in some way. Given my own institutional location in Australia, it would be remiss of me not to draw attention to the role played by racist eugenics in motivating the policy of removing Aboriginal children from their parents and families over a period that is now known as the 'Stolen Generations'.[28] Eugenic ideas also played a key role in informing the treatment of First Nations peoples in the United States and Canada.[29]

[25] Paul, 'Eugenics and the Left'.

[26] Kevles, *In the Name of Eugenics*.

[27] Kuhl, *The Nazi Connection*; Robert Proctor, *Racial Hygiene: Medicine Under the Nazis* (Cambridge, MA: Harvard University Press, 1988).

[28] Human Rights and Equal Opportunity Commission, *Bringing Them Home: Report of the National Inquiry into the Separation of Aboriginal and Torres Strait Islander Children from their Families* (Canberra: Human Rights and Equal Opportunity Commission, 1997).

[29] Robert A. Wilson, *The Eugenic Mind Project* (Cambridge, MA: MIT Press, 2018).

This history of the mistreatment of Indigenous peoples highlights the fact that eugenic ideas continued to have currency and play a role in public policy even after the end of World War II and despite their association with fascism.[30] It was not until the civil rights movement of the late fifties and early sixties made open expressions of racism politically unacceptable that the idea of eugenics achieved the negative valence that it—for the most part—retains today. Even then, the idea that certain people should be discouraged, or even prevented, from reproducing continued to play a role in shaping policies regarding various 'deviant' populations.[31]

Regardless of whether they were based on racist ideas or respected reproductive liberty, the eugenics of the nineteenth and the first half of the twentieth century were also vitiated by the fact that they were premised on bad science. Enthusiasts for eugenics were mistaken about the genetics of race, and arguably about the nature of the category. They had a poor understanding of the role played by the environment in determining the interaction between genotype and phenotype and often mistook human differences as a result of environmental influences for variation owing to genetic differences. They radically underestimated the extent to which particular human phenotypes are influenced by multiple genes and the importance of gene-gene interactions, with the consequence that they had an inflated sense of the power of individual genes. Thus, even if the policies they supported had not involved and/or led to such grotesque violations of human rights, the project would have been both doomed and unethical by virtue of its dubious scientific foundations.

As I argue below, concerns about the accuracy of the empirical claims underpinning enthusiasm for eugenics remain pressing today. However, it is also true that there have been vast improvements in scientific understanding of human genetics since the 1940s. In particular, researchers have now identified genetic variants associated with a large number of inheritable diseases and disorders and, increasingly, with other phenotypic traits in human beings.[32] They also have a much better understanding of gene-environment interactions. Combined with advances in reproductive, and other biomedical, technologies, these advances have led to the creation of a powerful suite of methods to provide prospective parents with information about the genes and likely phenotype of

[30] Comfort, *The Science of Human Perfection*; Stern, *Eugenic Nation*; Wilson, *The Eugenic Mind Project*.

[31] Nils Roll-Hansen and Gunnar Broberg, eds., *Eugenics and the Welfare State: Norway, Sweden, Denmark, and Finland* (East Lansing: Michigan State University Press, 2005); Wilson, *The Eugenic Mind Project*.

[32] McKusick-Nathans Institute of Genetic Medicine, Johns Hopkins University, *OMIM®: An Online Catalog of Human Genes and Genetic Disorders, 2024*, https://www.omim.org/, 2024; Tom Strachan, *Human Molecular Genetics*, 5th ed. (New York: Garland Science, 2018).

their future children.[33] In particular, where (prospective) parents have already created an embryo, technologies now exist to determine the likelihood that the child who might develop from the embryo will suffer from various genetic diseases and disorders. For the most part, the information made available by these technologies is used to inform parents' decisions about possible termination of a pregnancy or about which embryos to proceed to implant in the context of *in vitro* fertilization. However, the mere existence of these technologies, and the choices they make available, places pressure on the idea that parents should pay no mind to whether their children will be well born. It also exposes parents to social and economic coercion to make the 'right' choice. Finally, it requires the state to consider the public health implications of these new genetic choices. Critics of—and, to an extent, even advocates for—these technologies therefore refer to the use of these technologies, and especially the routinization of these technologies as part of antenatal care, as a 'new' eugenics.[34] The development of the CRISPR/Cas9 gene editing system—and other similar systems—has opened the door to the possibility that parents might directly edit the genomes of their offspring *in vitro* in order to try to enhance their prospects of leading a flourishing life.[35]

I will discuss the precise extent to which—and how—the new eugenics might differ from the old in the next section. Nevertheless, it is clear that there is an enormous gap between contemporary enthusiasm for the use of genetic testing (and other reproductive technologies) to try to ensure the birth of a healthy child, and Nazi eugenics. Moreover, the ethics of the former is an open question in a way that the ethics of the latter is not. When philosophers and others defend the moral permissibility of—or even advocate—eugenics today, it is (almost) always the new eugenics that they have in mind.

What's Wrong with Eugenics?

To this point I have left open the question of the normative status of eugenics. In particular, I have not built the idea that eugenics is morally wrong into the definition of eugenics. Nevertheless, I *do* think that eugenics is morally problematic for reasons that I shall explore briefly in this section.

[33] These technologies include: amniocentesis, chorionic villus sampling, pre-implantation genetic diagnosis, and non-invasive prenatal testing.

[34] Philip Kitcher, *The Lives to Come: The Genetic Revolution and Human Possibilities* (New York: Simon & Schuster, 1996); Wikler, 'Can We Learn from Eugenics?'.

[35] Neal Baer, ed., *The Promise and Peril of CRISPR* (Baltimore: Johns Hopkins University Press, 2024); National Academies of Sciences, Engineering, and Medicine, *Human Genome Editing*; Regalado, 'Engineering the Perfect Baby'.

The nature and extent of the human rights violations that were perpetuated in the course of Nazi eugenics, as well as a result of many of the other policies of the late nineteenth and early twentieth centuries motivated by eugenic ideas, means that these programs would have been evil regardless of their motivation. However, there are several other features of the 'old' eugenics that critics have pointed to as rendering it morally problematic.[36] These programs were typically directed, and carried out, by governments and were characterized by the arrogance, authoritarianism, and insensitivity that is often associated with bureaucracy. Even where they did not involve murder or forced sterilization, they were often coercive and involved violations of the civil rights, and especially the reproductive liberty, of vulnerable minorities. They were perfectionist in the sense that they had a distinct idea of the nature of the human good, which denied the plurality of ways in which human lives may flourish. Finally, they were oriented towards the good of the 'race' or the 'nation' rather than of individuals and characterized by a willingness to sacrifice the latter for the former.

On paper at least, the 'new' eugenics shares none of these features.[37] The technologies are often made available by, and accessed via a contractual arrangement with, private companies rather than by the state. Where programs of genetic testing are supported, or carried out by states, they respect reproductive liberty and are justified by reference to providing potential parents with knowledge to help them make an informed choice about their reproductive options. In particular, the profession of genetic counselling, which plays a key role in the new eugenics, aspires to an ideal of non-directive counselling, which aims to provide information to parents without encouraging or discouraging them to make any particular decision. To the extent that there is an idea guiding the deployment and use of these technologies, it is the child's 'open future', rather than a particular idea of what makes a human life go well.[38] Finally, both the development and use of these technologies are typically justified with reference to the health (or openness of future) of individuals rather than of populations.

Inevitably, these distinctions are overdrawn.

Insofar as the choices available to prospective parents are the products of a series of investment decisions, and an institutional and regulatory context, sustained by the broader society, the new eugenics is more state-centred and more

[36] Wikler, 'Can We Learn from Eugenics?'.

[37] Agar, *Liberal Eugenics: In Defence of Human Enhancement*; Anomaly, 'Defending Eugenics'.

[38] Agar, *Liberal Eugenics: In Defence of Human Enhancement*; Dena S. Davis, *Genetic Dilemmas: Reproductive Technology, Parental Choices, and Children's Futures* (New York: Routledge, 2001); Joel Feinberg, 'The Child's Right to an Open Future', in William Aiken and Hugh LaFollette, eds., *Whose Child? Children's Rights, Parental Authority, and State Power* (Totowa: Littlefield, Adams & Co., 1980), 124–153.

perfectionist than its advocates allow.[39] The ideas about 'health' and the images of what counts as a 'perfect baby' used to advertise and promote the new eugenics look remarkably like those that sat at the heart of the old eugenics.[40] Arguments about public health and the economic benefits of reducing the burden of disease on communities have a habit of creeping back into the justification for public funding of technologies of genetic selection.[41] Moreover, where the use of such technologies might generate problematic aggregate impacts—as in the case of sex selection, for instance—reference to these population-level effects in arguments about the wisdom and ethics of making the technology available seems both inevitable and defensible.[42] A key question about the ethics of the so-called new eugenics, then, is whether it is possible to offer parents choices in relation to the genetics of their children without the state assuming a dominant role and promoting a perfectionist vision of 'better babies'.

Similarly, the social context in which choices about reproduction occur provides ample grounds for concerns about the potential for coercion of parents in relation to these choices.[43] The new eugenics undermines social solidarity by increasing the responsibility of parents for the character of their children, which in turn exacerbates its coercive elements.[44] If I see my neighbour's children's disabilities as stemming from my neighbour's choices, rather than from brute luck, then I may be less willing to see my taxes used to subsidize care or accommodations for their children. If I know that I will bear the entire cost of raising my children, then I may be much less willing—or able—to risk them being born with an impairment. If I know that my children will have to compete with other children whose genes have been selected to maximize their talents and capacities, then I may feel compelled to ensure that my children also have the 'best' available genes. Both Lee Silver and George Annas have argued that this latter dynamic may even lead to what Annas calls 'species

[39] Inmaculada de Melo-Martín, *Rethinking Reprogenetics: Enhancing Ethical Analyses of Reprogenetic Technologies* (Oxford: Oxford University Press, 2017); Nancy Press, 'Assessing the Expressive Character of Prenatal Testing: The Choices Made or the Choices Made Available?', in Erik Parens and Adrienne Asch, eds., *Prenatal Testing and Disability Rights* (Washington, DC: Georgetown University Press, 2000), 214–233.

[40] Robert Sparrow, 'Ethics, Eugenics, and Politics', in Akira Akayabashi, ed., *The Future of Bioethics: International Dialogues* (Oxford: Oxford University Press, 2014), 139–153.

[41] Jonathan Anomaly and Garett Jones, 'Cognitive Enhancement and Network Effects: How Individual Prosperity Depends on Group Traits', *Philosophia*, 48/5 (2020), 1753–1768; Chris Gyngell and Thomas Douglas, 'Stocking the Genetic Supermarket: Reproductive Genetic Technologies and Collective Action Problems', *Bioethics*, 29/4 (2015), 241–250.

[42] Robert Sparrow, 'Imposing Genetic Diversity', *American Journal of Bioethics*, 15/6 (2015), 2–10; Robert Sparrow, 'Human Enhancement for Whom?', in Steve Clarke et al., eds., *The Ethics of Human Enhancement: Understanding the Debate* (Oxford: Oxford University Press, 2016), 127–142.

[43] de Melo-Martín, *Rethinking Reprogenetics*.

[44] Robert Sparrow, 'A Not-So-New Eugenics: Harris and Savulescu on Human Enhancement', *Hastings Center Report* 41/1 (2011), 32–42.

splitting', wherein the human race divides into two different species, one being the descendants of the poor with the same distribution of genes as exists today, and one being the descendants of the wealthy with a radically enhanced suite of genes.[45] Species splitting would only happen if powerful technologies of genetic selection or modification become available and are used over multiple generations. In the meantime, parents are increasingly routinely being encouraged, or even coerced, to practice negative selection by acting to terminate pregnancies on the basis of a diagnosis of a genetic disease or congenital disorder and to avoid transferring embryos with genes associated with disease in the context of IVF.

Finally, although the scientific underpinnings of selection against chromosomal disorders, and single genes strongly associated with genetic disorders, which currently constitutes a majority of the new eugenics, are firm, our understanding of the genetic determinants of human phenotype more generally is much more primitive: most human traits are the product of networks of multiple genes interacting with each other as well as with the environment. In particular, enthusiasm for human enhancement in popular culture—and, to an extent, in the philosophical and bioethical literature—over-emphasizes the role played by genetics and under-emphasizes the role played by the environment. Similarly, it gets very wrong the relative ease with which we can intervene into the genetic and the environmental determinants of phenotype. By encouraging the thought that we could make 'better' human beings via genetic selection and modification, advocates of human enhancement encourage people to think that the existing distribution of talents and opportunities is the product of genetics—and therefore merited—rather than of economic inequality and injustice. Like the old eugenics, the new eugenics both draws support from, and promotes, Social Darwinism. Given that, in most contemporary societies, social and economic inequalities have a strong racial component, this also means that advocating for human enhancement risks reinvigorating discredited ideas about the genetics of race—and, indeed, racism.

In practice, then, the new eugenics shares more features with the old eugenics than its advocates like to admit.[46] As I have argued at length elsewhere, I think that advocates of the new eugenics are naïve about the extent to which respect for human rights is embedded in public culture today and, consequently, about the level of risk that government programs that violate reproductive liberty, forcibly sterilize, or even murder people might emerge again as a result of public

[45] Lee M. Silver, *Remaking Eden: Cloning, Genetic Engineering and the Future of Humankind?* (London: Phoenix, 1999); George J. Annas, *American Bioethics: Crossing Human Rights and Health Law Boundaries* (New York: Oxford University Press, 2009).

[46] Sparrow, 'A Not-So-New Eugenics'; Garland-Thomson, 'Velvet Eugenics'.

support for eugenics.[47] In particular, social institutions that played a key role in resisting fascism and protecting human rights—the trade unions and communist parties—are much weaker now than they were in the 1920s to 1950s. Given what happened in the past, then, we should be extremely cautious about reinvigorating, or trying to rehabilitate, eugenics.

This line of criticism of eugenics points to its historical, and thus likely future, consequences. Like all consequentialist arguments, it is vulnerable to objections that contest the empirical assumptions that underpin the purported connection between actions—or, in this case, ideas—and outcomes.

It is, therefore, worth emphasizing that there are also powerful non-consequentialist criticisms of eugenics. In particular, to many, eugenics appears profoundly arrogant—and immoral because of that.[48] Who is to say what counts as being 'born well'? This line of criticism is often glossed as a concern about 'playing God', but it is important to recognize that it need not rest on theism.[49] For instance, virtue ethics—and especially an 'agent-based' virtue ethics—provide a plausible account of why a willingness to judge some human lives as being better than others is morally reprehensible.[50] A willingness to make such judgments—and to act upon them—also offends against egalitarian sentiments that are widespread in liberal democratic societies. The claim that some people are 'born better' or have 'better genes' than others is, at the very least, in tension with the idea that all people are fundamentally equal by virtue of their membership of the human species.[51] Disability critics have argued that the choice to use, and—even more plausibly—the choice of governments to fund, technologies to prevent the birth of children with disabilities 'expresses' or 'sends a message' that the lives of people with disabilities have less value than the lives of 'normal' people.[52] Moreover, in making choices about what sorts of people come into

[47] Sparrow, 'Ethics, Eugenics, and Politics'; Robert Sparrow, 'The Real Force of "Procreative Beneficence"', in Akira Akayabashi, ed., *The Future of Bioethics: International Dialogues* (Oxford: Oxford University Press, 2014), 183–192.

[48] Garland-Thomson, 'Velvet Eugenics'; Eva Feder Kittay, 'How Not to Argue for Selective Reproductive Procedures', *Kennedy Institute of Ethics Journal*, 27/2 (2017), 185–215.

[49] For an example of interpretation of criticism of technologies of genetic selection and modification as a concern about the ethics of 'playing God', see Harris, *Enhancing Evolution*.

[50] Michael J. Sandel, *The Case Against Perfection: Ethics in the Age of Genetic Engineering*, 1st ed. (Cambridge, MA: Belknap Press of Harvard University Press, 2007); Michael Slote, 'Agent-Based Virtue Ethics', *Midwest Studies in Philosophy*, 20/1 (1995), 83–101.

[51] Garland-Thomson, 'Velvet Eugenics'.

[52] Adrienne Asch, 'Reproductive Technology and Disability', in Sherrill Cohen and Nadine Taub, eds., *Reproductive Laws for the 1990s* (Clifton: Humana Press, 1988), 69–124; Adrienne Asch, 'Why I Haven't Changed My Mind about Prenatal Diagnosis: Reflections and Refinements', in Erik Parens and Adrienne Asch, eds., *Prenatal Testing and Disability Rights* (Washington, DC: Georgetown University Press, 2000), 234–258; Deborah Kaplan, 'Prenatal Screening and Its Impact on Persons with Disabilities', *Clinical Obstetrics and Gynecology*, 36/3 (1993), 605–612; Marsha Saxton, 'Disability Rights and Selective Abortion', in Rickie Solinger, eds., *Abortion Wars: A Half Century of Struggle* (Berkeley: University of California Press, 1997), 374–394; Susan Wendell, *The Rejected Body* (New York, Routledge, 1996).

existence in the next generation, or in future generations, the current generation asserts its rights over future generations in a manner that also seems hard to reconcile with the idea of political equality.[53] Finally, in judging that some genetic endowments are better than others, policymakers must—as must parents—make reference to ideas about the nature of a good human life that are contested in a liberal society. The perfectionism involved in eugenics therefore offends against liberal accounts of the proper role of government that emphasize state neutrality.[54]

The obvious riposte to all these lines of criticism is to point out that we can't avoid relying on ideas about the good life in thinking about the future, especially in relation to our children, but even in relation to our own lives. What critics of eugenics paint as arrogance is unavoidable and, thus, unremarkable. That parents already, necessarily, rely on a conception of what makes a human life go well is particularly evident when they act to try to prevent a particular phenotype (e.g., blindness or loss of a limb) developing as a result of an environmental circumstance (e.g., pollution, traumatic injury, or disease) in the early life of a child.[55] More generally, in making decisions about the physical and social environment in which their children grow up—and especially in relation to education—parents are guided by their sense of what matters when it comes to the lives and futures of their children.[56] Parents—and even policymakers—can act on these beliefs to promote the welfare of their children as they see it without communicating, or implying, anything about the amount of respect owed to existing persons.[57] It is also true of policy in relation to the use of the reproductive technologies that enable the new eugenics that banning them, or regulating the uses to which they may be put, as critics often suggest we should, seems to rely on ideas about the good and to violate rights to reproductive liberty just as much as the use of these technologies does.

The best contemporary critics of the new eugenics have developed, and are continuing to develop, responses to these defences of the new eugenics: reasons of space prevent me from pursuing these any further here.[58] However, I do want

[53] Jurgen Habermas, *The Future of Human Nature* (Cambridge: Polity Press, 2003).

[54] Sparrow, 'Liberalism and Eugenics'.

[55] Allen Buchanan, 'Choosing Who Will Be Disabled: Genetic Intervention and the Morality of Inclusion', *Social Philosophy and Policy*, 13/2 (1996), 18–46.

[56] Claudia Mills, 'The Child's Right to an Open Future?', *Journal of Social Philosophy*, 34/4 (2003), 499–509.

[57] Buchanan, 'Choosing Who Will Be Disabled'; James Lindemann Nelson, 'The Meaning of the Act: Reflections on the Expressive Force of Reproductive Decision Making and Policies', *Kennedy Institute of Ethics Journal*, 8/2 (1998), 165–182.

[58] Elizabeth Barnes, *The Minority Body: A Theory of Disability* (Oxford: Oxford University Press, 2016); de Melo-Martín, *Rethinking Reprogenetics*; Kittay, 'How Not to Argue for Selective Reproductive Procedures'; Matthew J. Barker and Robert A Wilson, 'Well-Being, Disability, and Choosing Children', *Mind*, 128/510 (2019), 305–328; Osagie K. Obasogie and Marcy Darnovsky, eds., *Beyond Bioethics: Toward a New Biopolitics* (Oakland: University of California Press, 2018).

to highlight—and, to an extent at least, endorse—one final line of criticism of the project of trying to shape the genes of our children. The colloquial description of the (mostly hypothetical) results of this project as 'designer babies' captures, I believe, something important about it, which is that it undercuts the distinction between 'the begotten' and 'the made'.[59] Although, today, this concern is directed towards the 'new eugenics'—and especially towards hypothetical germline gene editing of humans—it also arises, and was raised in relation to, the state-directed programs intended to improve the genes of the next generation that were characteristic of the old eugenics. The process of consciously trying to determine the phenotype of a child by determining its genes—the process of design—changes the relationship between the designers and the designed and places the child alongside other artefacts that have their properties because a designer has chosen them. The designers become responsible for the design choices that they have made: the child's life will be shaped by those choices. The relationship between designers and the designed is, therefore, deeply asymmetrical. Those individuals who have been designed will be able to trace the intentions of their designers in the shape of their own lives. As a consequence, they may struggle to feel themselves to be—and, indeed, may not even entirely be—the authors of their own lives.[60] These effects will be more present—and more significant—the more power parents have to determine the genomes of their children. Thus, they will be more pertinent if it becomes possible for parents to make multiple changes to their child's genome via germline genome editing.

Of course, parents have always wanted things for their children. Moreover, they have always had wishes for, and power over, their children, including some capacity to shape them to their own ends, and have exercised this by modifying their environment. Finally, no child comes into existence 'without' a genome and thus without their genome having some influence on their character and talents and thus the life options available to them. That is to say, whatever sense in which it is possible for individuals to be the authors of their own lives must be compatible with their coming into existence with a genome that they have not chosen. Children born with 'designed' genomes may even have more options available to them—a more 'open' future—by virtue of being born with genes that promote the development of talents and capacities that will be useful in the pursuit of multiple different ends.[61]

[59] Hans Jonas, 'Biological Engineering—-A Preview', in Hans Jonas, *Philosophical Essays: From Ancient Creed to Technological Man* (Englewood Cliffs: Prentice Hall, 1974), 141–167; Clive S. Lewis, *The Abolition of Man; or, Reflections on Education with Special Reference to the Teaching of English in the Upper Forms of Schools* (New York: Harper One, 2001); The President's Council on Bioethics, *Human Cloning and Human Dignity: An Ethical Inquiry* (Washington, DC: The President's Council on Bioethics, 2002); Sandel, *The Case Against Perfection.*

[60] Habermas, *The Future of Human Nature.*

[61] Agar, *Liberal Eugenics: In Defence of Human Enhancement*; Sparrow, 'A Child's Right to a Decent Future?'.

What is new and different about technologies of genetic selection and—especially—modification, though, is the way they require parents to relate to their (future) children in a 'technical' rather than a 'communicative' or 'deliberative' mode. As the metaphor of 'design' suggests, parents (and the medical team that supports them in their efforts) who want to determine the genes of their future child confront a problem that is essentially a matter of selecting the appropriate means to bring about the end they desire—that is to say, it is a problem of engineering: they relate to the embryo (or embryos) that will become their child as a lump of matter that might resist their efforts to reshape it to satisfy their own desires. This relationship is very different from the relationship that parents have to a child when they are trying to get their child to do something, or to develop a certain character, by manipulating the child's environment. In this latter case, the parents relate to the child as someone who might demand a justification for the way in which they are being treated: the child is interpellated as someone who is, at least potentially, a member of the 'kingdom of ends'.

The fact that the traits of a designer baby have been determined by their designers relating to them in this technical/instrumental mode is what makes them akin to other artefacts. Moreover, the political relationship between designers and the designed is very different in the case where parents have shaped their children via genetic rather than environmental interventions. The instrumental mode is fundamentally asymmetric.[62] By contrast, the communicative/deliberative mode has an egalitarian moment, insofar as it allows that the child has a right to an account of the reasons for the parents' choices, which in turn implies that the child might contest the reasons that are offered.[63] What threatens the freedom of designer babies is not that they necessarily have fewer life options but that there is someone standing in the relationship of their designer to them.[64] The sense in which designer babies are less free than those conceived naturally is a political one, rather than a question of how many life options they have.

Eugenics—old *and* new—is morally problematic, then, both because of the potential of ideas about superior genes to lend support to policies that violate human rights *and* because of the relationship it establishes between designers and designed.

[62] Anca Gheaus, 'Parental Genetic Shaping and Parental Environmental Shaping', *The Philosophical Quarterly*, 67/267 (2017), 263–281.

[63] Sparrow, 'Yesterday's Child'; Erik Malmqvist, 'Analysing Our Qualms about "Designing" Future Persons: Autonomy, Freedom of Choice, and Interfering with Nature', *Medicine, Health Care and Philosophy*, 10/4 (2007), 407–416.

[64] Lewis, *The Abolition of Man*.

'Iterated Embryo Selection'? Or '*In Vitro* Eugenics'?

With this account of the nature and normative status of eugenics in hand, we are now better placed to assess whether it is appropriate to describe iterated embryo selection as '*in vitro* eugenics'. Insofar as the motives of those carrying out IES are to produce a final generation of embryos who would be 'well born' by virtue of having 'better' genes, this process is quintessentially eugenic according to the 'thin' definition of eugenics. In order to achieve the desired genotype in the final generation, multiple decisions about which genes are desirable, as well as which embryos to 'mate' and which to discard, would need to be made. These decisions would necessarily be informed by a controversial account of human flourishing—of what a 'good' human life looks like. The process would require a guiding will, with a clear goal in mind, over an extended period of time: the will of the person guiding the process plays a role akin to that of the state in the old eugenics. Moreover, the process of IVE necessarily involves being concerned with the genetics of a *population* and not just the genetics of individuals. That is, even if the goal is to bring only a single individual, with some particular combination of genes, into existence at the end of the program, in order to achieve this researchers would need to ensure that they had all the necessary genes in the population of embryos used to permit the selection and combination of the desired genes, as well as sufficient genetic diversity to prevent other deleterious combinations of genes emerging in the context of the history of this process. Not only is IES eugenic, then, even according to the more substantive definition of eugenics that I have argued is more historically accurate, but in many ways it looks more similar to the old eugenics than the new!

Granted, as I discuss further below, the project need involve no violations of human rights. In particular, it does not involve forcing anyone to have children or preventing anyone from having children. However, as I argued above, while such human rights violations occurred in the course of policies motivated by eugenics in Nazi Germany and elsewhere, they are best understood as a product of eugenics rather than a condition of a policy, or idea, being eugenic. Thus, the absence of human rights violations in the course of IVE does not mean that it is not eugenic.

However, my account of the nature of eugenics does imply that there could be programs of IES that are less obviously eugenic than others. If IES were to be carried out purely for research purposes to understand human genetics without any ambition to bring children with 'superior' genomes into existence, then it might be argued that this did not count as a form of eugenics. This is arguable, rather than straightforward, because many of the features of eugenics would remain present even in this case. The process would still require a guiding will and would still involve a concern for the genetics of a population of embryos and not just individual embryos. Moreover, if the purpose of the research was to identify—even if not to create—'superior' genomes, then I believe this use of IES would still be a form of IVE. That is, IES may be eugenic by virtue of its place in a larger project of trying to improve the 'quality' of human births even if the process itself is not intended to lead to live births.

The Ethics of *In Vitro* Eugenics

As discussed above, that IES is a form of eugenics does not necessarily mean, even if it strongly implies, that it is morally wrong. IVE is a useful test case in this regard because it involves no direct harm to any individuals. Unless one believes that embryos have the moral status of persons, the technology does not require killing anyone. Nor, as long as researchers secure the permission of the individuals whose embryos (or gametes) are used to begin the process, as well as the consent of the women who might bring the final generation of embryos to term, need it involve any violation of reproductive liberty.[65]

Nevertheless, the criticisms of eugenics set out above apply, with no little force, to IVE. The goal of creating children with superior genes arguably reflects a morally problematic arrogance and is in tension with the ideal of political equality. Like genetic engineering—including genome editing—more generally, the attempt to shape the genomes of children via IVE would undercut the distinction between the begotten and the made. The children who would come into existence at the end of this process would be the result of an extended process of 'manufacture' conducted in a very technical mode. The person(s) who designed them would stand in a relationship to them that was entirely asymmetric. By virtue of the historical trace of the exercise of power by their designers over their character and traits, children born as the result of IVE would be radically less free—at least in the political sense described above—than children conceived naturally.

Perhaps more importantly—given that the number of people born as a result of IVE is likely to be small—the willingness to assert the superiority of some genomes, and the emphasis on the genetic, rather than the environmental, influences on life outcomes, involved in IVE risks encouraging false beliefs about the causal origins of the existing (unjust) social order. The project would need to be pursued over a number of years, perhaps as long as a decade, and

[65] There are, however, valid grounds for concerns about whether the decision of women to be implanted with the embryos created by IVE, and to give birth to the children who would develop from these embryos, is likely to be fully informed and/or whether the institutional setting in which they consent to do these things might involve coercion, especially insofar as these women would be encouraged to believe that they would give birth to 'better' babies.

would require significant institutional support. IVE is therefore highly 'public' in comparison to other reproductive choices, and concerns about what IVE communicates or expresses are, correspondingly, especially salient when it comes to its ethics. The message IVE sends resonates with, and lends support to, Social Darwinism. Given the historical associations between eugenics, Social Darwinism, and human rights abuses, there are ample reasons to worry that enthusiasm for IVE might contribute to the development of a political culture in which restrictions of the reproductive liberty of those judged to have 'inferior' genes are held to be justified. History has shown that, when such a culture emerges, 'bioethical' arguments based on concerns about the 'quality' of persons and their genes may be mobilized in defence of forced sterilization and murder. Insofar as the size of this risk depends on the social and political context in which IVE is being pursued or promoted, it would be premature to conclude that these concerns render the pursuit of IVE unethical. Nevertheless, I believe that they should weigh heavily in discussions of its all-things-considered moral permissibility.

Lessons for Other Reproductive Technologies ... and for Bioethicists

It may be that IVE remains an intriguing possibility that is never attempted: the technical barriers to safe iteration of gametogenesis may not be overcome, the project might take too long, or it might prove easier to make multiple changes to the genomes of embryos via direct genome editing. Even if this is the case, I do not believe that the time I have spent here considering the relationship between this hypothetical technology and the history of eugenics will have been wasted. As should be obvious, the ethical and political concerns I have surveyed here also arise regarding the use of genome editing, as well as existing technologies, such as pre-implantation genetic testing, to try to shape the genes of future children to ensure that they are 'well born'. Placing eugenics under the microscope helps us to see that the ethics of the new eugenics is a product of its politics.[66] It is what this project expresses in the public sphere, the relationship that it establishes between the designer and the designed, and the risk that it will reinvigorate false ideas about the origins of social and economic inequality that makes it morally problematic.

Importantly, two of these political concerns arise regarding discussions of the new eugenics, including philosophical and bioethical discussion thereof.[67]

[66] Sparrow, 'Ethics, Eugenics, and Politics'.
[67] Sparrow, 'Ethics, Eugenics, and Politics'.

That is to say, in advocating for the project of improving the genes of future generations—and even just in taking this possibility seriously—philosophers and bioethicists risk encouraging their readership to believe that genetics is destiny and that the wealthy are wealthy and the poor, poor because of their genes. Perhaps the most challenging implication of my discussion here, then, is that critics, like myself, as well as advocates of the new eugenics, have a moral obligation to consider whether the risks that they will contribute to a resurgence of the old eugenics is justified by the light they shed on the new.

Bibliography

Agar, Nicholas, 'Liberal Eugenics', *Public Affairs Quarterly*, 12/2 (1998), 137–155.

Agar, Nicholas, *Liberal Eugenics: In Defence of Human Enhancement* (Oxford: Blackwell, 2004).

Annas, George J., *American Bioethics: Crossing Human Rights and Health Law Boundaries* (New York: Oxford University Press, 2009).

Anomaly, Jonathan. 'Defending Eugenics', *Monash Bioethics Review*, 35/1 (2018), 24–35.

Anomaly, Jonathan, and Jones, Garett, 'Cognitive Enhancement and Network Effects: How Individual Prosperity Depends on Group Traits', *Philosophia*, 48/5 (2020), 1753–1768.

Asch, Adrienne, 'Reproductive Technology and Disability', in Sherrill Cohen and Nadine Taub, eds., *Reproductive Laws for the 1990s* (Clifton: Humana Press, 1988), 69–124.

Asch, Adrienne, 'Why I Haven't Changed My Mind about Prenatal Diagnosis: Reflections and Refinements', in Erik Parens and Adrienne Asch, eds., *Prenatal Testing and Disability Rights* (Washington, DC: Georgetown University Press, 2000), 234–258.

Baer, Neal, ed., *The Promise and Peril of CRISPR* (Baltimore: Johns Hopkins University Press, 2024).

Barker, Matthew J., and Wilson, Robert A., 'Well-Being, Disability, and Choosing Children', *Mind*, 128/510 (2019), 305–328.

Barnes, Elizabeth, *The Minority Body: A Theory of Disability* (Oxford: Oxford University Press, 2016).

Bashford, Alison, and Levine, Philippa, eds., *The Oxford Handbook of the History of Eugenics* (Oxford: Oxford University Press, 2010).

Buchanan, Allen, 'Choosing Who Will Be Disabled: Genetic Intervention and the Morality of Inclusion', *Social Philosophy and Policy*, 13/2 (1996), 18–46.

Comfort, Nathaniel, *The Science of Human Perfection: How Genes Became the Heart of American Medicine* (New Haven: Yale University Press, 2014).

da Fonseca, Flávio Guimarães, Ribeiro, Daniel Mendes, Carvalho, Nara Pereira, and Stancioli, Brunello, 'Human In Vitro Eugenics: Close, Yet Far Away', *Journal of Medical Ethics*, 40/11 (2014), 738–739.

Daniels, Norman, 'Normal Functioning and the Treatment-Enhancement Distinction', *Cambridge Quarterly of Healthcare Ethics*, 9/3 (2000), 309–322.

Davis, Dena S., *Genetic Dilemmas: Reproductive Technology, Parental Choices, and Children's Futures* (New York: Routledge, 2001).

de Melo-Martín, Inmaculada. *Rethinking Reprogenetics: Enhancing Ethical Analyses of Reprogenetic Technologies* (Oxford: Oxford University Press, 2017).

Douglas, Thomas, 'Stem Cell-Derived Gametes, Iterated In Vitro Reproduction, and Genetic Parenthood', *Journal of Medical Ethics*, 40/11 (2014), 723–724.

Feinberg, Joel, 'The Child's Right to an Open Future', in William Aiken and Hugh LaFollette, eds., *Whose Child? Children's Rights, Parental Authority, and State Power* (Totowa: Littlefield, Adams & Co., 1980), 124–153.

Fujita, Misao, Yashiro, Yoshimi, and Suzuki, Mika, 'Throwing the Baby Out with the Bathwater: A Critique of Sparrow's Inclusive Definition of the Term "In Vitro Eugenics"', *Journal of Medical Ethics*, 40/11 (2014), 735–736.

Galton, Francis, *Inquiries into Human Faculty and its Development* (London: J M Dent and Sons, 1883).

Garland-Thomson, Rosemarie, 'Velvet Eugenics: In the Best Interests of Our Future Children?', in Neal Baer, ed., *The Promise and Peril of CRISPR* (Baltimore: Johns Hopkins University Press, 2024), 151–171.

Gheaus, Anca, 'Parental Genetic Shaping and Parental Environmental Shaping', *The Philosophical Quarterly*, 67/267 (2017), 263–281.

Gyngell, Chris, and Douglas, Thomas, 'Stocking the Genetic Supermarket: Reproductive Genetic Technologies and Collective Action Problems', *Bioethics*, 29/4 (2015), 241–250.

Habermas, Jurgen, *The Future of Human Nature* (Cambridge: Polity Press, 2003).

Harris, John, *Enhancing Evolution: The Ethical Case for Making Better People* (Princeton: Princeton University Press, 2007).

Human Rights and Equal Opportunity Commission, *Bringing Them Home: Report of the National Inquiry into the Separation of Aboriginal and Torres Strait Islander Children from their Families* (Canberra: Human Rights and Equal Opportunity Commission, 1997).

Jonas, Hans, 'Biological Engineering—A Preview', in Hans Jonas, *Philosophical Essays: From Ancient Creed to Technological Man* (Englewood Cliffs: Prentice Hall, 1974), 141–167.

Kaplan, Deborah, 'Prenatal Screening and Its Impact on Persons with Disabilities', *Clinical Obstetrics and Gynecology*, 36/3 (1993), 605–612.

Kevles, Daniel J., *In the Name of Eugenics: Genetics and the Uses of Human Heredity* (Cambridge, MA: Harvard University Press, 1995).

Kitcher, Philip, *The Lives to Come: The Genetic Revolution and Human Possibilities* (New York: Simon & Schuster, 1996).

Kittay, Eva Feder, 'How Not to Argue for Selective Reproductive Procedures', *Kennedy Institute of Ethics Journal*, 27/2 (2017), 185–215.

Kuhl, Stefan, *The Nazi Connection: Eugenics, American Racism, and German National Socialism* (Oxford: Oxford University Press, 2002).

Lewis, Clive S., *The Abolition of Man; or, Reflections on Education with Special Reference to the Teaching of English in the Upper Forms of Schools* (New York: Harper One, 2001).

Macpherson, Ignacio, Roqué, María Victoria, and Segarra, Ignacio, 'Ethical Challenges of Germline Genetic Enhancement', *Frontiers in Genetics*, 10 (2019), article 767.

Malmqvist, Erik, 'Analysing Our Qualms about "Designing" Future Persons: Autonomy, Freedom of Choice, and Interfering with Nature', *Medicine, Health Care and Philosophy*, 10/4 (2007), 407–416.

Mathews, Debra J. H., 'Language Matters', *Journal of Medical Ethics*, 40/11 (2014), 733–734.

McKusick-Nathans Institute of Genetic Medicine, Johns Hopkins University, *OMIM®: An Online Catalog of Human Genes and Genetic Disorders, 2024*, https://www.omim.org/, 2024.

Mertes, Heidi, 'A Moratorium on Breeding Better Babies', *Journal of Medical Ethics*, 40/11 (2014), 734–735.

Mill, John Stuart, 'Nature', in John Stuart Mill, *Three Essays on Religion* (Amherst: Prometheus Books, 1998), 3–68.

Mills, Claudia, 'The Child's Right to an Open Future?', *Journal of Social Philosophy*, 34/4, (2003), 499–509.

National Academies of Sciences, Engineering, and Medicine, *Human Genome Editing: Science, Ethics, and Governance* (Washington, DC: The National Academies Press, 2017).

Nelson, James Lindemann, 'The Meaning of the Act: Reflections on the Expressive Force of Reproductive Decision Making and Policies', *Kennedy Institute of Ethics Journal*, 8/2 (1998), 165–182.

Obasogie, Osagie K., and Darnovsky, Marcy, eds., *Beyond Bioethics: Toward a New Biopolitics* (Oakland: University of California Press, 2018).

Palacios-González, Cesar, Harris, John, and Testa, Giuseppe, 'Multiplex Parenting: IVG and the Generations to Come', *Journal of Medical Ethics*, 40/11 (2014), 752–758.

Paul, Diane, 'Eugenics and the Left', *Journal of the History of Ideas*, 45/4 (1984), 567–590.

The President's Council on Bioethics, *Human Cloning and Human Dignity: An Ethical Inquiry* (Washington, DC: The President's Council on Bioethics, 2002).

Press, Nancy, 'Assessing the Expressive Character of Prenatal Testing: The Choices Made or the Choices Made Available?', in Erik Parens and Adrienne Asch, eds., *Prenatal Testing and Disability Rights* (Washington, DC: Georgetown University Press, 2000), 214–233.

Proctor, Robert, *Racial Hygiene: Medicine Under the Nazis* (Cambridge, MA: Harvard University Press, 1988).

Regalado, Antonio, 'Engineering the Perfect Baby', *MIT Technology Review*, 118/3 (2015), 26–33.

Roll-Hansen, Nils, and Broberg, Gunnar, eds., *Eugenics and the Welfare State: Norway, Sweden, Denmark, and Finland* (East Lansing: Michigan State University Press, 2005).

Russell, Wendy, and Sparrow, Robert, 'The Case for Regulating Intragenic GMOs', *Journal of Agricultural and Environmental Ethics*, 21/2 (2008), 153–181.

Sandel, Michael J., *The Case Against Perfection: Ethics in the Age of Genetic Engineering*, 1st ed. (Cambridge, MA: Belknap Press of Harvard University Press, 2007).

Savulescu, Julian, and Kahane, Guy, 'The Moral Obligation to Create Children with the Best Chance of the Best Life', *Bioethics*, 23/5 (2009), 274–290.

Saxton, Marsha, 'Disability Rights and Selective Abortion', in Rickie Solinger, eds., *Abortion Wars: A Half Century of Struggle* (Berkeley: University of California Press, 1997), 374–394.

Shulman, Carl, and Bostrom, Nick, 'Embryo Selection for Cognitive Enhancement: Curiosity or Game-Changer?', *Global Policy*, 5/1 (2014), 85–92.

Silver, Lee M., *Remaking Eden: Cloning, Genetic Engineering and the Future of Humankind?* (London: Phoenix, 1999).

Slote, Michael, 'Agent-Based Virtue Ethics', *Midwest Studies in Philosophy*, 20/1 (1995), 83–101.

Sparrow, Robert, 'A Child's Right to a Decent Future? Regulating Human Genetic Enhancement in Multicultural Societies', *Asian Bioethics Review*, 4/4 (2012), 355–373.

Sparrow, Robert, 'A Not-So-New Eugenics: Harris and Savulescu on Human Enhancement', *Hastings Center Report* 41/1 (2011), 32–42.

Sparrow, Robert, 'Ethics, Eugenics, and Politics', in Akira Akayabashi, ed., *The Future of Bioethics: International Dialogues* (Oxford: Oxford University Press, 2014), 139–153.

Sparrow, Robert, 'Human Enhancement for Whom?', in Steve Clarke, Julian Savulescu, C. A. J. Coady, Alberto Giubilini, and Sagar Sanyal, eds., *The Ethics of Human Enhancement: Understanding the Debate* (Oxford: Oxford University Press, 2016), 127–142.

Sparrow, Robert, 'Imposing Genetic Diversity', *American Journal of Bioethics*, 15/6 (2015), 2–10.

Sparrow, Robert, 'In Vitro Eugenics', *Journal of Medical Ethics*, 40/11 (2014), 725–731.

Sparrow, Robert, 'Liberalism and Eugenics', *Australasian Journal of Philosophy*, 89/3 (2011), 499–517.

Sparrow, Robert, 'Orphaned at Conception: The Uncanny Offspring of Embryos', *Bioethics*, 26/4 (2012), 173–181.

Sparrow, Robert, 'The Real Force of "Procreative Beneficence"', in Akira Akayabashi, ed., *The Future of Bioethics: International Dialogues* (Oxford: Oxford University Press, 2014), 183–192.

Sparrow, Robert, 'Yesterday's Child: How Gene Editing for Enhancement Will Produce Obsolescence—and Why It Matters', *American Journal of Bioethics*, 19/7 (2019), 6–15.

Sparrow, Robert, Koplin, Julian, and Mills, Catherine, 'Mitochondrial Replacement Techniques: A Critical Review of the Ethical Issues', in Diana Bowman, Karinne Ludlow, and Walter Johnson, eds., *Reproduction Reborn: How Science, Ethics and Law Shape Mitochondrial Replacement Therapies* (Oxford: Oxford University Press, 2023), 32–61.

Stepan, Nancy Leys, *The Hour of Eugenics: Race, Gender, and Nation in Latin America* (Ithaca: Cornell University Press, 1996).

Stern, Alexandra Minna, *Eugenic Nation: Faults and Frontiers of Better Breeding in Modern America*, 2nd ed. (Oakland: University of California Press, 2016).

Strachan, Tom, *Human Molecular Genetics*, 5th ed. (New York: Garland Science, 2018).

ten Have, Henk, and do Céu Patrão Neves, Maria, 'Eugenics', in Henk ten Have and Maria do Céu Patrão Neves, *Dictionary of Global Bioethics* (Cham: Springer Nature, 2021), 487–488.

Watt, Helen, 'Ancestor Embryos: Embryonic Gametes and Genetic Parenthood', *Journal of Medical Ethics*, 40/11 (2014), 759–761.

Wendell, Susan, *The Rejected Body* (New York: Routledge, 1996).

Wikler, Daniel, 'Can We Learn from Eugenics?', *Journal of Medical Ethics*, 25/2 (1999), 183–194.

Wilson, Robert A., *The Eugenic Mind Project* (Cambridge, MA: MIT Press, 2018).

Wilson, Robert A., 'Eugenics Undefended', *Monash Bioethics Review*, 37/1 (2019), 68–75.

6

Respecting Offspring; Identifying Parents

When Is *In Vitro* Gametogenesis (Truly) Parental?

Helen Watt

Respectful Parental Conception

Having a child to whom one is biologically related is valued by many, albeit in different ways.[1] Certainly, the existence of a child and the love between child and parents—genetic, gestational, and/or social—seem valuable in themselves, even in very adverse situations. Moreover, the child's moral dignity—which refers to, but does not reduce to,[2] the value of her life—is of course in no way affected by the way in which the child was conceived.

However, none of this means that arranging conception as one wishes is always morally defensible. Nor does it mean that the state may assist one's arrangements or even, in every case, permit them. There is a vast spectrum of cases between the loving, respectful conception of children in a way protective of their welfare and the most egregious ways in which children are disrespected by their coming-to-be. Even mid-spectrum, parents themselves may come to wish—however impossible this is in terms of identity[3]—that their dearly loved child was otherwise conceived. A one-night stand, for example, might be regretted, even if conception was the aim.

True, the state often stays in the background, letting people make their own conception choices though hopefully ready to pick up the pieces should the family need support. But there are limits to what the state will (or should) tolerate: putting embryos or gametes up on eBay, for example, might be frowned on. The activities of fertility clinics and laboratories, in particular, are not a

[1] Michal Pruski, 'The Relationship of Gametes to Those Who Procreate and Its Impact on Artificially Generated Gamete Technologies', *Ethics & Medicine*, 33/1 (2017), 28.

[2] Helen Watt, 'The Dignity of Human Life: Sketching Out an "Equal Worth" Approach', *Ethics and Medicine*, 31/1 (2020), 7–17.

[3] It is not necessary that one be 'worse off' than one would otherwise have been in order to be wronged. People can undergo wrongs—have their dignity impinged on—despite the fact that these wrongs are inextricable from their (valuable) existence, as when parents conceive a child with the aim of selling him into slavery. The conception for that purpose, not just the subsequent sale and enslavement, wrongs the child.

Helen Watt, *Respecting Offspring; Identifying Parents* In: *Revolutionizing Reproduction.* Edited by: Christopher Gyngell, Julian Savulescu, and Tsutomu Sawai, Oxford University Press. © Oxford University Press 2026.
DOI: 10.1093/9780197810415.003.0006

private domain but one where prohibitions in contexts including human reproduction are often accepted. Children's welfare matters, and the state should be concerned to protect it even in the face of alluring promises of reprotechnology and heartfelt responses of would-be parents.

This chapter will focus on three ways in which *in vitro* gametogenesis (IVG) might be used: IVG with autologous transplantation, the mass-production and screening of IVG embryos, and 'multiplex' reproduction by multiple people. There are many other issues worth exploring, not least potential adverse physical effects of IVG[4] on the child and later generations: the result of errors that are not merely passed on, as in natural conception, but introduced by the techniques themselves. Irrespective of such issues, I will argue that there are ethical objections to some forms of IVG discussed in this chapter, and also doubts about whether some forms would really help IVG users to achieve genetic parenthood. In the case of variants where genetic parenthood *is* achieved and reproduction is 'parental' as opposed to 'nonparental', the question still arises whether parents are invited to act 'unparentally'[5] towards their offspring.

Restoring Reproductive Health: IVG with Autologous Transplantation

Before turning to the second of these issues, the mass-production and screening of IVG embryos, I will begin with a form of IVG that would *not* involve these features if employed. This is IVG—call it 'IVG with autologous transplantation' (IVG-AT)—of a kind that also has the best claim to achieve not just genetic parenthood but reproductive health. Like a similar intervention following *in vitro* maturation (IVM), it would restore reproductive function to the infertile person, creating reproductive cells/tissue from his/her own cells not for external use but for transplantation so that full fertility is restored. What is transplanted becomes again part of the person, potentially allowing him or her to conceive a

[4] Seppe Segers, Guido Pennings, Wybo Dondorp, Guido De Wert, and Heidi Mertes, 'In Vitro Gametogenesis and Reproductive Cloning: Can We Allow One While Banning the Other?', *Bioethics*, 33/1 (2019), 68–75; Barbara Advena-Regnery, Hans-Georg Dederer, Franziska Enghofer, Tobias Cantz, and Thomas Heinemann, 'Framing the Ethical and Legal Issues of Human Artificial Gametes in Research, Therapy, and Assisted Reproduction: A German perspective', *Bioethics*, 32/5 (2018), 315.

[5] We will return to this below. The word 'unparental' is intended to suggest not that one is not a genetic parent but that one has not acted as a (morally) good one—so 'unparental' can be contrasted with 'nonparental' in the same way as 'unprofessional' is contrasted with 'nonprofessional'. Good parents accept children unconditionally, care about their welfare, and respect them as moral equals—though a child who was once seen as subordinate and even as property may later become unconditionally loved and respected. If the child seems happy and unaffected by his history (perhaps he never learns that he was conceived to be sold into slavery; see footnote 3), this does not mean that the parents did not act 'unparentally' at the time although they may be good parents now.

child with his/her partner in a way reflecting reproductive health and not merely a reproductive outcome.[6] Whatever other objections there may be to IVG-AT—for example, in terms of safety—there are some moral issues that simply will not arise if the patient is enabled to conceive 'naturally'.

I should stress that this is not about favouring 'nature' in the crude generic sense of what is 'out there'—good or bad, functional or dysfunctional—apart from or after some procedure. If that is all we mean by 'nature', all technology, medical or otherwise, is 'unnatural'. Rather, the point or points in focus are that IVG-AT enables *sexual* conception—something itself potentially of value—by promoting the person's *reproductive health*, unlike IVG with externalized conception. If 'health' is understood as an already-evaluative term referring to human functionality as an aspect of human flourishing, then the fact that IVG-AT would promote reproductive health is at least something in its favour. One would be a biological parent in the full sense, since the gametes or their precursors originate from one's own cells (intervened on to *make* them gametes), and conception would be the outcome of a holistic sexual act. At the most committed level, an IVG-AT pregnancy could express a lifelong bond in the archetypal way: unlike laboratory conception, marital sex can embody that bond[7] quite apart from any aim to conceive. That said, at this point in the chapter we are more concerned with the fact that 'natural' conception via IVG-AT will avoid some ethical issues in regard to mass-production, screening, and disposal of IVG embryos, to which we now turn.

Mass-Production and Polygenic Screening

Although mass-production and screening of embryos is a separate issue from IVG, they are worth examining together due to their not unlikely association.[8] IVG will enable mass-production of gametes and embryos, while mass-production of embryos would likely be combined with polygenic screening, including for non-medical reasons.[9] This outcome is regarded with satisfaction

[6] To illustrate the point: use of sperm in IVF does not treat the sperm provider in the sense of improving his reproductive health—bearing in mind that he could even be deceased if his sperm is used post-mortem. Nor does it treat his female partner, whether she is fertile (and therefore does not need treatment) or infertile (in which case, the intervention again provides a reproductive *opportunity* but does not make her body healthier in reproductive terms).

[7] Helen Watt, *The Ethics of Pregnancy, Abortion and Childbirth: Exploring Moral Choices in Childbearing* (New York: Springer, 2016), 114–118, 120–121; Pruski, 'The Relationship of Gametes to Those Who Procreate'.

[8] Seppe Segers, Guido Pennings, Wybo Dondorp, Guido De Wert, and Heidi Mertes, 'In Vitro Gametogenesis and the Creation of 'Designer Babies', *Cambridge Quarterly of Healthcare Ethics*, 28/3 (2019), 504–505.

[9] 'Medical' is arguably a misnomer for screening out embryos for disability: this neither prevents disability (the affected embryos already exist) nor improves the health of the parents, at least directly. Theories of medicine should not pass too quickly from personal preferences to what can be offered

by some commentators: in the words of Palacios-González and colleagues, 'Finally prospective parents will be able to choose among a hitherto unimaginable variety of potential children'.[10] Polygenic and other embryo screening, including for sex, is already on offer in some clinics in the context of *in vitro* fertilization (IVF), while screening of egg or sperm donors for features desired by intended parents is already practiced in many countries.[11] As with other innovative reproductive technologies, defences of IVG often stress the fact that current assisted reproductive technologies (ARTs), including the screening and/or discarding of IVF embryos, are acceptable or, at any rate, accepted.

My aim is in no way to defend existing ARTs, which should not, I would argue, be assumed to be benign—even if things can always deteriorate further. Structurally, existing ARTs like IVF have 'production-like' or 'consumerist'[12] connotations: they bring together not whole bodily persons but their disembodied parts (gametes) in what becomes an act of manufacture. The connotations are then played out in practice in both standard and more ambitious plans to select, control, or dispose of the human 'product'. Particularly where the embryo is conceived outside the maternal body, the production process lends itself, for technicians and (would-be) parents, to a recognizably producer-like mentality hospitable to quality control and the discarding of 'defective' or 'surplus' stock. Embryos are frequently produced in such numbers that they cannot be immediately transferred to the woman, and even those frozen for potential future 'use' (the word is significant) are often not 'used' in fact.

Unparental Parenthood

It is true that many IVF parents do have qualms about their frozen embryos, not unreasonably connecting them with siblings already born whose lives they can see unfolding. Many experience gnawing doubts about their surviving embryos[13] and delay a decision on their future year on year, even if storage costs must be paid.

as 'treatment'—even if satisfying a preference may have a knock-on effect of improving the person's psychophysical functioning. (If someone is a racist, evicting his Black neighbours may well improve his blood pressure and mental health. That does not make such eviction medicine: those wishing to defend it must do so in other terms.)

[10] Cesar Palacios-González, John Harris, and Giuseppe Testa, 'Multiplex Parenting: IVG and the Generations to Come', *Journal of Medical Ethics*, 40/11 (2014), 757.

[11] Segers et al., 'In Vitro Gametogenesis and the Creation of 'Designer Babies', 504.

[12] Ruth Padawer, 'The Two-Minus-One Pregnancy', *New York Times*, 10 August 2011, https://www.nytimes.com/2011/08/14/magazine/the-two-minus-one-pregnancy.html, accessed 29 January 2026. For discussion, see Watt, 'The Ethics of Pregnancy', 105–106.

[13] Valerie Shaikly, 'When to Stop Storage: Improving Conversations About Unused Embryos', *BioNews*, 1181, 6 March 2023, progress.org.uk; Celyn Harding-Jones, 'IVF Is Painful. Will I Leave My

For those who see these 'stalled' human lives as of serious moral weight, including some IVF parents unable or unwilling to bring them to term,[14] the situation is already very concerning. For those in the future who choose IVG and mass-production, they will perhaps have fewer qualms about the fate of any embryos who survive the general purge. (On the other hand, a change of heart about these embryos and their much more numerous siblings already discarded may be even more distressing for these parents.) Any embryos kept in storage will have survived a process where, say, 100 or even 1,000 embryos were created at a time, with commissioners invited to pick their preferred embryos, discarding, say, 95 or 995 in the process. Unless they meet extremely specific expectations, promoted via polygenic screening, embryos commissioned as part of the mass-production/screening package will have scant prospect of being chosen. This contrasts with standard forms of IVF today,[15] which do not involve mass-production, pre-implantation genetic diagnosis (PGD), and polygenic screening—even if, to some degree, these phenomena are already in our midst. For parents as well as scientists the sheer availability of IVG embryos in unimaginable numbers may make embryos less valued,[16] while not affecting their intrinsic moral status. The vast expansion in numbers may make it difficult to 'feel' that embryos—even surviving embryos who could still be brought to term—are morally considerable, let alone offspring of whom one is a parent, though one may have acted 'unparentally'.

Respect for Embryos

At this point, some readers will reply that an embryo is not offspring and in no way demands respect of a kind that would preclude mass-production and destruction, being a mere clump of cells or, at best, a 'potential human being'. This is not the place to discuss this question at length; however, a quick response

Final Embryo Frozen Forever?', *Guardian*, 15 July 2017, https://www.theguardian.com/commentisfree/2017/jul/15/ivf-painful-embryo-frozen-forever; Liza Mundy, 'Souls On Ice: America's Embryo Glut and the Wasted Promise of Stem Cell Research', *Mother Jones Magazine*, July–August 2006, https://www.motherjones.com/politics/2006/07/souls-ice-americas-embryo-glut-and-wasted-promise-stem-cell-research/; Sheryl de Lacey, 'Parent Identity and "Virtual" Children: Why Patients Discard Rather than Donate Unused Embryos', *Human Reproduction*, 20/6 (2005), 1661–1669.

[14] *Embryos on Ice: Jennie's Story* (online video), Center for Bioethics and Culture, 25 October 2022, https://www.youtube.com/watch?v=8kmEasV8SpE, accessed 29 January 2026.

[15] Currently, some IVF parents will have had no intention or expectation that more embryos (or, at least, more 'viable' embryos) would be produced than would eventually be transferred.

[16] Barbara Advena-Regnery, Hans-Georg Dederer, Franziska Enghofer, Tobias Cantz, and Thomas Heinemann, 'Framing the Ethical and Legal Issues of Human Artificial Gametes in Research, Therapy, and Assisted Reproduction: A German perspective', *Bioethics*, 32/5 (2018), 325.

would be that we are all made of cells, while the embryo is already the same individual human organism as the adult into which it grows. Leaving aside identical twinning, no embryo dies in producing a *separate* individual who is then born.[17] Human thoughts and mental abilities come and go, and are of course entirely lacking in the embryo; what we are essentially is not currently, formerly, or proximately thinking beings, but beings of a thinking *kind*.[18] If we begin as embryos, then the way is open to recognize significant human interests already at that stage. Despite the extreme immaturity of embryos, the idea that all of us have morally considerable interests from conception is very possible to defend if by 'interest' we mean simply an objective stake in our own well-being.[19]

Personal status, on this view, is not acquired but innate: there are no 'subpersonal' or 'subhuman' human beings. Moreover, even those who see embryos as having a lower moral status than older and already rational or conscious humans may still see embryos as deserving more respect than they receive with current ARTs. At any rate, many will recoil at the thought of embryos being mass-produced and mass-tested in vastly greater numbers than at present. As Sonia Suter comments, discussing the combination of IVG and PGD, 'One feature of IVG/PGD that might limit its widespread use, at least to some extent, is the fact that it would involve the creation and destruction of not just a handful of embryos, but literally thousands. For anyone who takes the position that the embryo is a person, any form of PGD would be problematic with or without IVG. But even for those who do not find the creation and destruction of embryos troubling under current uses of PGD, the exponential increase in embryo creation and destruction, as noted above, might be problematic'.[20]

Some people, including those planning a family, will be reluctant to countenance mass-creation and destruction of IVG embryos. However, others who see their neighbours mass-creating and selecting embryos—say, for cognitive ability—may feel pressure to do likewise[21] and may begin to ask if their own

[17] Alexander Pruss, 'I Was Once a Fetus: That Is Why Abortion Is Wrong', in Stephen Napier, ed., *Persons, Moral Worth, and Embryos* (Dordrecht: Springer, 2011), 19–42.

[18] Watt, *Ethics of Pregnancy*, 8–34.

[19] Christopher Tollefsen, 'Fetal Interests, Fetal Persons, and Human Goods', in Stephen Napier, ed., *Persons, Moral Worth, and Embryos* (Dordrecht: Springer, 2011), 163–184; Watt, *Ethics of Pregnancy*, 14–15, 26–28; Helen Watt, 'Do Fetuses Have the Same Interests as Their Mothers?', in Nicholas Colgrove, Bruce Blackshaw, and Daniel Rodger, eds., *Agency, Pregnancy and Persons* (New York: Routledge, 2022), 105–123.

[20] Sonia Suter, 'In Vitro Gametogenesis: Just Another Way to Have a Baby?', *Journal of Law and the Biosciences*, 3/1 (2016), 116–117.

[21] Marcelo de Araujo comments:

> If [embryo selection and gene editing] become commercially available within the next decades, prospective parents may suffer under the financial and emotional pressure to enhance their children, even if they would otherwise prefer not to enhance them ... But if, on the other hand, most parents opt for [genetic cognitive enhancement], they may believe that a non-enhanced child will lag behind in the competition for goods such as university

moral qualms on the subject are justified. We are all influenced by our moral environment, and changes of moral opinion and/or practice can be for the worse as well as the better. Progression to Gattaca-style selection would not be moral advancement but would surely testify to a very disturbing view of parental 'control'—a radical forgetting of what parenthood is all about.

Unconditional Acceptance

Parenthood, as many still believe, calls for unconditional acceptance of children as morally equal family members, though it does not preclude relinquishing a child to adoptive parents if one cannot meet the child's needs. Polygenic embryo screening for desired features is poor preparation for the unconditional acceptance of babies and older children, who may well disappoint in the very features for which parents were selecting. While this is also true of screening for disability, at least in some such cases, the motive will be child-centric: there may be a desire, however misguided, to spare a child a life of suffering, as opposed to simply wanting a different child. Parental concern is less the issue with 'designer' screening or screening for non-severe ailments, since parents would know that de-selected embryos would also have had a reasonable life—assuming they accept that embryos *grow into* and are not *replaced by* a child who is born. Among parents who see embryos as more like gametes (that is, as mere ingredients for a separate human being), there may be a generic concern for the best interests of a future child and a desire to have an 'even better-off' child. However, in the words of Sonia Suter, who also notes the effects of IVG screening on born disabled people and those with unwanted traits, 'we could imagine a gradual, but potentially notable, cultural shift in attitudes, which would probably involve a complicated mix of concerns about the best interest of the child (procreative beneficence) and a growing desire for perfection that goes beyond best interest concerns . . . If societal norms shift toward viewing IVG/PGD and IVG/gamete selection as mechanisms to fine-tune reproduction, rather than just to prevent

> admittance or better positions in the job market. . . . One of the surveys recently conducted by the SIENNA Project revealed, for example, that, on average, 79% of people across 11 countries believe that prospective parents will feel 'pressured to have genetic testing done on their unborn baby', if genetic testing becomes increasingly common . . . This . . . raises questions of social justice, as many prospective parents may suffer emotionally or financially under the pressure to have their unborn children genetically tested. Some genetic testing may, for instance, reveal an underlying condition that affects not only the unborn baby, but also the living siblings (or the mother, or the father), who might perhaps prefer not to be informed about that particular condition.

Marcelo de Araujo, 'The Ethics of Genetic Cognitive Enhancement: Gene Editing or Embryo Selection?, *Philosophies*, 5/3 (2020), 11–12.

suffering, the concerns about potential discrimination—including discrimination against the disabled and discrimination based on disfavored traits—also grow'.[22]

The point has been raised by Segers et al. that even the desire for genetically related children is, after all, a desire for children of a certain kind—that is, related to ourselves and perhaps resembling ourselves.[23] To this it can be replied that such desires mutate significantly and become 'turbo-charged' in a context of ART production and selection and, especially, mass-production and selection, which IVG would enable. In contrast, couples who conceive sexually—assuming a consensual, fully respectful encounter—are not conceiving in a 'manufacturer-like' way.[24] Their bodily act is interpersonal: unreserved, mutual, geared to express a lasting bond. Children are received as a *further* gift, and the couple are predisposed to accept 'whatever comes', even if they know and welcome the fact that 'whatever comes' will be related to them. The attitude of domination, of 'making it happen' and rejecting substandard 'products', is far from the couple's minds, even if their sexual act follows measures chosen to promote their child's health as best they can.[25]

As Smajdor et al. reflect: 'Artificial gametes change our relationship with our offspring. Children conceived with artificial gametes might query their parents' choices and decisions in new ways. Parents themselves might be regarded as occupying a role closer to that of designers or creators than hitherto . . . Artificial gametes change us. They change the nature of our choices and the scope of our moral responsibility. Not only this, but they pose new questions in areas where our existing abilities are already stretched and contentious'.[26]

If would-be IVG parents were multiplied in 'multiplex' reproduction, the topic of the next section, then multiple people—whether polyamorous partners or simply members of a 'parental co-op'—would contribute screening preferences to the mix. Envisaging a world in which a 'reproductive committee chairperson'

[22] Suter, 'In Vitro Gametogenesis', 118.

[23] Segers et al., 'In Vitro Gametogenesis and the Creation of "Designer Babies"', 506.

[24] Watt, *Ethics of Pregnancy*, 105–109, 117–118. Note that even with natural accidents such as an embryo splitting, resulting in a new twin/clone embryo or embryos arising from the first, no act of manufacture interposes between the couple's interaction and any child's conception.

[25] The health of actual or future children can of course be promoted before birth or conception by parents conceiving sexually; for example, they may take folic acid, avoid exposure to radiation, even screen out fellow Tay-Sachs carriers in their dating/matrimonial choices. If children are still 'received' at the time of conception from a sexual act of unconditional giving and receiving, the child-friendly structure that favours unconditional acceptance of any child conceived can still be preserved. Indeed, this could be the case even in a scenario—difficult to envisage without unethical prior experimentation—where germ-line therapy known to be safe was combined with IVG with autologous transplantation to prevent the transmission of a genetic condition. At the point of conception, sex could still involve holistic acceptance of one's partner and his/her fertility, paving the way for respectful acceptance of any child conceived.

[26] Anna Smajdor, Daniela Cutas, and Tuija Takala, 'Artificial Gametes, the Unnatural and the Artefactual', *Journal of Medical Ethics*, 44/6 (2018), 407.

negotiates with a smiling saleswoman on selecting a handsome and clever child, Christopher Kaczor comments: 'This brave new world, even if populated with only beautiful people, is an ugly world in which the gift of a child becomes a consumer product for committee evaluation and consumption'.[27]

Multiplex Reproduction

'Multiplex parenting', also known as 'polyparenting', has, however, been seen with approval by some writers, with Palacios-Gonzalez et al. observing that 'IVG would . . . allow the genome to be distributed and shared equally, bending the temporal necessity of genetic generations to the social and cultural preferences of our times'.[28] Palacios-González et al. further describe multiplex reproduction as 'taming genetic kinship for parenting preferences' and 'blurring the distinction between biological and social generations'.

How might this be achieved? If (for example) four people were involved, two IVF embryos could be created with paired gametes and used to harvest cells from which new gametes would be derived. Embryos would be created from those gametes and any child born would be related—parentally, some would argue—to the four adult commissioners.

But *are* they parentally related—or is this generation filled by the embryos used to create their successors? We should remember here that genetic parenthood is not social parenthood: *feeling* at any time that one is a genetic parent is neither necessary nor sufficient to *be* one. Embryos, like infants (who would also be unaware of their parenthood), can be made parents non-consensually. Like infants, they appear to be the *same living individual* as the older child and the adult, and it is implausible to suggest that the attribution of genetic parenthood must wait for a later stage. Imagine that an embryo was biopsied and the cells extracted used to create IVG gametes and then embryos, while the 'source' embryo was transferred to a woman's body and eventually born. Did not this individual become a genetic parent at the point the new embryo was created? Or are we claiming that a *third* individual would somehow come into existence already a parent?

In all likelihood, of course, the 'source' embryo would not be biopsied but destroyed in taking its cells. Such an embryo would become a genetic parent, no less than a man—including a brain-damaged man who had never been conscious—whose sperm was used post-mortem. The embryo existed, was

[27] Christopher Kaczor, 'Philosophy and Theology', *National Catholic Bioethics Quarterly*, 15/1 (2015), 174.
[28] Palacios-González et al., 'Multiplex Parenting', 757.

created solely for its parts, and became a genetic parent following its death: a real genetic parent, although this is a parody of what such parenthood should mean. The embryo's unawareness does not prevent, but grounds the wrong done to it: if rights apply throughout the human lifespan (admittedly a contested view), then it seems that no human being, of any age, should be made to conceive unaware, let alone destroyed in the process.

To the objection that these embryos would not be real genetic parents as they lack a biography,[29] it can be said that they have a biography of sorts: a brief and grim biography involving the mining of their cells to form new offspring at their own parents' behest. What is missing is precisely the *adult* biography of someone given time, space, and freedom to form a relationship suited to the conception of new human beings. Human conception is of great personal and social importance for parents, the child, and those around them; it deserves to be achieved by humane, conscious acts of a kind that signify unique, existing bonds.

Genetic Parents and Nonparents

Genetically, 'multiplex parenting' is a misnomer, as is 'polyparenting': in our example the four commissioners would be genetic grandparents[30] of a child whose genetic parents had been destroyed. 'Taming genetic kinship' in this way is a spin on genetic grandparenthood and not a benign one. So much did the commissioners want a child to reflect their 'parenting preferences' that they went to the length of assembling embryos not to keep but to press into reproductive service. Unlike most IVF embryos—even mass-produced IVG embryos—these embryos' survival was never envisaged: from the outset, they were treated as mere source material for the later embryo/child. It is hard to imagine a worse preparation for social parenthood than such thoroughly destructive reproduction—which could also, of course, be combined with mass-production and polygenic testing of the new IVG embryos. Moreover, 'polyparenting' promotes the fertility of no one: it does not change the body of anyone in a fertile direction; nor does it seem correct to say that we are all infertile because, without technical help, we cannot co-produce children with multiple people.[31] To be infertile, one

[29] Robert Sparrow, 'Orphaned at Conception: The Uncanny Offspring of Embryos', *Bioethics*, 26/4 (2012), 177.

[30] Palacios-González et al., 'Multiplex Parenting', 756.

[31] On the potential of IVG to create perceptions of infertility and expectations of treatment for those not previously considered infertile, see Anna Smajdor and Daniela Cutas, 'Will Artificial Gametes End Infertility?', *Health Care Analysis*, 23/2 (2015), 134–147; Daniela Cutas and Anna Smajdor, '"I am Your Mother and Your Father!" In Vitro Derived Gametes and the Ethics of Solo Reproduction', *Health Care Analysis* 25 (2017), 354–369; Lauren Notini, Christopher Gyngell, and Julian Savulescu, 'Drawing the Line on In Vitro Gametogenesis', *Bioethics*, 34 (2020), 123–134.

must have a blocked reproductive orientation, while no one has any such bodily orientation to source and use gametes from destroyed embryos.

Social Parents and Nonparents

Expansions of the 'four-parent' version of multiplex reproduction, as in the 'parental co-op' mentioned above, would further dilute both the genetic and the social connection of the commissioners to the offspring. As more and more people joined the co-op to reproduce, it would become more doubtful whether anything recognizably parental in social terms was involved. Parental attention cannot be indefinitely diluted, and it is in everyone's interest that there be no ambiguity as to who will guide and care for any child. A village can certainly *help* parents raise a child, without usurping parents' roles. However, if a village *literally* raises a child, that does not give the child a village-shaped number of parents. It gives the child no parents.

Sonia Suter, while accepting multiplex reproduction in lower numbers, suggests that 'At some point, as the number of intended "parents" increases, the social connections inevitably thin and intimacy diminishes, making multiplex reproduction more like the creation of "clans" and less like the creation of parent–child relationships. This is inconsistent with relational autonomy for the child, who would lose the benefits of the close relationship of a true social parent . . . The burden . . . should be on those seeking large multiplex parenting arrangements to show why this approach is consistent with relational autonomy as opposed to being driven by socially threatening, clan-like motivations'.[32]

Multiplex IVG Offspring

How would the offspring of multiplex IVG be affected by their method of conception? Leaving aside clan-sized 'parent' groups, there is surely a real potential for genealogical bewilderment and indeed survivor guilt, though some commentators are cautiously optimistic. In the words of Segers et al., 'having knowledge that an embryo had to be intentionally destroyed for the resulting child to live might be a source of discomfort . . . much will depend on how these issues will be framed and explained'.[33]

One can imagine a burgeoning cottage industry producing children's books like those parents currently buy for donor-conceived or surrogate-born children.

[32] Suter, 'In Vitro Gametogenesis', 114.
[33] Segers et al., 'In Vitro Gametogenesis and Reproductive Cloning', 73.

However, even those offspring told early and with reassuring gloss may come in later life, if not before, to question whether their origin is really so simple and benign. Perhaps they will be wholly protected from identity issues and survivor guilt in relation to their embryonic parents (and/or myriads of destroyed IVG siblings)—but perhaps they will not be so protected. And whether or not the offspring 'feel' it, is there really nothing to object to?

Outcomes for/Reactions of IVG and Other ART Offspring

Reading accounts of multiplex and/or mass-production IVG, it is difficult not to be struck with the insouciant approach of some commentators to outcomes for the offspring. There is a tendency to think of children—and the adults into which they grow—as endlessly adaptable, endlessly obliging in fitting around their parents' choices. However, this is not necessarily the case even with current ARTs, as the voices of, in particular, donor-conceived offspring are increasingly showing. This is despite what their parents were assured at various times by professionals about children not caring or not needing to know their donor or even (as many were once told) their donor status. While the research on outcomes for donor-conceived offspring is mixed,[34] there are many anecdotal accounts of those who, even if loved by and loving their social parents, perceived something unsettling in their environment and did not react well to knowledge of their donor conception.

'They created me in the same way as they breed pigs', recounted one man. 'All I know and am allowed to know about my father is that he masturbated his "sample" for a sum. Yes you could say I'm angry'.[35] While donor conception (here, paid, anonymous donation) would not be the issue with IVG—which would indeed be an *alternative* to donor conception—the agricultural connotations of mass-breeding and selecting the 'best' offspring via IVG, culling less favoured offspring, should not go unnoticed.

Donor-conceived offspring are frequently told that they should be grateful to be alive and loved,[36] but some reply that it is not that simple: they want to live but do not accept parental narratives as easily as their parents may wish. Not every

[34] Guido Pennings, 'The Moral Obligation to Have Genetically Related Children', *Journal of Medical Ethics*, 51/4 (2025), 274–277.

[35] Alexina McWhinnie, 'Afterword', in Joanna Rose, Christine Whipp, and Louise Jamieson, eds., *Who Am I? Experiences of Donor Conception* (Leamington Spa: Idreos Education Trust, 2006), 52.

[36] Donor Conceived UK, 'What NOT to say to a DCP', https://donorconceiveduk.org.uk/what-not-to-say-to-a-dcp/a, accessed 26 August 2024; Bethany K., 'Why I'm Angry about Being Donor-Conceived', https://www.wearedonorconceived.com/personal-stories/why-im-angry-about-being-donor-conceived/ accessed 26 August 2024; contributors in McWhinnie et al., *Who Am I?*, 6–7, 12, 22, 42.

form of human conception reflects the human dignity of what is produced. As we continue to overhaul genetic and social ties of immemorial standing and profound personal and social meaning, the voices of the more critical among donor-conceived and other ART offspring may be worth considering.

Gametes are parts, but important ones: they help to generate whole human beings and should arguably represent[37] whole adult human beings and whole, undivided relationships. Preparing for conception is preparing for parenthood, and future generations are not our genetic and social experiment, to do with as we wish.

In the words of one woman conceived by donor insemination (DI), with which I will end: 'As a DI adult, I want to sound a warning bell. The babies who "solve" today's infertility problems will have their own, long lives to lead. I foresee a correlation between the tangling of their family lines, and later confusion, pain or dysfunction, not necessarily during childhood, adolescence or even early adulthood, but almost certainly in later life. As part of the first "wave" of assisted conception, I feel I bear a responsibility to speak for those who are the products of newer techniques, and to urge greater circumspection on the part of the fertility industry and policymakers'.[38]

Bibliography

Advena-Regnery, Barbara, Hans-Georg Dederer, Franziska Enghofer, Tobias Cantz, and Thomas Heinemann, 'Framing the Ethical and Legal Issues of Human Artificial Gametes in Research, Therapy, and Assisted Reproduction: A German Perspective', *Bioethics*, 32/5 (2018), 314–326.

Cutas, Daniela, and Anna Smajdor, '"I am Your Mother and Your Father!" In Vitro Derived Gametes and the Ethics of Solo Reproduction', *Health Care Analysis*, 25 (2017), 354–369.

de Araujo, Marcelo, 'The Ethics of Genetic Cognitive Enhancement: Gene Editing or Embryo Selection?', *Philosophies*, 5/3 (2020), 1–20.

de Lacey, Sheryl, 'Parent Identity and "Virtual" Children: Why Patients Discard Rather than Donate Unused Embryos', *Human Reproduction*, 20/6 (2005), 1661–1669.

Embryos on Ice: Jennie's Story, online video, Center for Bioethics and Culture, 25 October 2022, https://www.youtube.com/watch?v=8kmEasV8SpE, accessed 29 January 2026.

Harding-Jones, Celyn, 'IVF Is Painful. Will I Leave My Final Embryo Frozen Forever?', *Guardian*, 15 July 2017, https://www.theguardian.com/commentisfree/2017/jul/15/ivf-painful-embryo-frozen-forever.

Jamieson, Louise, 'The DI Journey: Pain, Loss and Discovery', in Alexina McWhinnie, Joanna Rose, Christine Whipp, and Louise Jamieson, eds., *Who Am I? Experiences of Donor Conception* (Leamington Spa: Idreos Education Trust, 2006), 30–45.

[37] Calum Mackellar, 'Representative Aspects of Some Synthetic Gametes', *The New Bioethics*, 21/2 (2015), 105–116; Pruski, 'The Relationship of Gametes to Those Who Procreate'.

[38] Louise Jamieson, 'The DI Journey: Pain, Loss and Discovery', in Alexina McWhinnie, Joanna Rose, Christine Whipp, and Louise Jamieson, eds., *Who Am I? Experiences of Donor Conception* (Leamington Spa: Idreos Education Trust, 2006) 38.

Kaczor, Christopher, 'Philosophy and Theology', *National Catholic Bioethics Quarterly*, 15/1 (2015), 169–174.

Mackellar, Calum, 'Representative Aspects of Some Synthetic Gametes', *The New Bioethics*, 21, 2 (2015), 105–116.

McWhinnie, Alexina, 'Afterword', in Alexina McWhinnie, Joanna Rose, Christine Whipp, and Louise Jamieson, eds., *Who Am I? Experiences of Donor Conception* (Leamington Spa: Idreos Education Trust, 2006), 46–61.

McWhinnie, Alexina, Joanna Rose, Christine Whipp, and Louise Jamieson, *Who Am I? Experiences of Donor Conception* (Leamington Spa: Idreos Education Trust, 2006).

Mundy, Liza, 'Souls On Ice: America's Embryo Glut and the Wasted Promise of Stem Cell Research', *Mother Jones Magazine*, July–August 2006, https://www.motherjones.com/politics/2006/07/souls-ice-americas-embryo-glut-and-wasted-promise-stem-cell-research/.

Notini, Lauren, Christopher Gyngell, and Julian Savulescu, 'Drawing the Line on In Vitro Gametogenesis', *Bioethics*, 34 (2020), 123–134.

Padawer, Ruth, 'The Two-Minus-One Pregnancy', *New York Times*, 10 August 2011, https://www.nytimes.com/2011/08/14/magazine/the-two-minus-one-pregnancy.html, accessed 29 January 2026.

Palacios-González, Cesar, Harris, John, and Testa, Giuseppe, 'Multiplex Parenting: IVG and the Generations to Come', *Journal of Medical Ethics*, 40/11 (2014), 752–758.

Pennings, Guido, 'The Moral Obligation to Have Genetically Related Children', *Journal of Medical Ethics*, 51/4 (2025), 274–277.

Pruski, Michal, 'The Relationship of Gametes to Those Who Procreate and Its Impact on Artificially Generated Gamete Technologies', *Ethics & Medicine*, 33/1 (2017), 27–41.

Pruss, Alexander, 'I Was Once a Fetus: That Is Why Abortion Is Wrong', in S. Napier, ed., *Persons, Moral Worth, and Embryos* (Dordrecht: Springer, 2011), 19–42.

Segers, Seppe, Guido Pennings, Wybo Dondorp, Guido De Wert, and Heidi Mertes, 'In Vitro Gametogenesis and the Creation of 'Designer Babies', *Cambridge Quarterly of Healthcare Ethics*, 28/3 (2019), 499–508.

Segers, Seppe, Guido Pennings, Wybo Dondorp, Guido De Wert, and Heidi Hertes, 'In Vitro Gametogenesis and Reproductive Cloning: Can We Allow One while Banning the Other?', *Bioethics*, 33/1 (2019), 68–75.

Shaikly, Valerie, 'When to Stop Storage: Improving Conversations About Unused Embryos', *BioNews*, 1181, 6 March 2023, progress.org.uk.

Smajdor, Anna, and Daniela Cutas, 'Will Artificial Gametes End Infertility?', *Health Care Analysis*, 23/2 (2015), 134–147.

Smajdor, Anna, Daniela Cutas, and Tuija Takala, 'Artificial Gametes, the Unnatural and the Artefactual', *Journal of Medical Ethics*, 44/6 (2018), 404–408.

Sparrow, Robert, 'Orphaned at Conception: The Uncanny Offspring of Embryos', *Bioethics*, 26/4 (2012), 173–181.

Suter, Sonia, 'In Vitro Gametogenesis: Just Another Way to Have a Baby?', *Journal of Law and the Biosciences*, 3/1 (2016), 116–117.

Tollefsen, Christopher, 'Fetal Interests, Fetal Persons, and Human Goods', in Stephen Napier, ed., *Persons, Moral Worth, and Embryos* (Dordrecht: Springer, 2011), 163–184.

Watt, Helen, 'The Dignity of Human Life: Sketching Out an "Equal Worth" Approach', *Ethics and Medicine*, 31/1 (2020), 7–17.

Watt, Helen, 'Do Fetuses Have the Same Interests as Their Mothers?', in Nicolas Colgrove, Bruce Blackshaw, and Daniel Rodger, eds., *Agency, Pregnancy and Persons* (New York: Routledge, 2022), 105–123.

Watt, Helen, *The Ethics of Pregnancy, Abortion and Childbirth: Exploring Moral Choices in Childbearing* (New York: Routledge, 2016).

7

Multiplex Parenting Revised

César Palacios-González

The Generation of Gametes *In Vitro*

In vitro gametogenesis (IVG) research sparked a philosophical discussion that has been going on for the past twenty years.[1,2] A particularly salient moment in such research was the derivation of female and male functional gametes, or gamete-like cells, capable of generating viable offspring in non-human animals.[3,4] This development raised the prospect of human *in vitro*-derived gametes being used for reproductive purposes, which in turn raised the question of when, if ever, would it be ethical to use these types of gametes for reproduction. It would not come as a surprise, to anyone working in the reproductive ethics space, that the catalogue of ethical issues that have been discussed in relation to IVG includes consent of the future child, harm to the future child, long-term follow-up of the child, identity of the child, reproductive autonomy, harm to future generations, use and destruction of human embryos, slippery slope considerations, available reproductive/parenting alternatives, consent of prospective parents, and harm to prospective parents.[5,6]

In my consideration, the one finding in IVG research that has generated the most interesting philosophical debate is the ability to derive what we can call 'cross-sex gametes'.[7] Philosophers have been particularly concerned with what would be the ethical implications of generating and using such cross-sex gametes

[1] G. Testa and J. Harris, 'Ethical Aspects of ES Cell-Derived Gametes', *Science*, 305/5691 (2004), 1719.

[2] G. Testa and J. Harris, 'Ethics and Synthetic Gametes', *Bioethics*, 19/2 (2005), 146–166.

[3] K. Hayashi et al., 'Offspring from Oocytes Derived from In Vitro Primordial Germ Cell-like Cells in Mice', *Science*, 338/6109 (2012), 971–975.

[4] K. Nayernia et al., 'In Vitro-Differentiated Embryonic Stem Cells Give Rise to Male Gametes that Can Generate Offspring Mice', *Developmental Cell*, 11/1 (2006), 125–132.

[5] A. Smajdor and D. Cutas, *Background Paper on Artificial Gametes*, Nuffield Council on Bioethics, London, December 2015.

[6] H. T. Greely, *The End of Sex and the Future of Human Reproduction*, 1st ed. (Cambridge, MA: Harvard University Press, 2016).

[7] T. Douglas et al., 'Stem Cell Research and Same-Sex Reproduction', in M. Quigley, S. Chan, and J. Harris, eds., *Stem Cells: New Frontiers in Science and Ethics* (Singapore: World Scientific, 2012), 207–228.

César Palacios-González, *Multiplex Parenting Revised* In: *Revolutionizing Reproduction.* Edited by: Christopher Gyngell, Julian Savulescu, and Tsutomu Sawai, Oxford University Press.
DOI: 10.1093/9780197810415.003.0007

for human reproduction. The generation of cross-sex gametes entails two phenomena. First, the ability to use stem cells, either embryonic or induced pluripotent, that originate from a biological male to produce a functional oocyte—that is, an oocyte that can be fertilized by a naturally occurring, or *in vitro*-derived, sperm and can produce a live offspring. Second, the ability to use stem cells, either embryonic or induced pluripotent, that originate from a biological female to produce a cell that can be used in conjunction with a naturally occurring, or *in vitro*-derived, oocyte to generate viable offspring. In the case of female cells, the derivation of the sperm-proper is more difficult, since human cells with the XX chromosomes lack a necessary gene for sperm formation (the SRY gene).[8] For a recent review on the science of IVG, see Chapter 2, in this volume.[9]

One perennial issue within the IVG debate, broadly understood, is safety. When will it be safe to use *in vitro*-derived gametes in general, and cross-sex gametes in particular, for human reproduction?[10,11] Safety issues arise in at least three ways: safety in regard to the child, safety in regard to the parents, and safety in regard to society. When talking about bringing a child into existence, the point that gets discussed the most is whether the use of (cross-sex) *in vitro*-derived gametes would give rise to a healthy or unhealthy child. This is so because there is an underlying assumption that in all instances parents want to have a healthy child, and also because some people have the intuition that we can harm individuals by bringing them into existence.

Biotechnologies that can be used to generate new individuals take us to the fraught land of population ethics, and more specifically, to the non-identity problem bogland. In David Boonin's words, 'The non-identity problem arises from the fact that in some situations, our choices determine not only the quality of life that people will enjoy in the future, but the identity of the people who will exist to enjoy it'.[12] It is appropriate to, in metaphorical terms, present the non-identity problem as a bogland because, as with the physical place, one gets mesmerized by the beauty of what is in front of us, but a wrong step may mean that we might be stuck there for the foreseeable future. In a talk about the ethical and legal aspects of mitochondrial replacement techniques, at Harvard

[8] R. H. Lovell-Badge and A. Hacker, 'The Molecular Genetics of Sry and Its Role in Mammalian Sex Determination', *Philosophical Transactions of the Royal Society of London, Series B: Biological Sciences*, 350/1333 (1995), 205–214.

[9] K. Hayashi, 'State of the Science', in *In Vitro–Derived Human Gametes as a Reproductive Technology: Scientific, Ethical, and Regulatory Implications*, workshop proceedings (Washington, DC: National Academies Press, 2023), and M. Saitou and K. Hayashi, 'Mammalian In Vitro Gametogenesis', *Science*, 374/6563 (2021), eaaz6830.

[10] S. Segers et al., 'Using Stem Cell-Derived Gametes for Same-Sex Reproduction: An Alternative Scenario', *Journal of Medical Ethics*, 43/10 (2017), 688–691.

[11] S. Segers et al., 'Balancing Ethical Pros and Cons of Stem Cell Derived Gametes', *Annals of Biomedical Engineering*, 54/7 (2017), 1620–1632.

[12] D. Boonin, 'How to Solve the Non-Identity Problem', *Public Affairs Quarterly*, 22/2 (2008), 129–159.

University, Hank Greely gave a nice illustration of what is it to deal with the non-identity problem. There he stated that the non-identity problem 'makes my head hurt'.

Whether we can indeed harm or benefit individuals by bringing them into existence is a matter of ongoing controversy, and no definitive answer has been provided. This matter is of relevance since if we cannot conclusively ascertain whether we harm or benefit someone by bringing them into existence, then it is also impossible to fully address the matter of safety, from a person-affecting account of morality. It is true that if we talk about safety in impersonal terms then we could say something cogent about whether *in vitro*-derived cross-sex gametes are safe, or not. However, an impersonal account of morality has its own set of complications. And furthermore, a two-tier view, as that recently defended by McMahan and Savulescu, also has problems in terms, for example, of whether it is better to create someone healthy or someone with some disease that then we go on to cure.[13] For the most recent systematic treatment of this issue see David Boonin.[14]

From the above, a person unfamiliar with the analytic philosophical tradition might conclude that we cannot tackle other matters unless we answer the safety question first, and that in order to do so, we need to solve the non-identity problem. This is not quite right. The philosophical discussion on IVG follows the venerable tradition of bracketing finicky issues and moving on. In this case, we can bracket the problems related to safety and assume that children born from *in vitro*-derived gametes, and cross-sex *in vitro*-derived gametes, would be as healthy as children born via sexual intercourse. It is true that bracketing the matter of safety in this way does not do away with the more general problem of whether we might benefit or harm individuals by bringing them into existence. However, in making the safety assumption, and further supposing that it is morally permissible to bring children into existence, *pace* Benatar,[15] then we can move forward and engage with other philosophical matters concerning the use of (cross-sex) *in vitro*-derived gametes for human reproduction. Another way of thinking about what I have described here, humorously put by a colleague, is that we sprinkle some philosophical magic dust and make this problem disappear, for the time being.[16]

The reproductive options that *in vitro*-derived gametogenesis open are the following. Heterosexual couples could use *in vitro*-derived gametes to have

[13] J. McMahan and J. Savulescu, 'Reasons and Reproduction: Gene Editing and Genetic Selection', *The American Journal of Bioethics*, 24/8 (2024), 9–19.

[14] D. Boonin, *The Non-Identity Problem and the Ethics of Future People* (Oxford: Oxford University Press, 2014).

[15] D. Benatar, *Better Never to Have Been: The Harm of Coming into Existence* (Oxford: Oxford University Press, 2006).

[16] Matthias Steiner (personal communication, 22 August 2024).

a genetically related child. Same-sex couples could also use *in vitro*-derived gametes to have a genetically related child. More radically, individuals could use *in vitro*-derived gametes to have a genetically related child, 'solo' reproduction. In this they would use their naturally occurring gamete plus a cross-sex *in vitro*-derived one to produce a child. And even more radically, more than two individuals could employ (cross-sex) *in vitro*-derived gametes to have a child that is genetically related to all of them. This is not an exhaustive list. This chapter will focus on the latter case.

After this necessary excursus on IVG, cross-sex gametes, the non-identity problem, and safety, I can explain what I am going to do in this chapter. In the next section, I will present how more than two individuals could have a child that is genetically related to all of them—what, in the case of *in vitro*-derived gametes, colleagues and I have called 'multiplex parenting' in a previous publication. In the third section, I will present a challenge that could be raised to the use of *in vitro*-derived gametes for achieving multiplex parenting—that such children would not be biologically related to the intending parents in the desired way. In the fourth section, I respond to this challenge and argue that multiplex parenting fares no worse than certain cases of naturally occurring reproduction. In the fifth section, I present an objection to the argument advanced in the previous section. This is that only a reduced number of multiplex parenting cases are similar to those that occur in nature. In the conclusion, I respond to such objections by further specifying what is what people intend to achieve when engaging in multiplex parenting.

Natural and Biotechnological Pathways to Multiplex Genetic Parenthood

In 2014 John Harris, Giuseppe Testa, and I presented and explored what we called multiplex parenting.[17] In that publication we asserted, 'the most paradigm-shifting application of IVG [*in vitro* gametogenesis] could be a radical expansion of reproductive autonomy that allowed more than two persons to engage simultaneously in genetic parenting'. In that same paper we noted that there is at least another option for more than two individuals to be genetically related to a child: mitochondrial replacement techniques. Here two women and a man would be able to have a child that is genetically related to all of them. I will not explore how mitochondrial replacement techniques would achieve this. People interested in mitochondrial replacement techniques

[17] C. Palacios-González, C., J. Harris, and G. Testa, 'Multiplex Parenting: IVG and the Generations to Come', *Journal of Medical Ethics*, 40/11 (2014), 752–758.

and genetic parenthood can refer to Françoise Baylis[18] and Cavaliere and Palacios-González.[19]

What I will instead do is mention another phenomenon that would end up in three individuals being genetically related to a child. In order to understand it, the first that we need to know is that there is a phenomenon called tetragametic chimerism.[20] Tetragametic chimerism is a phenomenon that attains in humans, in nature, in the following way. A woman produces two oocytes during a menstrual cycle, rather than just one. The woman engages in sexual intercourse with a fertile man. Afterwards, the two oocytes get fertilized by two different sperm, from the same man, ending up in two zygotes. Finally, the two zygotes fuse due to being in close proximity. The resulting organism has two full different sets of DNA. Once we are clear about tetragametic chimerism, we need to be aware of another rare phenomenon that also occurs in humans, in nature: heteropaternal superfecundation.[21] Here the woman produces two oocytes during a menstrual cycle. The woman engages, during her fertile period, in sexual intercourse with two different fertile men. The result is twins being born where the twins are, in lay terms, half siblings. They have the same genetic mother but different genetic fathers.

Now, the phenomenon that I am about to present, to the best of my knowledge, has not been described in the academic literature and thus lacks a proper term. Thus, we could call it: heteropaternal tetragametic chimerism. Here the woman produces two oocytes during a menstrual cycle. She engages in sexual intercourse with two different men. One sperm from each man fertilizes one egg, forming two zygotes (that are half siblings). Finally, the two zygotes fuse into one single organism. As in the tetragametic chimerism case, this new organism has cells that contain the DNA from the woman and one of the men, and cells that contain the DNA from the woman and the other man. The child has two distinct sets of DNA, and we can trace back the DNA to three different individuals. Let us now turn back to the matter at hand.

Multiplex parenting can be understood as the phenomenon where more than two individuals rely on IVG to create a child who is genetically related to all of them. Here, I will not define what being 'genetically related' means, but I will do so in the next section. Now, imagine a polyamorous relationship, a quadruple,

[18] F. Baylis, 'The Ethics of Creating Children with Three Genetic Parents', *Reproductive Biomedicine Online*, 26/6 (2013), 531–534.

[19] G. Cavaliere and C. Palacios-González, 'Lesbian Motherhood and Mitochondrial Replacement Techniques: Reproductive Freedom and Genetic Kinship', *Journal of Medical Ethics*, 44/12 (2018), 835–842.

[20] K. Madan, 'Natural Human Chimeras: A Review', *European Journal of Medical Genetics*, 63/9 (2020), 103971.

[21] N. L. Segal, Craig, and M. P. Umstad, 'Challenge to the Assumed Rarity of Heteropaternal Superfecundation: Findings from a Case Report', *Australian Journal of Forensic Sciences*, 52/5 (2020), 547–552.

composed of two men, M1 and M2, and two women, F1 and F2. Further imagine that they want to have a child that is genetically related to all of them. In this case they would oppose having a child that is only genetically related to some of them.

What they could do is the following. First, they would resort to IVF to create two zygotes, in this case one of the zygotes would be created with a sperm from M1 and an oöcyte from F1. The second zygote would be created with a sperm from M2 and an oocyte from F2. Once the zygotes become embryos, two human embryonic stem cell lines would be established. We would then use these embryonic stem cell lines to obtain cells that would be then derived into complementary gametes. Here we can also assume, for simplicity, that both embryos are destroyed; however, this does not necessarily need to be the case. For further simplicity, let us imagine that the embryo that originated from M1 and F1, let us call it E1, had the sexual chromosomes XX, and that the embryo that originated from M2 and F2, let us call it E2, had the sexual chromosomes XY. Via IVG we would derive an oocyte from E1 and a sperm from E2. Finally, we would fuse the oocyte obtained from E1 and the sperm obtained from E2 to produce a third zygote. This third zygote will turn into a third embryo, let us call it E3, which will be then transferred into one of the women, the gestational mother. In the end we have a child that is genetically related to the four adults.

How a particular zygote, in terms of its genetic makeup, gets generated will depend on the number of individuals who belong to the group, and what pathway is taken to generate the embryo. However, regardless of the specific DNA percentage that a child inherits from each intending parent, all instances of multiplex parenting have the effect of temporally compressing at least one generation. For example, in the case of the group of four individuals we temporally compress one generation. Some of these generations will exist only for a few days, as I will explain in more detail in the next section.

Parenthood through Multiplex Parenting

Someone might accept that multiplex parenting would temporally compress generations, and that such children would be genetically related to what we have called the intending parents. However, they would argue that whether any particular adult within a group, in a multiplex parenting scenario, is in fact a *genetic parent* would depend on the specific pathway that is taken to generate the child. Many adults engaging in multiplex parenting would be, in fact, *genetic grandparents* rather than *genetic parents*. In order to see this, let us re-examine the quadruple case. In the quadruple case, the child inherits 25% of his DNA from each one of the adults, via an intermediate generation, putting them into an equivalent position to naturally occurring genetic grandparents.

The former assumes that we are certain about the account of human reproduction that we are employing, and what are the necessary conditions for this to attain. Even if we do not adopt any particular account, contemporary accounts of reproduction have as one of their central elements that reproduction is the creation of new organisms. As Peter Godfrey-Smith puts it, reproduction is making a new individual, in the sense of creating a numerically distinct individual.[22] And creation, in its reproductive sense, should be understood as proximately materially causing the existence of a new individual.

Our interlocutor would further argue that what people want from IVG, in general, is to be a *genetic parent*, and not only to be genetically related to a child—as in the way in which genetic grandparents are to their genetic grandchildren, or how the child of someone who is an identical twin is genetically related to his 'identical' uncle. Genetic relatedness can thus be understood, following Avery Kolers,[23] in an informational sense or a physical one. According to the former, A and B are genetically related if it is the case that certain percentage of their DNA's information is the same. On the latter sense, A and B are genetically related if it is the case that some of the DNA, in its material sense, have been transmitted from A to B, either directly or via an intermediary.

It seems to me that when the objector says that some people resorting to IVG would be genetically related to the resulting child as grandparents are genetically related to their genetic grandchildren, what they have in mind is genetic relatedness in its material sense. As per what I have said above, A and B are in a genetic parent offspring relation if A was a proximate material cause of B's existence, and B is (partially) made out material parts from A, his DNA. In his work on the nature of reproduction James Griesmer adds the condition that the created individuals should be able to, in turn, create individuals that can start again the process of material transfer.[24] 'Reproduction is neither mere multiplication of individuals nor mere copying. It requires that multiplication involve material overlap and it bounds the specification of development'.[25] Development, here, should be understood as the material overlap resulting in offspring 'with the capacity to acquire the capacity to reproduce'.[26]

If it is indeed the case that our interlocutor is right, which is not a given since this empirical claim requires empirical support, then we would have to conclude

[22] P. Godfrey-Smith, 'Darwinian Individuals', in F. Bouchard and P. Huneman, eds., *From Groups to Individuals: Perspectives on Biological Associations and Emerging Individuality* (Cambridge, MA: MIT Press, 2012), 17–36.

[23] A. Kolers, 'Cloning and Genetic Parenthood', *Cambridge Quarterly of Healthcare Ethics*, 12/4 (2003), 401–410.

[24] J. Griesemer, 'Reproduction and the Reduction of Genetics', in P. Beurton, R. Falk, and H. J. Rheinberger, eds., *The Concept of the Gene in Development and Evolution: Historical and Epistemological Perspectives* (Cambridge University Press, 2000), 240–285.

[25] Griesemer, 'Reproduction and the Reduction of Genetics', 240–285.

[26] J. Griesemer, 'The Units of Evolutionary Transition', *Selection*, 1/1–3 (2000), 67–80.

that multiplex parenting would not, in most instances, deliver what people want from it, and thus we would not have strong reasons to pursue it. Let us suppose, for the sake of argument, that our interlocutor is correct in his assessment. To illustrate this point further let me flesh out the quadruple case. In the case of the quadruple, what happens is that we first generate two embryos, E1 and E2, and then we use them to generate a third one, E3. The issue is, according to our interlocutor, that M1 and F1 are the genetic parents of embryo E1, and M2 and F2 are the genetic parents of E2. Thus, M1, F1, M2, and F2 are the genetic grandparents of E3. As I noted before, I am supposing that E1 and E2 would be destroyed to generate M3. Here we have temporally compressed a generation, in the sense that we are able to obtain functional gametes from these embryos (E1 and E2) rather than the embryos developing normally to become children, and then these children start producing gametes once teenagers. It must be obvious that the point raised by our interlocutor only works if we accept that early embryos are organisms, following for example Mathew Liao's position.[27] Let us accept this view, since this would be the way in which the objection to multiplex parenting would be the strongest, or if not the strongest at least it would be a robust one.

From Parent to Grandparent

In our original publication my colleagues and I noted that members of a quadruple engaging in multiplex parenting would in fact be the genetic grandparents of the resulting child: 'The resulting embryo would be genetically related to all four prospective parents, who would technically be the child's genetic grandparents'.[28] This point was also echoed by Sonia Suter, who writes: 'the genetic relationship between each member of the quadruple and the child would be like that of grandparent and grandchild, rather than parent and child'.[29]

One way in which we can deal with the argument advanced by our interlocutor, that multiplex parenting does not deliver on its promise, is to bite the bullet, whilst emphasizing that the fact that such individuals would be the genetic grandparents of the created children is not particularly worrisome, neither practically nor morally. In moral terms, what we would need to show is that even when these human embryos are organisms, they have no moral status, and thus destroying them is morally neutral in itself. We could here espouse an argument similar to that advanced by Mary Anne Warren.[30] A second option

[27] S. M. Liao, 'The Organism View Defended', *The Monist*, 89/3 (2006), 334–350.
[28] Palacios-González, 'Multiplex Parenting', 752–758.
[29] S. M. Suter, 'In Vitro Gametogenesis: Just Another Way to Have a Baby?', *Journal of Law and the Biosciences*, 3/1 (2016), 87–119.
[30] M. W. Warren, 'On the Moral and Legal Status of Abortion', *The Monist*, 57/1 (1973), 43–61.

would be to defend that early human embryos have moral status, but very little of it, as defended by Rosamund Scott.[31] In this case the reproductive goals of the quadruple would trump such moral status. Of both possibilities the one that I favour is the first one. I consider that the philosophical arguments that have been presented for why early human embryos do not have moral status are convincing. So rather than rehashing old arguments here, I am going to assume that human early embryos do not have moral status, and examine what I think is a more interesting point. However, before doing so it is important to note that if human embryos have the same moral status as human persons, then IVG would only be morally permissible via the iPSC route—or when the derivation of gametes from embryonic stem cell lines did not destroy the embryo (as advocated by Mathew Liao[32]) and further, that the cell that would be coaxed into turning into a gamete was also not a single-cell embryo.

Now, what I meant when I said that in practical terms multiplex parenting is not so radical is that if we pay attention to what already occurs in nature, we find at least two similar phenomena. By this I mean that we have instances where who we think are the genetic parents seem rather to be the genetic grandparents. The first phenomenon, which I described above, is tetragametic chimerism. Here a woman and a man produce two embryos in one menstrual cycle, let us call them E4 and E5, and then these two embryos fuse forming a single organism, E6. If we accept that E4 and E5 are organisms, and that their fusion destroys both of them and creates a new organism, then we cannot claim that the members of such a couple are the genetic parents of E6, but rather they seem to be in a position similar to that of a genetic grandparent.

The second phenomenon is that of monozygotic twining.[33] Here a couple produces an embryo, E7, which splits into two, resulting in identical twins, E8 and E9. In this case E7 does not survive the fission process. Here E7 is the genetic parent of E8 and E9. And the couple who produced E7 are the genetic grandparents of E8 and E9.

I am unsure if what has happened in the tetragametic chimerism case would count as an instance of reproduction, but what has happened in the fission case is in fact a type of reproduction that happens in nature, asexual reproduction. For example, in binary fission, a bacterium grows to a specific size before splitting into two daughter cells of equal size, and '[d]uring the course of each doubling in the cell mass, this immense [DNA] molecule must be faithfully replicated so that

[31] R. Scott, *Choosing between Possible Lives: Law and Ethics of Prenatal and Preimplantation Genetic Diagnosis*, 1st ed. (London: Hart Publishing, 2007).

[32] S. M. Liao, 'Rescuing Human Embryonic Stem Cell Research: The Blastocyst Transfer Method', *American Journal of Bioethics*, 5/6 (2005), 8–16.

[33] J. G. Hall, 'Twinning', *The Lancet*, 362/9385 (2003), 735–743.

there will be one complete copy for each daughter cell'.[34] Here someone might object that humans do not reproduces asexually. However, the problem for such an objector is to explain why this particular type of fission in humans is not an instance of reproduction, when we accept that the fission of bacteria is indeed reproduction. The objector could reply that it is part of human nature to reproduce sexually, but this would just beg the question, and miss that species are historical individuals.[35]

Let me conclude this section by drawing two implications of biting the bullet. First, multiplex parenting is not as a radical development as first thought, in that there are already individuals who engage in the parenting, as a specific type of caring, of their genetic grandchildren, unbeknown to them. Multiplex parenting would not deliver, *in terms of genealogical relations*, something that does not already occur in nature. Second, what is radical about multiplex parenting is that we would be able to achieve the temporal compression of human generations intentionally, rather than it happening fortuitously.

Extended Lineages in Quintuple Multiplex Parenting

At this point our interlocutor could accept our response, and level a second challenge. This second challenge is that whereas it is true that the case of the quadruple is, in genealogical terms, identical to that of monozygotic twinning and the tetragametic chimerism one, there are other cases that do not have a natural counterpart. For example, if a quintuple relied on multiplex parenting to have a child, some of them would be the child's genetic *great grandparents*, whereas another one might be the child's genetic parent. In order to appreciate this more clearly imagine the following.

Imagine that three members of a quintuple are men, let us call them V1, V2, and V3, and two are women, W1 and W2. In this case V1 and W1 generate an embryo, Z1. V2 and W2 also generate an embryo, Z2. Z1 has the sex chromosomes XX, and Z2 has the sex chromosomes XY. From Z1 we then derive an oocyte, and from Z2 we derive a sperm. Now we create a third embryo, Z3, with the sperm from Z2 and the oocyte from Z1. Afterwards we use Z3 to generate an egg. And again, for simplicity, let us assume that Z1, Z2, and Z3 are destroyed. Finally, we use the egg generated from Z3 and a *naturally occurring* sperm from V3 (i.e., a non-*in vitro*-generated one) to generate embryo Z4, which will be then transferred to either W1 or W2. Genealogically, what has happened here is the

[34] W. Margolin, 'Binary Fission in Bacteria', in John Wiley & Sons Ltd., eds., *Encyclopedia of Life Sciences* (New Jersey, 2001).

[35] D. L. Hull, 'A Matter of Individuality', *Philosophy of Science*, 45/3 (1978), 335–360.

following. V3 is the genetic parent of Z4. Z3, which gets destroyed, is also a genetic parent of Z4. Z1 and Z2, which are destroyed, are the genetic parents of Z3, and genetic grandparents of Z4. V1 and W1 are the genetic parents of Z1, and genetic grandparents of Z3. V2 and W2 are the genetic parents of Z2, and genetic grandparents of Z3. And here is the most important point, in terms of the quintuple: whereas V3 is a genetic parent of Z4, V1, V2, W1, and W2 are the genetic great grandparents of Z4.

It is true, as far as I can see, that in nature we do not have something similar to the quintuplet case. However, there are naturally occurring cases where someone might be the genetic great grandparent of a child, even when she thinks that she is the genetic parent. For example, in 2020 a story was reported in the US media about the birth of monochorionic quadruplets.[36] Here an embryo splits in half, forming two embryos; then these two embryos split in half again, forming two embryos each. If we accept, as we have done so far, that embryos are organisms then we have the following. A man, V4, and a woman, W3, produce embryo Z5. Embryo Z5 divides in two and produces embryos Z6 and Z7. Z6 divides in two, producing embryos Z8 and Z9; and Z7 also divides in two producing embryos Z10 and Z11. Here V4 and W3 are the genetic great grandparents of Z8, Z9, Z10, and Z11.

At this point our interlocutor might argue that this is an exercise in futility, since we can easily come up with a configuration of people who want to engage in multiplex parenting for which there is no natural counterpart. I accept this; however, what I think would be the most common cases of multiplex parenting would be those for which there is a natural counterpart.

The Significance of Genealogical Counterparts

A reader might accept that human embryos do not have moral status. She could also accept that most instances of multiplex parenting would have naturally occurring genealogical counterparts. However, she might not be clear about the significance of genealogical counterparts. And furthermore, she might also be worried that what has happened here is that we have, implicitly, introduced the idea that natural phenomena, and their counterparts, are morally good, and that this is a morally problematic assumption.

To further elaborate on the conclusion of the previous section, the thing that we gain in showing that such genealogical counterparts exist in nature is removing the sensationalist edge about what this use of *in vitro*-derived

[36] S. Page, 'Quartet of Marvels: A Woman Has Quadruplets without Fertility Treatments, and They Are Identical', *The Washington Post* (21 May 2020).

gametes entails. We remove it by showing that it is not the case that being a genetic grandparent, or great grandparent, of a child, to whom we think we are their genetic parents, is unheard of. Multiplex parenting just joins the list of phenomena where the intending parents might not be genetic parents, even when they are genetically related to the children in a material distant causal way. This list includes: tetragametic chimerism, monozygotic twinning, heteropaternal tetragametic chimerism, and mitochondrial replacement techniques.

In regard to the worry that I just introduced and accepted, the idea that natural phenomena and their human instantiated counterparts are morally good, I would respond the following. No. My point is actually the opposite. If we do not think that the items in the previous list are morally suspicious, in themselves, then in order to be consistent we need to assume that multiplex parenting, as a phenomenon that temporally compresses human generations, is also not morally suspicious. Our moral suspicion, if we have one, must come from somewhere else. Maybe we think that it is morally problematic for children to be parented by a group of individuals. Or maybe we think that it is morally problematic to engage in multiplex parenting because such children will be bullied at school. Or maybe we think that it is morally problematic to engage in multiplex parenting for some other reason. Now it is up to the objector to spell out any particular objection in detail.

Let me close the chapter with two observations. First, all of what I have said before, of course, will depend on how we fare in terms of safety and IVG research, or what level of risk we are willing to accept in order to achieve certain reproductive goals. The second observation is that more work needs to be done on whether there might be moral reasons for more than two people to be involved in a parenting project, and how this would impact our assessment of multiplex parenting. Daniela Cutas[37] and Kalle Grill[38] have argued, for example, that children might be better off if they had more than two parents. Grill says: 'There is some indication that two [parents] may be better than one and also that three or four may be better than two, but these indications are too weak to support a general social presumption in favor of any particular number or numbers below the psychological limit'.[39] It might be the case that whereas reproductive autonomy morally allows for multiplex parenting, that we have other moral considerations to engage in it.

[37] D. Cutas, 'On Triparenting: Is Having Three Committed Parents Better than Having Only Two?', *Journal of Medical Ethics*, 37/12 (2011), 735–738.

[38] K. Grill, 'How Many Parents Should There Be in a Family?', *Journal of Applied Philosophy*, 37/3 (2020), 467–484.

[39] Grill, 'How Many Parents Should There Be in a Family', 467–484.

Bibliography

Baylis, F., 'The Ethics of Creating Children with Three Genetic Parents', *Reproductive Biomedicine Online*, 26/6 (2013), 531–534.

Benatar, D., *Better Never to Have Been: The Harm of Coming into Existence* (Oxford: Oxford University Press, 2006).

Boonin, D., 'How to Solve the Non-Identity Problem', *Public Affairs Quarterly*, 22/2 (2008), 129–159.

Boonin, D., *The Non-Identity Problem and the Ethics of Future People* (Oxford: Oxford University Press, 2014).

Cavaliere, G., and Palacios-González, C., 'Lesbian Motherhood and Mitochondrial Replacement Techniques: Reproductive Freedom and Genetic Kinship', *Journal of Medical Ethics*, 44/12 (2018), 835–842.

Cutas, D., 'On Triparenting: Is Having Three Committed Parents Better than Having Only Two?', *Journal of Medical Ethics*, 37/12 (2011), 735–738.

Douglas, T., Harding, C., Bourne, H., and Savulescu, J., 'Stem Cell Research and Same-Sex Reproduction', in M. Quigley, S. Chan, and J. Harris, eds., *Stem Cells: New Frontiers in Science and Ethics* (Singapore: World Scientific, 2012), 207–228.

Godfrey-Smith, P., 'Darwinian Individuals', in F. Bouchard and P. Huneman, eds., *From Groups to Individuals: Perspectives on Biological Associations and Emerging Individuality* (Cambridge, MA: MIT Press, 2012), 17–36.

Greely, H. T., *The End of Sex and the Future of Human Reproduction*, 1st ed. (Cambridge, MA: Harvard University Press, 2016).

Griesemer, J., 'Reproduction and the Reduction of Genetics', in P. Beurton, R. Falk, and H. J. Rheinberger, eds., *The Concept of the Gene in Development and Evolution: Historical and Epistemological Perspectives* (Cambridge University Press, 2000), 240–285.

Griesemer, J., 'The Units of Evolutionary Transition', *Selection*, 1/1–3 (2000), 67–80.

Grill, K., 'How Many Parents Should There Be in a Family?', *Journal of Applied Philosophy*, 37/3 (2020), 467–484.

Hall, J. G., 'Twinning', *The Lancet*, 362/9385 (2003), 735–743.

Hayashi, K., 'State of the Science', in *In Vitro–Derived Human Gametes as a Reproductive Technology: Scientific, Ethical, and Regulatory Implications*', *Proceedings of a Workshop* (Washington, DC: National Academies Press, 2023).

Hayashi, K., Ogushi, S., Kurimoto, K., Shimamoto, S., Ohta, H., and Saitou, M., 'Offspring from Oocytes Derived from In Vitro Primordial Germ Cell-like Cells in Mice', *Science*, 338/6109 (2012), 971–975.

Hull, D. L., 'A Matter of Individuality', *Philosophy of Science*, 45/3 (1978), 335–360.

Kolers, A., 'Cloning and Genetic Parenthood', *Cambridge Quarterly of Healthcare Ethics*, 12/24 (2003), 401–410.

Liao, S. M., 'Rescuing Human Embryonic Stem Cell Research: The Blastocyst Transfer Method', *The American Journal of Bioethics*, 5/6 (2005), 8–16.

Liao, S. M., 'The Organism View Defended', *The Monist*, 89/3 (2006), 334–350.

Lovell-Badge, R. H., and Hacker, A., 'The Molecular Genetics of Sry and Its Role in Mammalian Sex Determination', *Philosophical Transactions of the Royal Society of London, Series B: Biological Sciences*, 350/1333 (1995), 205–214.

Madan, K., 'Natural Human Chimeras: A Review', *European Journal of Medical Genetics*, 63/9 (2020), 103971.

Margolin, W., 'Binary Fission in Bacteria', in John Wiley & Sons Ltd., ed., *Encyclopedia of Life Sciences* (New Jersey, 2001). https://doi.org/10.1002/9780470015902.a0001420.pub2.

McMahan, J., and Savulescu, J., 'Reasons and Reproduction: Gene Editing and Genetic Selection', *The American Journal of Bioethics*, 24/8 (2024), 9–19.

Nayernia, K., Nolte, J., Michelmann, H. W., Lee, J. H., Rathsack, K., Drusenheimer, N., et al., 'In Vitro-Differentiated Embryonic Stem Cells Give Rise to Male Gametes that Can Generate Offspring Mice', *Developmental Cell*, 11/1 (2006), 125–132.

Page, S., 'Quartet of Marvels: A Woman Has Quadruplets without Fertility Treatments, and They Are Identical', *The Washington Post* (21 May 2020), https://www.washingtonpost.com/lifestyle/2020/05/21/quartet-marvels-woman-had-quadruplets-without-fertility-treatments-they-are-identical, accessed 21 March 2025.

Palacios-González, C., Harris, J., and Testa, G., 'Multiplex Parenting: IVG and the Generations to Come', *Journal of Medical Ethics*, 40/11 (2014), 752–758.

Saitou, M., and Hayashi, K., 'Mammalian In Vitro Gametogenesis', *Science*, 374/6563 (2021), eaaz6830.

Scott, R., *Choosing between Possible Lives: Law and Ethics of Prenatal and Preimplantation Genetic Diagnosis*, 1st ed. (London: Hart Publishing, 2007).

Segal, N. L., Craig, J. M., and Umstad, M. P., 'Challenge to the Assumed Rarity of Heteropaternal Superfecundation: Findings from a Case Report', *Australian Journal of Forensic Sciences*, 52/5 (2020), 547–552.

Segers, S., Mertes, H., de Wert, G., Dondorp, W., and Pennings, G., 'Balancing Ethical Pros and Cons of Stem Cell Derived Gametes', *Annals of Biomedical Engineering*, 54/7 (2017), 1620–1632.

Segers, S., Mertes, H., Pennings, G., de Wert, G., and Dondorp, W., 'Using Stem Cell-Derived Gametes for Same-Sex Reproduction: An Alternative Scenario', *Journal of Medical Ethics*, 43/10 (2017), 688–691.

Smajdor, A., and Cutas, D., *Background Paper on Artificial Gametes*, Nuffield Council on Bioethics, December 2015, http://nuffieldbioethics.org/wp-content/uploads/Background-paper-2016-Artificial-gametes.pdf, accessed 21 March 2025.

Suter, S. M., 'In Vitro Gametogenesis: Just Another Way to Have a Baby?', *Journal of Law and the Biosciences*, 3/1 (2016), 87–119.

Testa, G., and Harris, J., 'Ethical Aspects of ES Cell-Derived Gametes', *Science*, 305/5691 (2004), 1719.

Testa, G., and Harris, J., 'Ethics and Synthetic Gametes', *Bioethics*, 19/2 (2005), 146–166.

Warren, M. A., 'On the Moral and Legal Status of Abortion', *The Monist*, 57/1 (1973), 43–61.

8

The Ethics of Solo Reproduction

Anantharaman Muralidharan

In vitro Gametogenesis and Solo Reproduction

Advances in reprogramming technologies[1] and techniques for differentiating pluripotent stem cells have recently brought the possibility of *in vitro* gametogenesis (IVG) closer to reality. IVG is a technology that allows gametes to be produced from a person's somatic cells. It has the potential to address forms of infertility that current *in vitro* fertilization (IVF) techniques are unable to. Insofar as certain forms of infertility deprive some persons of the capacity to produce gametes, IVG can help such persons. IVG also promises to allow same-sex couples to have children who are approximately equally genetically related to both parents. The same technology that can help couples who are infertile for medical or social reasons can also allow a single individual to reproduce entirely by themselves (or at most with the help of a surrogate). In this entry, I will examine some of the objections people might have to solo reproduction and show that, with the exception of physiological risk, they do not work.

Since solo reproduction seems to have some intuitive parallels with cloning, Solo Reproduction versus Cloning will compare the two technologies. Strong Objections to Solo Reproduction will examine a number of strong objections that were traditionally offered against cloning that seemingly also apply to solo reproduction. These objections include concerns about physiological risk, social risk, dignity, the telos of human sexuality, and vice. If any of these objections succeed, solo reproduction is wrong. I will argue that these objections by and large (with one exception) do not succeed. Weak Objections examines some weak objections that try to show that solo reproduction should be deprioritized relative to other uses of IVG. I will argue that these weaker objections too do not succeed.

[1] Yusuke Murase et al., 'In Vitro Reconstitution of Epigenetic Reprogramming in the Human Germ Line', *Nature*, 631/8019 (2024), 170–178.

Anantharaman Muralidharan, *The Ethics of Solo Reproduction* In: *Revolutionizing Reproduction.* Edited by: Christopher Gyngell, Julian Savulescu, and Tsutomu Sawai, Oxford University Press.
 DOI: 10.1093/9780197810415.003.0008

Solo Reproduction versus Cloning

IVG is a process by which gametes are created from stem cells.[2] These stem cells can be stimulated to become Primordial Germ Cell-Like Cells (PGCLCs). Like Primordial Germ Cells, these cells will develop, eventually, into sperm and egg cells. In principle, if the technology is successful, scientists would be able to control whether the PGCLCs develop into sperms or eggs. This means that through this technology, a genetic male could give rise to eggs and a genetic female could give rise to sperm. For simplicity of expression, let us call the person who gives rise to the sperm the 'biological father' and the person who gives rise to the egg the 'biological mother'. With solo reproduction, the same person is both biological mother and father of a particular child. This is done by using IVG to produce one cross-sex gamete and combining that with another gamete from the same individual.[3] For ease of expression, let us call the child that results from solo reproduction a 'solone'.

Cloning, like solo reproduction, also starts with somatic cells and, if successful, may allow someone to reproduce without involving anyone else's genetic material. Nuclear DNA is extracted from the somatic cell. It is then transferred into an egg that has had its own nuclear genetic material removed. This means that without an IVG-produced egg, a genetic male or infertile female's clone would require someone else's mitochondrial genetic material. A fertile female may use her own egg. IVG could, in principle, allow anybody to clone themselves without an egg donor.

Solones are different from clones in the following way. Clones are approximately[4] genetically identical at the level of nuclear DNA to their 'parent'. Solones, on the other hand, are unlikely to be genetically identical. There is an approximately 100% chance that a solone's parent will have any of the solone's alleles. However, there is only a 75% chance (provided that the parent is heterozygous at that gene locus) that a solone will have at least one copy of any given allele of their parent's.

Solones are also different from clones in another way. While solones can legitimately point to a biological father and mother (who happens to be the same person), the relationship is less clear with a clone. Is someone's clone their twin

[2] This would include both embryonic stem cells as well as pluripotent stem cells induced from somatic cells. See Mitinori Saitou and Katsuhiko Hayashi, 'Mammalian In Vitro Gametogenesis', *Science*, 374/6563 (2021), eaaz6830.

[3] The other gamete can either be generated naturally or by IVG.

[4] Clones are only approximately and not perfectly identical because there is, inevitably, a small rate of random mutation. While some women would be able to use their own eggs and hence ensure that the mitochondrial DNA of the clone is approximately identical to her parent's, other people who cannot or are not willing to donate an egg will have to use a donor's egg. Hence, their clones will have different mitochondrial DNA as well.

or their offspring? Genetically, since a clone is not produced through the union of sperm and egg, a person's clone may be more like her twin than her offspring. To be fair, however, in other species that reproduce asexually, we regard the clones as children or progeny rather than as twins. In any case, clones, unlike solones, have less well-defined biological relationships.

Despite these differences, solo reproduction is, in many ways, morally similar to reproductive cloning. For one, cloning is one method by which a person may reproduce without involving anyone else's genetic material. Secondly, like cloning, some people might see it as a particularly 'unnatural' way of reproducing. While, unlike cloning, it involves the union of a sperm and an egg, either or both gametes are not generated 'naturally' (i.e., extracted from testes or ovaries). The same types of criticism that have been levelled at cloning, that it involves a promethean striving to exceed human limitations, that it evinces a desire for a specific genetic makeup for one's offspring, or that it involves a kind of narcissism can be directed towards solo reproduction.

Strong Objections to Solo Reproduction

We can classify strong objections to cloning and solo reproduction under five broad categories: concern for the physiological well-being of the resultant child, concerns about the social well-being of the resultant child, dignitarian concerns, concerns about the teleology of human sexuality, and concerns about whether cloning expresses vicious dispositions.

Physiological Risk

Physiological risk is a significant objection to solo reproduction. IVG technology is still in its infancy, and comprehensive safety data does not exist. Using the technology, when there is a risk of harmful consequences, would be irresponsible. This is true for many new technologies. However, there is reason to believe that solo reproduction may introduce additional levels of physiological risks. Fusing two gametes from the same individual vastly increases the risk of recessive genetic disorders being expressed. After all, repeated inbreeding in dog breeds has resulted in the accumulation of deleterious alleles. For instance, cavalier King Charles spaniels, like a few other breeds, are especially prone to certain kinds of congenital heart diseases.[5] Likewise, generations of inbreeding

[5] Erik Axelsson et al., 'The Genetic Consequences of Dog Breed Formation—Accumulation of Deleterious Genetic Variation and Fixation of Mutations Associated with Myxomatous Mitral

among the Habsburgs notoriously resulted in many of the dynasty's members suffering from a number of genetic diseases. This is caused by the accumulation of deleterious mutations in a small gene pool. To get some idea of the numbers, in Bradford, risk of severe birth defects is about 6% for the offspring of first cousins against background risks of 3% in Bradford and about 1% in the United Kingdom.[6] In another study[7] in Prague, children of incestuous pairings were six times[8] as likely to have serious congenital anomalies as children of non-incestuous pairings. The risk of severe congenital anomalies for solones could be very high (upwards of 25%, perhaps between 50 and 75%) even for someone who had a typical genetic load.[9] It would arguably be wrong to produce children via a procedure that has such high chance of resulting in severe birth defects.

To be sure, this argument is most effective when such birth defects are so severe that the resultant child's life is not worth living. In these kinds of cases, it is clearly wrong to create a being whose life is guaranteed to contain more suffering than happiness. In cases where the solone has a good chance of having a life worth living, things get more complicated. Suppose we were to subscribe to a strong form of procreative beneficence according to which we ought to, among the children we could possibly have, choose to have the child(ren) with the best expected life prospects. Even if such a principle were true, this could count against solo reproduction only if the person who wanted to solo reproduce was willing to have children by other means too. If, for her own reasons, she wanted to have children only by solo reproduction, then all her potential children would be solones. One might argue that she could, in principle, choose to have children by IVF or naturally, but this interpretation is too strong. Consider a couple, one of whom is homozygous for the allele that confers Huntington's Disease. Their child is guaranteed to have Huntington's Disease. The person who does not have the disease could, in principle, choose to have a child with someone else, but they are unwilling to because they are committed to their marriage. Yet, few, except for the most fervent anti-natalist, would regard the couple as morally obligated to not have children.

Despite the high levels of risk of genetic defects, so long as the child has a good chance of having a life worth living, even the strong principle of procreative

Valve Disease in Cavalier King Charles Spaniels', ed. Gregory S. Barsh, *PLOS Genetics*, 17/9 (2021), e1009726.

[6] 'Born in Bradford: Consanguinity and Birth Defects', *Archives of Disease in Childhood*, 99 (2014), 38.

[7] Eva Seemanová, 'A Study of Children of Incestuous Matings', *Human Heredity*, 21/2 (1971), 108–128.

[8] 52% risk for parent-child or sibling incest versus a background risk of 8.7%

[9] The background risk of birth defects from non-consanguineous or non-incestuous matings were 3% and 8.7%, respectively, in the Bradford and Prague studies. This is much higher than a more typical risk of about 1%.

beneficence does not rule out solo reproduction. This means that even if the increased homozygosity did not result in any specific genetic disease but reduced overall health due to loss of hybrid vigour, so long as the child had a life worth living and the parent was unwilling to have a child any other way, solo reproduction would be permissible. Suppose that we were to instead adopt an even weaker principle whereby we were not obligated to choose the child with the best life, but only to choose any child whose life was good enough. On this principle, if the solone could have a good enough life despite any congenital defects, then even if the parent was willing to have children by other means, solo reproduction would still be permissible for her.

However, if we use whole genome sequencing as pre-implantation genetic testing (PGT), we could select for embryos which do not have the birth defect-causing alleles. Given that IVG can be used with somatic cell-derived pluripotent stem cells as a starting point, it may be possible to generate solone embryos until we obtain at least a few that have not inherited significantly deleterious alleles from their parent. If the use of PGT to screen away embryos with congenital abnormalities is permissible,[10] then worries about the physiological health of the solone can be assuaged. There is a caveat, namely that this works only on the assumption that we have identified all genetic diseases and can screen for them. There are, as of the writing of this entry, many congenital diseases that we do not understand the genetics of. Conducting solo reproduction when such knowledge is incomplete will increase the risk of rare diseases that we are unable to screen for. As such, until these technical problems are solved, more caution will be required.

Risk to Social Well-Being

Apart from physiological risk, there is also a worry that solones will be at higher risk of oppression of various forms.[11] For instance, a solone's parent or society may place unreasonable expectations on her or may think it all but certain that she will follow in her parent's footsteps. For instance, they may expect the solone of a professional football player to also be especially good at football. Teachers and family members might, perhaps unconsciously, steer or pressure her into pursuing football. The solone may, as a result, feel that her options are limited. Society may also discriminate against, denigrate, or otherwise exclude solones in virtue of them having only one biological parent.

[10] Objections to PGT are beyond the scope of this article.

[11] Michael J. Sandel, 'The Ethical Implications of Human Cloning', *Perspectives in Biology and Medicine* 48/2 (2005), 241–247.

Given the kinds of harms and wrongs that solones are likely to experience from society or their parents, or so the objection goes, solo reproduction would be wrong.

However, such an argument seems to prove too much. What is wrong here is bad parenting and an oppressive society. Society should not oppress and exclude those who have only one biological parent just as society in the past should not have discriminated against and excluded those who were born out of wedlock. Likewise parents should not over-parent and instrumentalize their children. Many parents instrumentalize their naturally conceived children and place unreasonable expectations on them. While this is wrong, it does not make their prior decision to have children in the first place wrong too. Likewise, many racial minorities in racist societies face significant oppression. It does not follow that those minorities should not have children.

Dignity

According to the dignity objection, cloning, and by extension solo reproduction, violates the dignity of the clone or solone. However, concerns about human dignity are controversial. On some views, dignity, as a concept, should be done away with because it reduces to either respect for the capacity to reason (i.e., Kantian autonomy), vague restatements of other moral positions, or a covert appeal to religious dogma.[12] Even defenders of the concept concede that there is significant ambiguity in the usage of the term.[13] Schulman, for instance, distinguishes between stoic dignity, biblical dignity, Kantian autonomy, and the accounts of dignity found in twentieth-century international declarations and constitutions.[14] Malby distinguishes three strands: one pertaining to Kantian autonomy and the capacity to reason, the second pertaining to a conception of civilized behaviour, and the third pertaining to group identity and membership.[15] Likewise other commentators have also distinguished between accounts of dignity that relate

[12] Ruth Macklin, 'Dignity Is a Useless Concept', *British Medical Journal*, 327 (2003), 1419–1420; Timothy Caulfield, 'Human Cloning Laws, Human Dignity and the Poverty of the Policy Making Dialogue', *BMC Medical Ethics* 4/1 (2003), 3.

[13] Oscar Schachter, 'Human Dignity as a Normative Concept', *American Journal of International Law*, 77/4 (1983), 848–854; Steven Malby, 'Human Dignity and Human Reproductive Cloning', *Health and Human Rights* 6/1 (2002), 102; Adam Schulman, 'Bioethics and the Question of Human Dignity', in *Human Dignity and Bioethics*, ed. Thomas W. Merrill and Adam Schulman (Washington, DC: President's Council on Bioethics, 2008), 3–18; H. Rolston III, 'Human Uniqueness and Human Dignity: Persons in Nature and the Nature of Persons', in *Human Dignity and Bioethics*, ed. Thomas Merrill W. and Adam Schulman (Washington, DC: President's Council on Bioethics, 2008), 129–154.

[14] Schulman, 'Bioethics and the Question of Human Dignity'.

[15] Malby, 'Human Dignity and Human Reproductive Cloning'.

to respect for Kantian autonomy and conceptions pertaining to culture and civilization[16]or humiliation.[17] For our purposes, we can distinguish between two distinct conceptions of dignity: dignity as the capacity to reason and eudaimonic dignity. We will address each conception in turn.

Dignity as Respect for Kantian Autonomy

Consider the first conception of dignity. Critics of cloning contend that at least some of the purposes with which people might pursue cloning instrumentalizes the clone and hence disrespects their personhood. For instance, someone may wish to clone a child in order to replace a dead child or to select perceived favourable characteristics. The objection goes that in such cases, there is 'some intent to control how a new human should be'[18] and this uses the clone as a mere means to a personal, selfish end. A parallel argument can also be offered against solo reproduction. While solo reproduction may not likely be used to replace a dead child, it might be used because someone thinks very highly of their own genetic makeup and therefore wishes to produce offspring that is a product only of their own genetic heritage and who would thus emulate themselves as much as possible. Solo reproducing for such a narcissistic reason would, arguably, use the resultant child merely as a means to an end.

There are two problems with this argument. Firstly, not all of the motivations are objectionable. Secondly, it is not clear that reproductive acts are the type of act that can use the resultant child merely as a means to an end.

Consider, first, the question of motivation. Plausibly, there might be motivations where cloning and solo reproduction that do not instrumentalize the resultant child. For instance, a woman who has been abused by men all her life may wish to have a child without involving any man even by sperm donation. Or suppose someone has a genetic condition that makes it extremely risky for them to contribute genetically to the offspring. If they and their spouse wish to have offspring without getting genetic material from outside the marriage, they might pursue cloning or solo reproduction if either technology were viable. Likewise, consider a single person who wants to start a family without introducing genetic material from outside the family unit.[19] If such a motivation

[16] Deryck Beyleveld and Roger Brownsword, 'Human Dignity, Human Rights, and Human Genetics', *The Modern Law Review* 61/5 (1998), 661–680; Rolston III, 'Human Uniqueness and Human Dignity: Persons in Nature and the Nature of Persons'.

[17] Schachter, 'Human Dignity as a Normative Concept'.

[18] Malby, 'Human Dignity and Human Reproductive Cloning', 123.

[19] Sonia M. Suter, '*In Vitro* Gametogenesis: Just Another Way to Have a Baby?', *Journal of Law and the Biosciences*, 3/1 (2016), 87–119.

is legitimate for heterosexual couples who wish to pursue IVF and same-sex couples pursuing IVG, then that same motivation can also be legitimate for singles. What this means is that solo reproduction, if wrong, is not wrong *tout court*, but wrong only when performed for certain objectionable motives but not others.

Even if the motivation is objectionable, it is not clear whether cloning or solo reproduction could treat the resultant child as merely a means to an end. For instance, as Caulfield[20] notes, at the point at which the cloning takes place, the child does not yet exist. It is not clear how a non-existent being could be used merely as a means to an end. We might also think that the possibility that reproductive acts could use the resultant child merely as a means to an end has anti-natalist implications that opponents of cloning and solo reproduction may not be willing to endorse.

Consider that nobody ever consents to being conceived, carried to term, and then being born. Even for people who have really great lives, being alive involves experiencing some suffering or harm. As Shiffrin[21] argues, it is wrong to inflict a harm on someone in order to confer a greater benefit without their consent. Since reproduction involves the non-consensual infliction of some harm (i.e., all unavoidable pains of existence) on someone in order to confer a benefit[22] (i.e., all the joys of existence), reproduction, according to Shiffrin, is wrong.

In order to avoid the anti-natalist conclusion, we must either abandon the principle that non-consensually harming someone in order to benefit them is wrong, or suppose that the principle does not apply to reproductive acts. Yet, if we accept that it would be wrong to use people merely as a means to an end, it is not clear how non-consensually harming someone in order to benefit them would be permissible. If we do not think it is wrong to use people merely as a means to an end, then the instrumentalization objection does not succeed. Likewise, if the principle does not apply to reproductive acts, then this can only be because the duty to not use people merely as a means does not apply to reproductive acts. If this is right, the instrumentalization objection and a broad antinatalism rise and fall together. That is, if it is wrong to solo reproduce since that would instrumentalize the solone, then it would be wrong to reproduce because that would instrumentalize the child who never consents to being born. However, opponents of solo reproduction are rarely willing to adopt such an anti-natalist position.

[20] 'Human Cloning Laws, Human Dignity and the Poverty of the Policy Making Dialogue'.

[21] 'Wrongful Life, Procreative Responsibility, and the Significance of Harm', *Legal Theory*, 5 (1999), 117–148.

[22] Shiffrin allows that we may non-consensually harm someone in order to prevent a greater harm but not in order to confer a benefit.

Eudaimonic Dignity

The core of the eudaimonic conception of dignity is some conception or other about what it means to live a sufficiently flourishing life. Anything that causes someone's life to lack one or more of the elements found in that conception of flourishing is injurious to that person's dignity. For instance, on the stoic account of dignity, dignity is an achievement that one has when one lives by the stoic virtues (i.e., in a thoughtful and reflective way).[23] A person loses dignity if she fails to live up to these values. On what Malby calls 'the European conception of dignity',[24] living a flourishing life involves, among other things, not engaging in masochistic acts, and not performing in peep shows or dwarf throwing. On these accounts of the good life, performing such acts is demeaning, and people who perform such acts ought to feel humiliated. On Malby's account, this aspect of dignity pertains to what it means to be a civilized person. Malby also talks about another aspect of dignity that is more collective. Here violating the norms of a group may be demeaning to that group. Consider, for instance, that attempts to insist on Western norms of informed consent on certain Indigenous peoples are viewed by them as signalling distrust and hence humiliating[25] to those Indigenous groups. This consequently presupposes a particular conception of what it means for that group to flourish.

Opponents of cloning might claim that fundamental to what it means to be a human being is to be part of a species which has certain features. One feature is that each person has two genetic parents. Another is that each individual is the product of a genetic lottery. Yet another feature may be that nobody is the product of deliberate systematic scientific manipulation. The Catholic Church, for instance, claims that children ought to be the natural product of their parents' love for each other within a marriage.[26] Since cloning and solo reproduction result in children who have only one genetic parent, who are removed to some degree from the genetic lottery, and who are not the natural products of their parents' love within the confines of marriage, the act of producing such children supposedly offends the dignity of (depending on the

[23] Schulman, 'Bioethics and the Question of Human Dignity'.

[24] Malby, 'Human Dignity and Human Reproductive Cloning', 110.

[25] John Noel Montaño Viaña, 'Interrogating Sites of Knowledge Production: The Role of Journals, Institutions, and Professional Societies in Advancing Epistemic Justice in Bioethics', *The American Journal of Bioethics*, 24/4 (April 2024), 63–66, https://doi.org/10.1080/15265161.2024.2308144; Christopher Mayes, 'White Medicine, White Ethics: On the Historical Formation of Racism in Australian Healthcare', *Journal of Australian Studies*, 44/3 (2020), 287–302; Warwick Anderson, 'The Whiteness of Bioethics', *Journal of Bioethical Inquiry* 18/1 (2021), 93–97.

[26] Second Vatican Council, *Declaration on Religious Freedom, Dignitis Humanae on the Right of the Person and of Communities to Social and Civil Freedom in Matters Religious*, 1965, https://www.vatican.va/archive/hist_councils/ii_vatican_council/documents/vat-ii_decl_19651207_dignitatis-humanae_en.html.

details of the view under consideration) that child, God, or even the human species as a whole.[27]

Suppose that there are coherent conceptions of the good that contain such elements as having two genetic parents, being subject to the genetic lottery, or being natural products of their parents' love within the confines of marriage. We might still question whether every coherent and attractive conception of human flourishing requires any of these elements at all. Apart from religious fiat, there seems to be no reason to think that human flourishing requires any of these elements, let alone all of them. Perhaps they are conventionally a part of our self-conception as human beings. But, that itself is not a reason to think that these elements *have* to be. People could reasonably have conceptions of human flourishing and what it means to be human that have nothing to do with whether one has two genetic parents or whether one is the product of scientific manipulation or the product of the genetic lottery. On these latter kinds of conceptions of human flourishing, cloning and solo reproduction would not violate human dignity.

Summing up, dignitarian arguments either fail to show that solo reproduction as such is wrong, rely on dubious premises, or commit us to an implausible version of antinatalism.

Teleology of Human Sexuality

Another objection to cloning and solo reproduction is that pursuing such artificial methods of reproduction divorces human sexuality from reproduction.[28] This is supposedly objectionable because the purpose or telos of human sexuality is reproduction and divorcing a capacity from its telos is wrong. On such often very conservative views, it is wrong to have forms of sexual intercourse that are guaranteed to never result in children or to have children without having sexual intercourse.

There are at least two problems with this argument. Firstly, it arguably proves too much. Procedures like IVF and artificial insemination, which most find unobjectionable, also divorce human sexuality from reproduction. That is to say, like solo reproduction, IVF and artificial insemination allow people to reproduce

[27] Malby, 'Human Dignity and Human Reproductive Cloning'.

[28] Second Vatican Council, *Dignitis Humanae*; Congregation of the Doctrine of the Faith, 'Donum Vitae: Instruction on Respect for Human Life in Its Origin and on the Dignity of Procreation', 1987, https://www.vatican.va/roman_curia/congregations/cfaith/documents/rc_con_cfaith_doc_19870222_respect-for-human-life_en.html; Robert P. Kraynak, 'Human Dignity and the Mystery of the Human Soul', in Thomas Merrill W. and Adam Schulman, eds., *Human Dignity and Bioethics* (Washington, DC: President's Council on Bioethics, 2008), 61–82; Leon Kass, 'The Wisdom of Repugnance', *The New Republic*, June 1997.

without having sex intercourse. At least some of those who have made the teleology objection against cloning have explicitly considered IVF as objectionable for the same reason.[29] Insofar as these procedures are permissible, the mere fact that cloning and solo reproduction allow people to have children without having sex does not make them wrong.

Secondly, such the objection faces a dilemma. Depending on how we understand the term 'telos', the claim that 'the telos of human sexuality is reproduction' is either plausible but has no moral import, or has moral import but presupposes some religious or otherwise controversial account of the good life.

One way we could interpret 'telos' is as biological function. On this view, the proper biological function of sexuality is sexual reproduction because human sexuality is a trait that evolved by facilitating sexual reproduction[30] or because human sexuality currently contributes to biological fitness by facilitating reproduction.[31] However, this has no moral import because biological fitness and evolutionary success are merely a matter of the proliferation of genes, and do not necessarily translate into well-being or any other morally relevant property.[32]

One morally thicker account of telos pertains to what God intends or desires. On this view, God intends or wants people to reproduce sexually and bypassing human sexuality to reproduce would contravene God's wishes and hence be sinful.[33]

However, the biblical argument proves too much. The same argument, that God ordained the world to be in a particular way (e.g., childbirth being painful or disease being punishment for sin), was used to argue against anaesthesia and vaccines.[34] Yet most do not find such arguments credible when applied to anaesthesia and vaccination.

On certain neo-Aristotelian views, the telos of a capacity is the thing it uniquely provides that contributes to the flourishing of the bearer of the capacity (or the species of which the bearer is a member).[35] On such an account, to say

[29] Second Vatican Council, *Dignitis Humanae*; Congregation of the Doctrine of the Faith, 'Donum Vitae'; Kraynak, 'Human Dignity and the Mystery of the Human Soul'; Kass, 'The Wisdom of Repugnance'; John M. Finnis, 'Public Reason, Abortion, and Cloning', *Valparaiso University Law Review*, 32 (1998), 361–382.

[30] William J. FitzPatrick, *Teleology and the Norms of Nature*, Studies in Philosophy (New York London: Garland, 2000); Kim Sterelny and Paul Griffiths, *Sex and Death: An Introduction to Philosophy of Biology*, Science and Its Conceptual Foundations (Chicago: University of Chicago Press, 1999).

[31] Sterelny and Griffiths, *Sex and Death*.

[32] FitzPatrick, *Teleology and the Norms of Nature*.

[33] Second Vatican Council, *Dignitis Humanae*; Congregation of the Doctrine of the Faith, 'Donum Vitae'; Kraynak, 'Human Dignity and the Mystery of the Human Soul'.

[34] Patricia Churchland, 'Human Dignity from a Neurophilosophical Perspective', in Thomas Merrill W. and Adam Schulman, eds., *Human Dignity and Bioethics* (Washington, DC: President's Council on Bioethics, 2008), 99–121.

[35] Philippa Foot, *Natural Goodness* (Oxford: Clarendon Press, 2001); William D. Casebeer, *Natural Ethical Facts: Evolution, Connectionism, and Moral Cognition* (Cambridge, MA: MIT Press, 2005), 1; Micah Lott, 'Have Elephant Seals Refuted Aristotle? Nature, Function, and Moral

that the telos of human sexuality is reproduction is to say that human sexuality uniquely and irreplaceably contributes to human flourishing by facilitating sexual reproduction and further that having children while bypassing one's sexuality is strictly worse than having children by having sex. However, there does not seem to be any good reason to think that this is indeed true.

Suppose that we grant that human sexuality contributes to human flourishing by facilitating sexual reproduction at least typically for cisgendered heterosexual persons. The teleological claim can only be vindicated if reproduction is required for flourishing. One might argue that there is no human flourishing unless members of the species *homo sapiens* exist. However, while reproduction is required at the species level, it seems harder to show that any given individual needs to reproduce in order to flourish. No doubt that for many, their children are a great source of joy in their lives. Yet, it is possible that there could be aspects of a childless life that would also contribute to flourishing. Even if having children is necessary for flourishing, it is not clear why having children by artificial methods is less conducive to flourishing.

All in all, teleological arguments are insufficient to show that solo reproduction is wrong. Conceptions of telos are either too thin to have normative import or too thick to be plausible.

The Vice Objection

One objection against cloning and solo reproduction that has received little if any attention in the literature is that they characteristically exemplify certain vices.[36] The objection is not that solo reproduction harms, instrumentalizes, or otherwise wrongs the solone. Rather, it is that even if solo reproduction did not harm or wrong anyone, it would be wrong simply because it expressed certain vices.

Cloning and solo reproduction supposedly express and reinforce a kind of promethean hubris already rife in modern societies because it is an attempt to control the genetic makeup of one's child. By forgoing more traditional modes of reproduction where the characteristics of one's child are more left up to chance, parents lose an opportunity to develop a sense of humility and openness to the natural world as given.[37] Some have also argued that cloning and, by extension,

Goodness', *Journal of Moral Philosophy*, 9/3 (2012), 353–375; Parisa Moosavi, 'From Biological Functions to Natural Goodness', 19/51 (2019).

[36] Finnis, 'Public Reason, Abortion, and Cloning'; Sandel, 'The Ethical Implications of Human Cloning'; Kraynak, 'Human Dignity and the Mystery of the Human Soul'.

[37] Sandel, 'The Ethical Implications of Human Cloning'.

solo reproduction may be narcissistic in most circumstances[38] because most motivations to clone oneself supposedly involve a certain (excessively) high degree of regard towards oneself and one's own genetic makeup.[39]

One problem with the narcissism argument is the same as with the instrumentalization worry. There are at least some legitimate motivations for pursuing solo reproduction. As previously mentioned, a single person who wants to start a family without using genetic material from outside the family unit would need solo reproduction to fulfil this desire. No other technology (apart from cloning) would do. Insofar as it is legitimate for couples to desire to maintain the genetic integrity of the family unit, it is also legitimate for singles to desire to maintain the genetic integrity of the family unit. Perhaps there are other, non-narcissistic motivations that would be better served by something other than solo reproduction.[40] An abuse survivor who wants to reproduce without input from any man could obtain sperm produced by IVG from another woman. However, this is not dispositive as it is an empirical question as to what motivations most people who pursue solo reproduction will have. We cannot decide in advance whether legitimate or objectionable motivations will predominate.

One problem with the hubris argument is that critics seem to be willing to level the charge of hubris against any attempt to use technology to achieve what was previously unachievable for humans. The story of Icarus, the classic morality tale about hubris, was about a person who was so consumed with hubris he flew too close to the sun. Yet no one now thinks that flying or, for that matter, space travel is hubristic. Perhaps the ancients would have, *ex-ante*, considered it hubristic to try to do all we have accomplished with modern technology and medicine. Yet, it is hard to deny that it is far better to be alive today than a hundred, let alone a few thousand, years ago. If this is right, then whatever the vice of hubris is, it cannot simply be a matter of defying the natural order of things or of trying to do what was previously impossible for humans to do. If this is right, then we have no reason to think that solo reproduction is hubristic.

Summing up, the vice argument seems to be, as of this moment, inadequate. At best, solo reproduction would actually need to be implemented in order ascertain whether objectionable motivations predominate and further whether there are any unforeseen consequences to allowing its use. Perhaps the objection could be refined further, but that is beyond the scope of this chapter.

[38] Kass, 'The Wisdom of Repugnance'; Malby, 'Human Dignity and Human Reproductive Cloning'; Sandel, 'The Ethical Implications of Human Cloning'.

[39] Malby, 'Human Dignity and Human Reproductive Cloning'.

[40] Malby, 'Human Dignity and Human Reproductive Cloning'.

Weak Objections

We have so far looked at objections to solo reproduction and seen that apart from questions of physiological risk to the solone, arguments that try to show that solo reproduction is wrong do not succeed. Even for concerns about physiological risk, these may be addressed with further technological advances. Solo reproduction, if wrong, is only contingently wrong. Let us now turn to weaker objections that do not quite argue that solo reproduction is wrong, but only that access to it is less valuable or less urgent than for other uses of IVG. If these arguments succeed, other uses of IVG may merit our attention and resources more than solo reproduction.

Distress

One argument against solo reproduction is that people experience less infertility-related distress for being unable to solo reproduce than for being unable to reproduce with their partner.[41] Hence, or so the argument goes, there is little positive reason to allow people access to IVG for the purpose of solo reproduction.

To be clear, while most people would indeed feel little distress at being unable to solo reproduce, someone who is willing to spend large amounts of money to solo reproduce may very well feel large amounts of distress if she is denied access to such technology. Importantly, while we find it difficult to imagine people feeling distress in the current context where the technology is not available, this need not be the case. Consider that prior to the invention of smartphones, people would not feel distressed if they did not have internet access all the time. However, now, we can easily imagine someone feeling distressed if they did not have internet access for an hour. Plausibly the question of whether someone might feel distress is going to depend on what options are live for her.

Justice

Notini and colleagues offer another argument against solo reproduction. They argue that same compared to using IVG for same-sex couples, people who intend to solo reproduce have less of a claim to use IVG technology. They argue that while singles, like same-sex couples, are entitled to compensatory justice and hence owed access to assistive reproductive technologies, this does not

[41] Lauren Notini et al., 'Drawing the Line on In Vitro Gametogenesis', *Bioethics* 34/1 (2020), 123–134.

include the use of IVG for solo reproduction. That is because singles can attain genetic parenthood (but not solo genetic parenthood) by other means.

The problem with this objection is that it presupposes that the option of being the sole genetic parent is not valuable. However, there might be some reason to think that being the sole genetic parent is a valuable option. After all, many people consider being genetically related to one's children to be valuable in and of itself. If there is some value in having a child who is only 50% related to oneself, it would, despite being unusual, not obviously be a mistake to assign even greater value to having a child who is related to an even greater degree. In addition, as previously mentioned, being able to preserve the genetic integrity of one's family unit is as legitimate a desire for singles as it is for couples. If this is right, then solo reproduction is just as valuable an option as any other use of IVG. Compensatory justice, therefore, does not require deprioritizing solo reproduction.

Conclusion

We have examined various arguments against solo reproduction. Of the arguments attempting to show that access to solo reproduction is less valuable than other uses of IVG, we have argued that neither succeed. Of the arguments attempting to show that solo reproduction is wrong, only the physiological risk objection succeeds. This objection would, however, become obsolete if the technology were to become sufficiently reliable that physiological risk of solo reproduction became on par with other permissible reproductive technologies like IVF.

Bibliography

Anderson, Warwick, 'The Whiteness of Bioethics', *Journal of Bioethical Inquiry*, 18/1 (2021), 93–97, https://doi.org/10.1007/s11673-020-10075-y.

Axelsson, Erik Ljungvall, Ingrid, Bhoumik, Priyasma, Conn, Laura Bas, Muren, Eva, Ohlsson, Asa et al., 'The Genetic Consequences of Dog Breed Formation—Accumulation of Deleterious Genetic Variation and Fixation of Mutations Associated with Myxomatous Mitral Valve Disease in Cavalier King Charles Spaniels', *PLoS Genetics*, 17/9 (2021), e1009726, https://doi.org/10.1371/journal.pgen.1009726.

Beyleveld, Deryck, and Brownsword, Roger, 'Human Dignity, Human Rights, and Human Genetics', *The Modern Law Review*, 61/5 (1998), 661–680, https://doi.org/10.1111/1468-2230.00172.

'Born in Bradford: Consanguinity and Birth Defects', *Archives of Disease in Childhood*, 99 (2014), 38, https://doi.org/10.1136/archdischild-2013-305613.

Casebeer, William D., *Natural Ethical Facts: Evolution, Connectionism, and Moral Cognition* (Cambridge, MA: MIT Press, 2005).

Caulfield, Timothy, 'Human Cloning Laws, Human Dignity and the Poverty of the Policy-Making Dialogue', *BMC Medical Ethics*, 4/1 (2003), 3, https://doi.org/10.1186/1472-6939-4-3.

Churchland, Patricia, 'Human Dignity from a Neurophilosophical Perspective', in T. W. Merrill and A. Schulman, eds., *Human Dignity and Bioethics* (Washington, DC: President's Council on Bioethics, 2008), 99–121.

Congregation for the Doctrine of the Faith, '*Donum Vitae*: Instruction on Respect for Human Life in Its Origin and on the Dignity of Procreation', 1987, https://www.vatican.va/roman_curia/congregations/cfaith/documents/rc_con_cfaith_doc_19870222_respect-for-human-life_en.html.

Finnis, John Mitchell, 'Public Reason, Abortion, and Cloning', *Valparaiso University Law Review*, 32 (1998), 361–382.

FitzPatrick, William J., *Teleology and the Norms of Nature* (New York: Garland, 2000).

Foot, Philippa, *Natural Goodness* (Oxford: Clarendon Press, 2001).

Kass, Leon, 'The Wisdom of Repugnance', *The New Republic*, 2 June 1997, 17–27.

Kraynak, Robert P., 'Human Dignity and the Mystery of the Human Soul', in T. W. Merrill and A. Schulman, eds., *Human Dignity and Bioethics* (Washington, DC: President's Council on Bioethics, 2008), 61–82.

Lott, Micah, 'Have Elephant Seals Refuted Aristotle? Nature, Function, and Moral Goodness', *Journal of Moral Philosophy*, 9/3 (2012), 353–375, https://doi.org/10.1163/174552412X625727.

Macklin, Ruth, 'Dignity Is a Useless Concept', *BMJ*, 327 (2003), 1419–1420.

Malby, Steven, 'Human Dignity and Human Reproductive Cloning', *Health and Human Rights*, 6/1 (2002), 102–114, https://doi.org/10.2307/4065316.

Mayes, Christopher, 'White Medicine, White Ethics: On the Historical Formation of Racism in Australian Healthcare', *Journal of Australian Studies*, 44/3, 2020, 287–302, https://doi.org/10.1080/14443058.2020.1796754.

Moosavi, Parisa, 'From Biological Functions to Natural Goodness', *Philosophia*, 19/51 (2019), 1–20.

Murase, Yusuke, Yokogawa, Ryuta, Yabuta, Yukihiro, Nagano, Masahiro, Katou, Yoshitaka, Mizuyama, Manami et al., 'In Vitro Reconstitution of Epigenetic Reprogramming in the Human Germ Line', *Nature*, 631/8019 (2024), 170–178, https://doi.org/10.1038/s41586-024-07526-6.

Notini, Lauren, Gyngell, Christopher, and Savulescu, 'Drawing the Line on In Vitro Gametogenesis', *Bioethics*, 34/1 (2020), 123–134, https://doi.org/10.1111/bioe.12679.

Rolston III, Holmes, 'Human Uniqueness and Human Dignity: Persons in Nature and the Nature of Persons', in T. W. Merrill and A. Schulman, eds., *Human Dignity and Bioethics* (Washington, DC: President's Council on Bioethics, 2008), 129–154.

Saitou, Mitinori, and Hayashi, Katsuhiko, 'Mammalian In Vitro Gametogenesis', *Science*, 374/6563 (2021), eaaz6830, https://doi.org/10.1126/science.aaz6830.

Sandel, Michael J., 'The Ethical Implications of Human Cloning', *Perspectives in Biology and Medicine*, 48/2 (2005), 241–247, https://doi.org/10.1353/pbm.2005.0063.

Schachter, Oscar, 'Human Dignity as a Normative Concept', *American Journal of International Law*, 77/4 (1983), 848–854, https://doi.org/10.2307/2202536.

Schulman, Adam, 'Bioethics and the Question of Human Dignity', in Thomas W. Merrill and Adam Schulman, eds., *Human Dignity and Bioethics* (Washington, DC: President's Council on Bioethics, 2008), 3–18.

Second Vatican Council, *Dignitatis humanae*: Declaration on Religious Freedom, 1965, https://www.vatican.va/archive/hist_councils/ii_vatican_council/documents/vat-ii_decl_19651207_dignitatis-humanae_en.html.

Seemanová, Eva, 'A Study of Children of Incestuous Matings', *Human Heredity*, 21/2 (1971), 108–128, https://doi.org/10.1159/000152391.

Shiffrin, Seana Valentine, 'Wrongful Life, Procreative Responsibility, and the Significance of Harm', *Legal Theory*, 5 (1999), 117–148.
Sterelny, Kim, and Griffiths, Paul, *Sex and Death: An Introduction to Philosophy of Biology* (Chicago: University of Chicago Press, 1999).
Suter, Sonia M., 'In Vitro Gametogenesis: Just Another Way to Have a Baby?', *Journal of Law and the Biosciences*, 3/1 (2016), 87–119, https://doi.org/10.1093/jlb/lsv057.
Viaña, John Noel Montaño, 'Interrogating Sites of Knowledge Production: The Role of Journals, Institutions, and Professional Societies in Advancing Epistemic Justice in Bioethics', *The American Journal of Bioethics*, 24/4 (2024), 63–66, https://doi.org/10.1080/15265161.2024.2308144.

9

In Vitro Gametogenesis and the Concept of Genetic Parenthood

Hilary Bowman-Smart

Introduction

This chapter will proceed in the following sections. First, I will explain what *in vitro* gametogenesis (IVG) is, and why people might be interested in using it. Second, I will explore the reasons why genetic parenthood in particular might be valuable in itself and/or valuable to understand as a concept; I will conclude that genetic parenthood has limited value and utility for questions of moral, legal, or social parenthood, but that nonetheless it may be valuable to have some sort of understanding of it as a concept. Third, I will explore a few different approaches to understanding genetic parenthood that exist in the literature—then, in turn, the implications that IVG might have for these definitions, and the project of defining genetic parenthood at all. Finally, I will argue that using the term 'genetic parenthood' causes us to mistake the value and meaning of some types of genetic relations; that it reinforces bionormativity; that IVG (perhaps inadvertently) highlights this fact; and, that we should divorce the conceptual and ethical debate from the paradigm of 'parenthood' to allow us to more clearly articulate what we mean when we talk about it. Instead, we will need to think of an umbrella concept that covers various types of genetic relation—we can refer to this as (genetic) 'progeneration'. This is important because IVG will change not just the social dynamics of parenthood, but also our understandings of what it means to reproduce.

In Vitro Gametogenesis—What Is It, and Why Are We Developing It?

Currently, in order to reproduce sexually—whether this is through intercourse, intrauterine insemination, *in vitro* fertilization (IVF), or some other assisted reproductive technology—we need to have at least two gametes (a sperm and

Hilary Bowman-Smart, In Vitro *Gametogenesis and the Concept of Genetic Parenthood* In: *Revolutionizing Reproduction.* Edited by: Christopher Gyngell, Julian Savulescu, and Tsutomu Sawai, Oxford University Press.
 DOI: 10.1093/9780197810415.003.0009

an egg). These gametes might be harvested or donated, and we may do some interventions in the process (e.g., mitochondrial replacement therapy). However, fundamentally, a sperm and an egg must be sourced from somewhere in order to create a fertilized embryo.

This presents challenges to prospective parents who are unable to produce, source, or use their preferred gametes. This can be in a wide range of cases. This can include people who do not produce any functional gametes of their own, perhaps because of developmental variation, injury to their reproductive organs, an illness such as cancer, or age.[1] In other cases, prospective parents may produce functional gametes, but these are incompatible with the gametes of their preferred partner or co-parent—for example, a cisgender same-sex couple (where both parties may produce eggs, or both produce sperm).[2]

IVG presents a potential solution to a lack of functional or compatible gametes by allowing us to create sperm and egg cells from other types of cells. For example, a skin cell could be 'reprogrammed' to become a sperm cell, which could then be used to fertilize a harvested egg to create an embryo.[3] This embryo could then be implanted and gestated, and result in viable offspring. IVG has already been demonstrated in mice, but is not yet available for clinical use in humans. Alongside the scientific and medical work that must be done, there also remain significant ethical, legal, and regulatory challenges.[4]

However, while IVG has frequently been described in the media as a solution for biological or social infertility, it offers significantly more possibilities than that. For example, IVG may become integrated into assisted reproductive technologies as a way to avoid egg harvesting (oocyte retrieval).[5] Egg harvesting is an invasive and physically challenging procedure, particularly when compared to the generally more straightforward process of retrieving sperm. It can involve

[1] Victoria G. Wesevich, Christopher Arkfeld, and David B. Seifer, 'In Vitro Gametogenesis in Oncofertility: A Review of Its Potential Use and Present-Day Challenges in Moving toward Fertility Preservation and Restoration', *Journal of Clinical Medicine*, 12/9 (2023), 3305; Engineering National Academies of Sciences et al., 'Potential Clinical Implications of IVG', in *In Vitro–Derived Human Gametes as a Reproductive Technology: Scientific, Ethical, and Regulatory Implications: Proceedings of a Workshop* (Washington, DC: National Academies Press, 2023), https://www.ncbi.nlm.nih.gov/books/NBK599667/.

[2] Eli Y. Adashi, Katsuhiko Hayashi, and I. Glenn Cohen, 'Ethical and Legal Challenges in Assisted Same-Sex Conception through in Vitro Gametogenesis', *Nature Medicine*, 30/2 (2024), 322–323; Anne Le Goff et al., 'Anticipating in Vitro Gametogenesis: Hopes and Concerns for IVG among Diverse Stakeholders', *Stem Cell Reports* 19/7 (2024), 933–945.

[3] Mitinori Saitou and Katsuhiko Hayashi, 'Mammalian in Vitro Gametogenesis', *Science*, 374/6563 (2021), eaaz6830; Maria Victori Romualdez-Tan, 'Modelling In Vitro Gametogenesis using Induced Pluripotent Stem Cells: A Review', *Cell Regeneration*, 12/1 (2023), 33.

[4] Engineering National Academies of Sciences et al., 'Social, Ethical, and Legal Considerations Raised by IVG', in *In Vitro–Derived Human Gametes as a Reproductive Technology: Scientific, Ethical, and Regulatory Implications: Proceedings of a Workshop* (Washington, DC: National Academies Press, 2023), https://www.ncbi.nlm.nih.gov/books/NBK599672/.

[5] Guido Pennings, 'Why We Need Stem-Cell Derived Gametes', *Reproductive BioMedicine Online*, 47/5 (2023), 103322.

regular injections of hormones and painful surgical procedures.[6] Currently, this is an essential step in the process of IVF, but IVG could allow prospective parents to bypass this. Participants in a qualitative study from the United States highlighted the ability to avoid egg retrieval as a key benefit of IVG, with one remarking, 'I'd rather have a biopsy [for IVG] any day than go in for monitoring five times a week, and inject myself with hormones three times a day. . . . I think honestly it just sounds pretty great'.[7]

In addition to providing options for those who experience infertility, and augmenting existing assisted reproductive technologies, IVG also opens the doors to more novel ways of bringing about offspring. Many of these have already been explored in the literature. One example is the concept of 'solo reproduction', whereby one individual creates a sperm and an egg cell from their own cells.[8] The resulting child would not be a clone, but they would be created using genetic material from only one individual. While this is speculative, and it is unclear how many, if any, people would be interested in this, it does raise some ethical concerns—for example, solo reproduction is likely to increase the probability of the child having a genetic condition or illness, as it is essentially 'an extreme case of inbreeding'.[9] This would greatly increase the risk of homozygosity for recessive conditions. Sparrow has also raised the idea of using repeated iterations of IVG to produce multiple generations *in vitro*. This could be used to produce cell lines for research, but could also be used as a form of selection for human enhancement.[10]

Another possibility is what Palacios-González et al have labelled 'multiplex parenting'.[11] This is a scenario where four individuals bring about a child that is the 'genetic grandchild' of all four, and the intervening generation is kept *in vitro*. Persons 1 and 2 create an embryo using IVG (Embryo A), and persons 3 and 4 create another embryo (Embryo B). Embryo A and Embryo B are kept *in vitro*, but IVG is used to in turn generate a sperm and egg cell from these embryos. These sperm and egg is used to generate another fertilized embryo—Embryo C—which is then gestated to bring about a live child. In this case, 'C' has parents who have never existed as persons, because they remain embryos, and the only living persons to whom they are genetically related are their 'grandparents' (persons 1, 2, 3, and 4).

[6] Diane Tober et al., 'Alignment between Expectations and Experiences of Egg Donors: What Does It Mean to Be Informed?', *Reproductive Biomedicine & Society Online*, 12 (2021), 1–13.

[7] Goff et al., 'Anticipating in Vitro Gametogenesis'.

[8] Daniela Cutas and Anna Smajdor, '"I Am Your Mother and Your Father!" In Vitro Derived Gametes and the Ethics of Solo Reproduction', *Health Care Analysis*, 25/4 (2017), 354–369.

[9] Lauren Notini, Christopher Gyngell, and Julian Savulescu, 'Drawing the Line on in Vitro Gametogenesis', *Bioethics*, 34/1 (2020), 123–134.

[10] Robert Sparrow, 'In Vitro Eugenics', *Journal of Medical Ethics*, 40/11 (2014), 725–731.

[11] César Palacios-González, John Harris, and Giuseppe Testa, 'Multiplex Parenting: IVG and the Generations to Come', *Journal of Medical Ethics*, 40/11 (2014), 752–758.

In addition, we could imagine that 'C' grows up, becomes interested in the idea of finding out what her genetic 'parents' would have been like, and chooses to gestate and give birth to both embryos A and B. She then becomes the 'gestational parent' of her 'genetic parents'. (This is illustrative of some of the problems with using the parenthood paradigm for this discussion, which will be examined further in later sections.) Although unlikely, we might even consider further hypothetical scenarios, including generating novel types of gametes or different ways of reproducing, such as the androgenesis or gynogenesis seen in aquatic animals. In this process, the presence of two gametes may be required, but the offspring only inherit genetic material from one of the progenitors.[12]

IVG therefore may lead to a range of non-traditional or unexpected approaches to reproducing and creating familial relationships. This illustrates a key challenge for the development and clinical application of IVG. It is frequently positioned or framed as a solution or useful tool for infertility, as a means to ensure that you can have genetically related children. However, the range of possibilities associated with the technology itself disrupts the concept of genetic parenthood, and in turn brings the value and social meaning of it into question. Seger refers to this as the 'paradox of genetic relatedness'.[13]

This chapter will advance the claim that IVG has the potential to accelerate the decoupling of different types of parental roles, rights, and duties from the concept of 'genetic' or 'biological' parenthood. When we examine the concept of 'genetic parenthood', the case of IVG highlights how we need to divorce the discussion of these types of genetic relation from the paradigm of parenthood.

A Preliminary Note on Language and Reframing Genetic Parenthood as Genetic Progeneration

As I will ultimately argue in this chapter, the language we are using is in fact the thing that makes this debate about the nature of 'genetic parenthood' so unclear and murky. Despite this view, in this chapter I will continue to refer to genetic parenthood where this is the term used in the literature or theories that I am describing.

There are other terms that I will also use, and that I suggest should be used in the debate moving forward—or at least something like them (see Table 9.1). Later in this chapter, I will explore the difficulties in providing a clear and useful definition of 'genetic parenthood'. Despite this—and indeed precisely because of

[12] Hans Komen and Gary H. Thorgaard, 'Androgenesis, Gynogenesis and the Production of Clones in Fishes: A Review', *Aquaculture* 269/1 (2007), 150–173.

[13] Seppe Segers, 'The IVG "Relatedness Paradox": Researchers Should Mind Speculation', *Trends in Biotechnology* 41/10 (2023), 1220–1222.

Table 9.1 Notes on terminology

Current or common terminology	Preferred terminology
Genetic parent (person) *qua* source of genetic material (general term)	Progenitor* (mitochondrial, nuclear, other)
Genetic parent (*in vitro* organism/non-person) *qua* source of genetic material	Precursor (mitochondrial, nuclear, other)
Gestational parent	Gestational progenitor
Other person(s) involved in some way in bringing about the child (general term)	Adjutor**
Child (at any stage of development, inc. embryonic)	Offspring
Genetic parenthood	Genetic progeneration Genetic transfer or inheritance Progenitor-offspring genetic relation

** The term 'progenitor' is used here because it is already a relatively common word, but we must emphasize the idea of 'source' or 'origin'. Therefore, progenitor is an umbrella term inclusive of those who have provided either mitochondrial or nuclear DNA (or other)—it can simply be modified by the appropriate adjective. The same can be done for 'precursor', referring to non-persons (e.g., in vitro embryos).*

***This might include, for example, doctors or technicians involved in reproductive technologies. As with above, this term is an umbrella term that can be modified or specified as required.*

this difficulty—I will suggest the term (genetic) 'progeneration' as a general umbrella term to describe those various genetic relations where a person or material is the source or origin of another's genetic material, in some shape or form. Specifications and details can be added as necessary, and stipulative definitions used where required. Other related terms I suggest we apply include 'progenitor' to refer to persons who have a relation to offspring. This term has the advantage of already being in use in many parts of the literature.

However, one thing that is important to pull out here is that much of the language around reproduction in the bioethics literature presumes that there has been a *person* involved in some form of action or engagement with the process of reproduction. IVG demonstrates that it is possible to have an immediate source of starting material that is *not a person*, such as *in vitro* embryos.

The term 'progenitor' may certainly be used in other fields in a broader sense to refer to *any* organism (e.g., plant, bacteria). However, in this particular debate, it is useful to have a term that picks out cases where the 'genetic parent' is not a *person.* This includes the *in vitro* embryos in the 'multiplex parenting' example described earlier. Therefore, we should also consider using the term 'precursor' where appropriate (i.e., for *in vitro* embryos), to emphasize their position as a form of 'starting material' rather than as a person.

Genetic Progeneration—Why Might It Matter?

Having explained what IVG is, and its potential uses, we now turn our attention to the key concept at the heart of this chapter, which is 'genetic parenthood'—or, as we will refer to it later, genetic progeneration, the progenitor-offspring genetic relation, or some variation or further specification thereof (see Table 9.1).

'Biological parenthood' is still widely considered valuable in many societies, and has been for a long time. Prior to the discovery of genes, the tie of biological parenthood was often conceptualized in terms of 'blood' and the 'blood relation', terms that are still used in everyday language.[14] Nowadays, however, 'biological parenthood' is often understood in terms of a genetic relation—and specifically, the transfer or inheritance of *nuclear* genetic material.[15]

Although other definitions of biological parenthood are not the focus of this chapter, we will acknowledge that there are other ways of understanding 'biological parenthood', such as through the gestational relationship. For example, Gheaus defends the gestational connection as being more important in the debate around the right to rear the child, due to the emotionally intimate relationship that forms between the gestational progenitor and the foetus.[16] However, the development of technologies like ectogenesis can disrupt the concept of 'gestational parenthood', just as IVG may disrupt the concept of 'genetic parenthood'.[17] In addition, the idea that gestation necessarily creates bonds between foetus and gestational progenitor is subject to significant critique. As Romanis argues, it is not universal to experience or feel some bond or emotionally intimate relationship with a foetus you might be gestating.[18] Furthermore, care work and parenting are relational activities. If you argue that the foetus is *part* of the gestational progenitor rather than a separate individual, then pregnancy itself is not parenting but rather a precursor to it.[19] I have taken this position elsewhere and argued that it has implications for how we understand claims over information generated through prenatal testing.[20]

[14] Hane Maung, 'Parenthood and the Concept of the Biological Tie', *DiGeSt—Journal of Diversity and Gender Studies*, 7/2 (2021), 7–19.

[15] Catherine Mills, 'Nuclear Families: Mitochondrial Replacement Techniques and the Regulation of Parenthood', *Science, Technology, & Human Values*, 46/3 (2021), 507–527.

[16] Anca Gheaus, 'Biological Parenthood: Gestational, Not Genetic', *Australasian Journal of Philosophy*, 96/2 (2018), 225–240.

[17] James J. Cordeiro, 'How Ectogestation Can Impact the Gestational versus Moral Parenthood Debate', *Journal of Medical Ethics*, 51 (2025), 98–99.

[18] Elizabeth Chloe Romanis, 'Parenthood', in *Biotechnology, Gestation, and the Law*, ed. Elizabeth Chloe Romanis (Oxford: Oxford University Press, 2025).

[19] Romanis, 'Parenthood'.

[20] Hilary Bowman-Smart and Michelle Taylor-Sands, 'Fetal Information as Shared Information: Using NIPT to Test for Adult-Onset Conditions', *Monash Bioethics Review* (2021).

Parenthood as a broader concept also has a complex relationship with the process of *reproduction*. For example, through one model, reproduction as a process can be understood as being comprised of three different stages: the begetting (conception), the bearing (gestation and birth), and the rearing (parenting and raising a child).[21] Biological parenthood is relevant to the begetting and the bearing, but is not necessarily connected to the rearing. IVG is a specific means of *begetting*, but this may be separate from *bearing* and/or *rearing*.

The fundamental challenge in this discussion is that the term 'parent' has many meanings, which themselves are often not fixed or clear. In addition to the idea of a 'biological parent', the term 'parent' also refers to those who assume moral, social, and legal parental roles. In the day-to-day of raising a child, there are many different people who may fulfil these various parental roles—a 'biological parent', a familial guardian who may or may not be genetically related, a foster carer, an adoptive parent, a step-parent, a sibling, or some other sort of mentor. It can also be that people who are legally assigned parental rights and responsibilities do not *act as* parent, and people who *act as* parent do not have legal rights or responsibilities (as is frequently the case with, e.g., step-parents).

Biological parenthood is often seen as valuable because it is presumed to have some normative value in assigning moral parenthood. Moral parenthood is the question of how we assign parental rights and responsibilities (which in practical terms might mean paying for new shoes, or making decisions about medical treatments). This chapter will not outline in detail accounts of moral parenthood; however, there is a significant amount of existing work in the literature that explores how moral parenthood can be accounted for. These accounts usually do not require or emphasize the existence of a particular kind of genetic relation between two individuals. For example, Millum advances an account that focusses on the nature and extent of the investment that a parent makes, and Ferracioli focusses on the role of the parent in creating the conditions for the child's flourishing.[22] Neither of these accounts requires the parent to be biologically or genetically related to the child in order to establish moral rights and responsibilities.

I will now explore five broad categories of reasons why genetic progeneration might be considered either valuable in and of itself, or why it might be considered valuable to *have some understanding of it* as a concept. We must separate out these two points, as it might be valuable to have some sort of understanding of a concept, while not viewing it as having much normative value or utility. The value of understanding the progenitor-offspring genetic relation may be considered

[21] Georgina Antonia Hall, 'Reconceiving Reproduction: Removing 'Rearing' From the Definition—and What This Means for ART', *Journal of Bioethical Inquiry*, 21/1 (2023), 117.

[22] Joseph Millum, *The Moral Foundations of Parenthood* (Oxford: Oxford University Press, 2017); Luara Ferracioli, *Parenting and the Goods of Childhood* (Oxford: Oxford University Press, 2023).

before the reproductive event, in the context of people seeking to have genetically related offspring, but it may also be considered in terms of seeking to determine who is the genetic progenitor *after* the reproductive event.

These five broad categories of reasons (which may overlap) are: (i) parental rights and responsibilities, (ii) connection with a broader group, (iii) personal identity, (iv) preferences and beliefs, and (v) practical reasons. The aim of this section is not to *endorse* these reasons, but rather to generally get the lay of the land.

First, genetic progeneration may be considered relevant to the assigning of parental rights and responsibilities. This may be in the sense of *moral* rights and responsibilities; while genetic progenitors may not necessarily be guaranteed moral parenthood, they are often privileged in the process of establishing moral parenthood.[23] Similarly, genetic progeneration is often considered valuable in terms of assigning *legal* rights and responsibilities. However, privileging genetic progeneration in the legal setting can present challenges where children have been conceived using assisted reproductive technologies and/or gamete donation, and for families that do not fit the traditional nuclear heterosexual family model.[24] Unintentional embryo mix-ups during IVF further highlight these complexities, with one recent case in Australia only being identified when the child was already a toddler, having been born to and being raised by the gestational progenitor and her partner.[25]

Second, information about the genetic progenitor may be considered instrumentally valuable in establishing a biological connection with a broader kinship or social group. For example, children in different families who have been conceived using the same sperm donor may seek each other out and establish a new kinship network.[26] Furthermore, children may value knowing who their genetic progenitor is insofar as that person functions as a 'genetic bridge' in their relationship with another family member, such as an aunt or grandparent. Someone might highly value the idea that they have a particular genetic relation with some other sort of family member (e.g., grandparent). If they found out that the person who they thought was their genetic progenitor had not, in fact, provided a gamete, they may experience distress more because of how this affects their understanding of the relationship with their *grandparent*. The

[23] Joseph Millum, 'How Do We Acquire Parental Rights?', *Social Theory and Practice*, 36/1 (2010), 112.

[24] Douglas Nejaime, 'The Nature of Parenthood', *The Yale Law Journal*, 126/8 (2017), 2260–2381.

[25] Sinead Prince, Andrew McGee, Hilary Bowman-Smart, and Julian Savulescu, 'In Vitro Fertilisation Mix-ups and Contested Parenthood', *Journal of Medical Ethics* (2026), https://doi.org/10.1136/jme-2025-111285.

[26] Rosanna Hertz and Margaret K. Nelson, *Random Families: Genetic Strangers, Sperm Donor Siblings, and the Creation of New Kin* (Oxford: Oxford University Press, 2018); Astrid Indekeu, Sophie Bolt, and A. Janneke B. M. Maas, 'Meeting Multiple Same-Donor Offspring: Psychosocial Challenges', *Human Fertility*, 25/4 (2022), 677–687.

genetic progenitor is, in this instance, fungible with any other member of their family (e.g., sibling, cousin) who would function as the same 'genetic bridge' to the grandparent.

Third, genetic progeneration may be valuable to understand as it relates to personal identity, for either the progenitor or the offspring. This may emerge from knowledge of genetic ancestry as much as the parent-child social relationship—for example, through the construction of racial or ethnic identity. A person may build their personal identity around the belief that they have a specific racial or ethnic ancestry, and the discovery that this is not the case may disrupt their self-concept.[27] An example is an American woman who learnt through a direct-to-consumer ancestry test that she was not the child of her father, who was African-American, but was rather the child of a man with Ashkenazi Jewish heritage. She then changed her name and converted to Judaism.[28] Ancestry tests such as these have been criticized as reinforcing racial essentialism and the concept of race as a genetic or biological category.[29] Similar to the previous case, the individual genetic progenitor is fungible with any individual with the same racial or ethnic background. However, genetic progeneration may also be relevant to constructing personal identity related to the specific progenitor-offspring relation ('I am my mother's son', 'I am my daughter's father'). In this case, the genetic progenitor is not fungible, and the construction of personal identity is related to a specific individual (offspring or progenitor). Furthermore, for those who see themselves *as* genetic progenitors, this may play a role in constructing their identity as it relates to being a parent. To discover that they are not the genetic progenitor of a child or offspring may impact their sense of self as a parent at all.

Fourth, genetic progeneration may be valuable more straightforwardly in terms of preferences and beliefs, regardless of whether these are justified, reasonable, or true. Prospective parents frequently value the opportunity to have genetically related offspring, to the degree they pay large sums of money for assisted reproductive technologies.[30] This does not necessarily mean that these preferences outweigh other considerations, such as a potential *pro tanto* duty to

[27] Bessie L. Lawton et al., 'Experiences of Misattributed Parentage Communities: Impacts of Discovering New Familial Kinships', *Journal of Family History*, 49/1 (2024), 75–97.

[28] Jenny Kleeman, 'The People Whose DNA Held a Secret', https://www.bbc.com/future/article/20231004-these-people-took-dna-tests-the-results-changed-their-lives, 5 October 2023, accessed 7 November 2024.

[29] Amina Zarrugh and Luis Romero, 'Genetic Racialization: Ancestry Tests and the Reification of Race', *Social Problems*, 28 November 2023, spad056; Wendy D. Roth and Biorn Ivemark, 'Genetic Options: The Impact of Genetic Ancestry Testing on Consumers' Racial and Ethnic Identities', *American Journal of Sociology*, 124/1 (2018), 150–184.

[30] Saskia Hendriks et al., 'The Relative Importance of Genetic Parenthood', *Reproductive BioMedicine Online* 39/1 (2019), 103–110; Felicia von Estorff et al., 'Driving Factors in Treatment Decision-Making of Patients Seeking Medical Assistance for Infertility: A Systematic Review', *Human Reproduction Update* 30/3 (2024), 341–354.

adopt a child.[31] However, these preferences do exist, and they frequently exist strongly; they are likely to be socially reinforced in turn. These preferences may additionally be produced by material drivers—having a child through assisted reproduction may be the more cost-effective or economical option compared to, for example, adoption.[32] Adoption has also been subject to significant critique, and may itself be a morally problematic practice.[33] A preference to have genetically related offspring may also be born out of a desire for your offspring to resemble you (whether in terms of genotype or phenotype).[34] There may be a psychological motive to have genetically related offspring in order to leave some sort of 'legacy' into the future, as a means of leaving an impression of oneself onto the world after death.[35] In addition to these reasons, people may hold these preferences and want a child that is 'theirs' without being able to articulate *why* this is important.[36]

Fifth, there may be some specific practical reasons to value genetic progeneration and/or information about it. One such reason may be medical—for example, the clinical utility of establishing a family history of a particular medical condition. A prospective parent may also seek to have genetically related offspring for their own medical reasons (e.g., to serve as an organ donor). Information about genetic progeneration can also have forensic utility, such as to identify the perpetrator of a sexual assault that resulted in the conception of a child.[37]

It is important to note that the overview of these reasons is not intended to serve as *endorsement* or *justification* of these reasons. Many of these reasons are generated by widespread bionormativity—where biological relations are privileged above others.[38] Someone may have strong preferences to know who their 'genetic parent' is—and from a preference satisfaction perspective, this may have

[31] Tina Rulli, 'Preferring a Genetically-Related Child', *Journal of Moral Philosophy*, 13/6 (2016), 669–698.

[32] I. Glenn Cohen and Daniel L. Chen, 'Trading-Off Reproductive Technology and Adoption: Does Subsidizing IVF Decrease Adoption Rates and Should It Matter?', *Minnesota Law Review*, 95/2 (2010), 485–577.

[33] JaeRan Kim and Melanie (JaeHee) Chung-Sherman, 'Critiques of Adoption', in *Encyclopedia of Social Work* (2024).

[34] Stina Isaksson et al., 'Managing Absence and Presence of Child–Parent Resemblance: A Challenge for Heterosexual Couples Following Sperm Donation', *Reproductive Biomedicine & Society Online*, 8 (2019), 38–46.

[35] Annedore Hoppe, Immo Fritsche, and Nicolas Koranyi, 'Self-Transcendence as a Psychological Parenthood Motive: When Mortality Salience Increases the Desire for Non-Biological Children', *European Journal of Social Psychology*, 47/4 (2017), 488–500.

[36] Heidi Mertes, 'Gamete Derivation from Stem Cells: Revisiting the Concept of Genetic Parenthood', *Journal of Medical Ethics* 40/11 (2014), 744–747.

[37] John S. Buckleton, Duncan Taylor, and Jo-Anne Bright, 'Parentage Testing', in *Forensic DNA Evidence Interpretation*, 2nd ed. (Boca Raton: CRC Press, 2018), 411–430.

[38] Katharine Baker, 'Bionormativity and the Construction of Parenthood', *Georgia Law Review*, 42 (2008), 649.

some moral weight, even if these preferences stem from mistaken beliefs or do not make them better off.[39] Some of these reasons may be important, but only in specific contexts or for specific situations. That these reasons exist does not *necessarily* mean that they should continue to exist or that we should give them weight, but it does mean that understanding the concepts we are talking about is important.

Genetic Progeneration—What Is It?

We have explored the reasons why genetic progeneration might be valuable in and of itself as a concept, and/or why it might be useful to have knowledge about the specific source of genetic material or information which someone has inherited or been generated from. We then turn our attention to how exactly we might define genetic progeneration—or 'genetic parenthood', as is the term currently used in much of this literature.

Genetic progeneration in most cases of unassisted sexual reproduction is fairly straightforward to define and establish (excepting some uncommon scenarios such as natural chimerism[40]). However, given the range of new techniques and technologies that have become part of the project of reproduction (including IVG, genome editing, mitochondrial replacement, etc.), we are increasingly faced with 'hard cases' where defining genetic progeneration and who counts as a genetic progenitor is more challenging. Indeed, these 'hard cases' demonstrate why finding a watertight definition of 'genetic parenthood' that aligns with our intuitions, or has utility in ethical or legal debates, is an endeavour likely to fail.

I have explored these definitions in a previous paper, in part to demonstrate various scenarios in which each of them is deficient.[41] There have been a range of approaches, but many of them involve similar criteria or generally align. As Palacios-González has highlighted, this debate has largely occurred in two disciplines within philosophy: bioethics and philosophy of biology.[42] Philosophers of biology generally have more interest in producing an account of reproduction (including non-human) to understand processes such as evolution and natural selection. On the other hand, bioethicists have focussed

[39] Philip Pettit, 'Preference, Deliberation and Satisfaction', *Royal Institute of Philosophy Supplements*, 59 (2006), 131–153.

[40] César Palacios-González, 'Human Dignity and the Creation of Human–Nonhuman Chimeras', *Medicine, Health Care and Philosophy*, 18/4 (2015), 487–499.

[41] Hilary Bowman-Smart, 'Orphans by Design: The Future of Genetic Parenthood', *Bioethics*, 35/1 (2021), 23–30.

[42] César Palacios-González, 'What Is to Reproduce? On the Overlap, Development, and Persistence Account of Reproduction', *The Philosophical Quarterly*, 74/4 (2024), 1376–1385.

more on how accounts of (human) reproduction can have implications for our understandings of parenthood, and what it means to have genetically related offspring—because of the potential connection with moral parenthood, or for other reasons with possible normative weight as outlined earlier.[43]

Bioethicists should, in part, turn to philosophy of biology for more neutral and specific language to pick out what we mean when we talk about this question—for example, referring to the 'genetic progenitor' rather than 'genetic parent'. However, it is also the case that bioethicists are focussed more on how to understand the ethical implications of genetic progeneration, rather than aiming to produce an account useful for population-level questions around evolution and natural selection. Bioethicists are ultimately more concerned with ethical, social, and legal matters of parenthood (a broader concept that may or may not involve a biological relation), whereas philosophers of biology are more concerned with specialized accounts of biological reproduction. This is an important distinction to keep in mind. One of the earlier questions that spurred this debate in bioethics is whether a clone would be the child of the individual it was cloned from, or a sort of 'delayed twin'. As will be discussed, an undercurrent of this debate has been the undesirability of any definition that produced the conclusion that it *would* be a parent-child relation.[44] Similarly, the debate around mitochondrial replacement has been characterized by the privileging of nuclear DNA over mitochondrial DNA in an effort to avoid having the mitochondrial donor legally 'count' as a parent.[45]

Initial definitions emerging from the philosophy of biology literature include Griesemer's account of reproduction.[46] This includes a requirement for material overlap from one generation to the next—the offspring must be 'made from parts of the parents'.[47] Furthermore, the process of reproduction must confer the capacity for reproduction in the next generation; this criterion is introduced because we would not say we had reproduced if we, for example, removed an appendix, or cut off an arm—the appendix or arm would not become a 'child' of which we are a 'parent'. Piotrowska has later augmented Griesemer's account of reproduction to clarify this second criterion, by stating that conferring the capacity for reproduction must be understood as 'restart[ing]

[43] Palacios-González, 'What Is to Reproduce? On the Overlap, Development, and Persistence Account of Reproduction'.

[44] Robert Sparrow, 'Cloning, Parenthood, and Genetic Relatedness', *Bioethics*, 20/6 (2006), 308–318.

[45] Mills, 'Nuclear Families: Mitochondrial Replacement Techniques and the Regulation of Parenthood'.

[46] James R. Griesemer, 'Reproduction and the Reduction of Genetics', in *The Concept of the Gene in Development and Evolution: Historical and Epistemological Perspectives*, ed. Hans-Jörg Rheinberger, Peter J. Beurton, and Raphael Falk, Cambridge Studies in Philosophy and Biology (Cambridge: Cambridge University Press, 2000), 240–285.

[47] J. Griesemer, 'The Units of Evolutionary Transition', *Selection*, 1/1–3 (22 July 2005), 67–80.

the process of material transfer'.[48] Piotrowska also adds another criterion, which is persistence—the parts that overlap and confer the capacity for reproduction must persist into the following generation.[49] However, the challenge for both Griesemer's and Piotrowska's accounts of reproduction is that they suggest that where the offspring is sterile, 'reproduction' has not successfully occurred. This may be less of a problem when the focus is on evolution and natural selection, but illustrates that these definitions have less utility when focussing on the debate around human reproductive biotechnologies. For example, if someone has a child with Turner or Klinefelter Syndrome, conditions that generally result in infertility or sterility, this would suggest that they have not reproduced and are not a genetic progenitor. Furthermore, if there is uncertainty about whether someone has the capacity for reproduction, or they lose the capacity for reproduction (as often happens with age), this position would suggest that their genetic progenitors' status as reproducers is either uncertain (in the former) or revoked (in the latter).

Mertes and Penning have also put forth a proposed definition of genetic parenthood, which involves three criteria.[50] Firstly, they state that genetic *information* must be passed from the progenitor to the offspring. Secondly, they introduce a criterion for (physical) causality: the progenitor must be a physical cause of the offspring's existence. Thirdly, when genes are transferred from the progenitor to the offspring, they must be reshuffled in the process. This third criterion is intended to address the question of cloning, but as Piotrowska points out, this excludes instances of asexual reproduction, which are quite common in various non-human organisms.[51] It is also important to note that Mertes has since argued that genetic parenthood is not a value-neutral concept and should be considered 'open to interpretation', rather than something that we can clearly define and fix.[52]

Douglas and Devolder have also developed an account of genetic parenthood that they call 'Modified Direct Proportionate Genetic Descent' (MDPGD).[53] Their account rejects the requirement that Piotrowska places on material overlap and persistence of parts, as they believe that informational overlap or transfer could be sufficient. Their definition is in part formulated to exclude clones from being the children of the individual from whom they are cloned, as

[48] Monika Piotrowska, 'Is "Assisted Reproduction" Reproduction?', *The Philosophical Quarterly*, 68/270 (1 January 2018): 138–157.

[49] Piotrowska, 'Is "Assisted Reproduction" Reproduction?'.

[50] Heidi Mertes and Guido Pennings, 'Embryonic Stem Cell-Derived Gametes and Genetic Parenthood: A Problematic Relationship', *Cambridge Quarterly of Healthcare Ethics: CQ: The International Journal of Healthcare Ethics Committees*, 17/1 (2008), 7–14.

[51] Piotrowska, 'Is "Assisted Reproduction" Reproduction?'.

[52] Mertes, 'Gamete Derivation from Stem Cells: Revisiting the Concept of Genetic Parenthood'.

[53] Thomas Douglas and Katrien Devolder, 'A Conception of Genetic Parenthood', *Bioethics*, 33/1 (2019), 54–59.

well as formulated to exclude mitochondrial donors from being considered genetic parents. Under MDPGD, someone qualifies as a genetic parent if some proportion (~20–80%) of the child's genes are derived from the parent's genes, and not through deriving from the genes of some third, intervening individual from whom that parent derived a proportion of *their* genes (a variety of proportions are possible, as long as it excludes ~100%). Previously, I have outlined a range of scenarios where, with the use of different reproductive biotechnologies, this definition may not hold and/or may not accord with our intuitions.[54] Palacios-Gonzalez has elsewhere also highlighted a key challenge for Douglas and Devolder's account, which is that it insufficiently integrates causation as a condition.[55] Critically, it is their use of the term 'derived' that introduces some ambiguity; they have explicitly excluded material overlap as necessary, but without sufficient integration of causation, informational overlap or transfer could result in a much wider range of people becoming eligible to be considered genetic parents (e.g., other family members).

From these definitions, we can pull out some common threads for conditions or criteria for understanding genetic progeneration or 'genetic parenthood'. See Table 9.2 for a list of these common components, particularly from the definitions of Piotrowska, Mertes and Penning, and Douglas and Devolder. Here, they are sorted into two general categories: (i) requirements relating to overlap, transfer, and resemblance; and (ii), requirements relating to causality, direction, and intention.

The first category covers the various conditions and criteria that require there be some sort of overlap, transfer, or resemblance. One is material overlap or transfer; however, material overlap or transfer could in fact be a reason to privilege gestational accounts of parenthood.[56] Informational overlap or transfer is preferred by Douglas and Devolder, but has challenges for similar reasons that conditions relating to resemblance are not sufficient. I have previously described how requiring a persistence of parts and conferring the capacity for reproduction would produce some unusual outcomes impacting its utility and applicability to many scenarios (i.e., sterile offspring); and, as is the key point of this chapter, our efforts to define 'genetic parenthood' are motivated largely by arguments for its utility and value as a concept.

Resemblance is not sufficient on its own, because while this may be enough to establish a genetic relation generally, it is ambiguous with regard to the direction

[54] Bowman-Smart, 'Orphans by Design'.

[55] César Palacios-González, 'Genetic Parenthood and Causation: An Objection to Douglas and Devolder's Modified Direct Proportionate Genetic Descent Account', *Bioethics*, 33/9 (2019), 1085–1090.

[56] Avery Kolers and Tim Bayne, '"Are You My Mommy?" On the Genetic Basis of Parenthood', *Journal of Applied Philosophy*, 18/3 (2001), 273–285.

Table 9.2 Components of different definitions of 'genetic parenthood'

Category	Component
Overlap, transfer, and resemblance	Material overlap or transfer (material from the progenitor overlaps with or is transferred through the process of reproduction to the offspring)
	Informational overlap or transfer (genetic information from the progenitor overlaps with or is transferred through the process of reproduction to the offspring, usually facilitated by material overlap, though not necessarily in some scenarios, e.g., synthetic embryos)
	Conferring capacity for reproduction (the parts that overlap materially or informationally confer the capacity for reproduction in the next generation)
	Persistence of parts (the parts that confer the capacity for reproduction in the next generation persist into that generation)
	Derivation (the genetic material of the offspring is 'derived' from the genetic material of the progenitor)
	Resemblance (the genotype of the offspring resembles the genotype of the progenitor; this may or may not result in phenotypic or physical resemblance (which we might term here 'phenotypic overlap'))
Causality, direction, and intention	Causality (physical) (the direct physical cause of the offspring's existence)
	Causality (other senses) (more broadly causing the offspring's existence)
	Specified direction of transfer (any transfer of material or information is from the progenitor TO the offspring, and not the other way around)
	Intention (the progenitor intends to bring about the child's existence)
	Consent (the progenitor consents to bring about the child's existence—while this is usually concomitant with intent, this is not necessarily always the case) [a]

[a] Grant Gillett, 'Intention, Autonomy, and Brain Events', *Bioethics*, 23/6 (2009), 330–339.

or nature of the relationship. For example, if my father has an identical twin, I would resemble him as much as my father; furthermore, without some specification of the direction of the relationship, I could on this account be the parent of my father. Resemblance alone does not capture the asymmetry (temporal, causal) of the progenitor-offspring relation. Therefore, adding the criterion of derivation is a next step, because this establishes the direction of the relationship (while both parties may resemble each other, one party is derived *from* the other).[57] Another consideration is that if parents use technologies such as genome editing, there may be material overlap or transfer, causality, and so on—but resemblance is lost in the process of editing the embryo.

This leads us to the second category of requirements. Causality is also a critical factor, though out of these three definitions, it is primarily Mertes and Penning that place it as a central condition. Determining and understanding causal relationships is a generally challenging task and remains so in the context of reproduction. If we are integrating causality into definitions of genetic progeneration, with the goal of producing a definition that has certain utility and/or aligns with our intuitions about who counts as a parent, we would need to do so in such a way that captures edge cases, without casting too wide a net. Causality seems important, but setting its scope has long been a challenge for the project of understanding the idea of parenthood more generally. Many people may be involved in bringing about a child's existence (e.g., doctors and laboratory staff), so invoking it too strongly as a criterion leads to the 'too many parents' problem.[58]

Some of these criteria are compatible and may go together. Others seem to produce definitions that have little use in the broader ethical, legal, and social debate around parenthood (e.g., requirements for conferring the capacity for reproduction). Critically, what this demonstrates is that the project of defining 'genetic parenthood' is largely dead in the water, for two reasons:

(a) 'genetic parent' is a vague term that leads to circular definitions—a fruitless chase of attempting to contort criteria to avoid particular situations (e.g., cloned person as parent of their clone) or to fit our intuitions about its connection to moral, legal, and social parenthood—which are the things that really matter;
(b) these intuitions become increasingly irrelevant anyway, as novel reproductive biotechnologies disrupt them by creating wholly unfamiliar social and reproductive dynamics.

The Implications of IVG

Having explored a range of different definitions of 'genetic parenthood' or genetic progeneration, we can see how IVG as a technology may result in 'hard cases'. Earlier in this chapter, we explored a few potential applications of IVG,

[57] Kolers and Bayne, 'Are You My Mommy?'.
[58] Jason Hanna, 'Causal Parenthood and the Ethics of Gamete Donation', *Bioethics*, 33/2 (2019), 267–273.

though it is likely that as the technology develops, more novel options will present themselves.

In these cases, IVG can still largely fulfil many criteria related to overlap, transfer, and resemblance. For example, it would still generally involve material and informational overlap or transfer—the provision of a skin cell can still be traced through to the embryo that is ultimately produced. Once the skin cell has undergone the gametogenesis process and a fertilized embryo has been created, the process of reproduction could remain largely similar to IVF as it is today.

If we are focussing on these criteria, IVG could be more easily positioned as an 'advancement' or 'next step' in assisted reproductive technologies, to allow those who cannot otherwise have biologically related children to do so. IVG is thus simply a technological solution to a problem (infertility). However, possibilities such as 'solo reproduction' (one person providing both the sperm and the egg) illustrate the difficulty in clinging to this framing of the technology.[59] The possibility for multi-generational *in vitro* embryos further complicates things. The distinction we have made between 'progenitors' and 'precursors' here is helpful. We can consider the multiplex parenting case—where gametes are sourced from two *in vitro* embryos, each created by two people.[60] Here, there is indeed material and informational overlap, but from a precursor, not a progenitor. This does not necessarily make it a challenge to assign 'genetic parenthood' (it goes to the *in vitro* embryo as precursor). However, it *does* disrupt the idea of IVG as simply another assisted reproductive technology like IVF for those who cannot reproduce.

For criteria related to causality, direction, and intention, IVG has the potential to be much more challenging. The most concerning case has been raised by Smajdor and Cutas—that of non-consensual or 'unwitting' parenthood.[61] IVG could allow someone to take a skin cell from another person, and use IVG to create a child from that skin cell. This is akin to existing practices such as sperm theft, but could widen the scope of possibilities.[62] In these cases, one of the progenitors has not consented or directly caused the existence of the resulting child. This could be damaging emotionally, but potentially also financially, if regulatory approaches to parental obligations are not reviewed. Furthermore, as will be discussed further below, reproduction is still largely positioned as resulting from some form of romantic or sexual relationship. Non-consensual parenthood can therefore be experienced as a sort of sexual violation, or at least

[59] Cutas and Smajdor, '"I Am Your Mother and Your Father!" In Vitro Derived Gametes and the Ethics of Solo Reproduction'.

[60] Palacios-González, Harris, and Testa, 'Multiplex Parenting'.

[61] A. Smajdor and D. Cutas, 'Artificial Gametes and the Ethics of Unwitting Parenthood', *Journal of Medical Ethics*, 40/11 (2014), 748–751, https://doi.org/10.1136/medethics-2013-101824.

[62] Smajdor and Cutas, 'Artificial Gametes and the Ethics of Unwitting Parenthood'.

a violation of bodily integrity. As will be discussed, the bionormative positioning of genetic progeneration as 'genetic parenthood' exacerbates this.

Direction of transfer is also another consideration. Earlier, we used an example of a case where someone gives birth to their own 'genetic parent', because they were derived from *in vitro* embryos as precursors. With IVG, it is hypothetically possible to temporally precede the existence of your 'genetic parent' as a *person*. The idea of this is completely incompatible and incoherent when we use the parenthood paradigm. This again also makes it helpful to distinguish somewhat between 'precursor' and 'progenitor'.

However, the real and most likely challenge that IVG poses for our understandings of genetic progeneration—and our vigorous attempts to define it—is through the way it can change the dynamics of reproduction and parenthood. A key element of IVG is that it facilitates the process of reproduction beyond the context of a sexual relationship. As Romanis and Brown state in their analysis of relevant law in the United Kingdom, 'The problem lies in the fact that legal parenthood is constructed around the notion of the nuclear and sexual family'.[63] Assisted reproductive technologies like IVF—or even artificial insemination—have allowed for reproduction without sexual intercourse for some time. However, they do not disrupt sexed roles in the reproductive process—sperm from father, egg from mother.[64] This encompasses not only sexed roles but also the presumption of a sexual and romantic relationship as default (i.e., the 'sexual family'). Technologies like ectogenesis are not covered here, but these have the potential to also challenge gendered concepts of gestation.[65]

With IVG, you can have children with your friends. Why should we not create families with people to whom we are not romantically or sexually connected? Of course, one can do this already in a sense (e.g., through gamete donation or artificial insemination). Historically, many gay men and lesbian women have had children together, often as friends, through informally negotiated reproductive relationships.[66]

However, as the anthropologist Emily Martin describes in her essay 'The Egg and the Sperm', cultural stereotypes based on the idea of sperm as masculine and eggs as feminine leads to the imagery of biological reproduction as a *romantic* relationship. Thus, 'in this microscopic "culture" a cellular "bride" (or femme

[63] Elizabeth Chloe Romanis and Alan Brown, 'Legal Parenthood, Novel Reproductive Practices, and the Disruption of Reproductive Biosex', *The Modern Law Review*, 88/1 (2025), 155–191, https://doi.org/10.1111/1468-2230.12914.

[64] Elizabeth Chloe Romanis and Alan Brown, 'Legal Parenthood, Novel Reproductive Practices, and the Disruption of Reproductive Biosex'.

[65] Claire Horn, 'Gender, Gestation and Ectogenesis: Self-Determination for Pregnant People Ahead of Artificial Wombs', *Journal of Medical Ethics*, 46/11 (2020), 787–788.

[66] Deborah Dempsey, 'Conceiving and Negotiating Reproductive Relationships: Lesbians and Gay Men Forming Families with Children', *Sociology*, 44/6 (2010), 1145–1162.

fatale) and a cellular "groom" (her victim) make a cellular "baby"'.[67] The widespread implicit framing of gametes fusing to create an embryo as some sort of *sexual relationship* has limited the idea of reproduction in non-romantic and non-sexual relationships to a rare and strange phenomenon, frequently with the undercurrent of tension. The 2011 romantic comedy film *Friends with Kids* opens with two heterosexual friends choosing to have a child together; by the end of the film, they have formed a romantic and sexual relationship.[68]

By allowing us to reproduce through a skin cell provided by biopsy—which is then used, unseen to us, to create an embryo—IVG can help to break down this framing of reproduction as a romantic and sexual relationship, which introduces new possibilities for a wide range of family structures to be seen as 'legitimate'. IVG also allows for you to have children with not just one friend, but perhaps many friends (as in the multiplex parenting case).

Dubious Definitions and Parenthood as the Mistaken Paradigm

If we continue to use the paradigm of parenthood, as I have somewhat hypocritically done throughout this chapter in describing the current state of affairs, we are bringing in mistaken intuitions. This is because it is a form of circular reasoning. We wish to examine these various cases presented by biotechnologies such as IVG in order to come to a conclusion about what type of biological or genetic relations accord with our existing normative view on *who should count* as a biological or genetic 'parent'.

Much of the debate around 'genetic parenthood' has focussed on attempting to parse *who* 'counts' as a genetic progenitor, particularly when there are multiple direct sources of genetic material used in the process of reproduction (e.g., in the case of mitochondrial replacement) or only one direct source (e.g., in the case of cloning). However, this approach implicitly assumes that (a) we can define 'genetic parent' in a way that is coherent, and (b) that the idea of biological or genetic 'parent' has a particular level of normative value or utility.

Conceptual projects in bioethics often risk being derailed by an implicit or unrecognised desire to shore up a particular normative claim. By this, I am not saying that we should therefore aspire to produce 'value-neutral' accounts of something like 'genetic parenthood'—rather the opposite. Debates in bioethics are frequently full of thick concepts that have both descriptive and evaluative

67 Emily Martin, 'The Egg and the Sperm: How Science Has Constructed a Romance Based on Stereotypical Male-Female Roles', *Signs: Journal of Women in Culture and Society*, 16/3 (1991), 485–501.

68 '*Friends with Kids*', in *Wikipedia*, 21 December 2024, https://en.wikipedia.org/w/index.php?title=Friends_with_Kids&oldid=1264405061.

elements, and we might want to think of a concept in different ways for different purposes.[69] However, it does not seem useful that a definition be contorted to meet a pre-existing position (e.g., that a mitochondrial donor is not a 'genetic parent') that has been arrived at for other reasons (e.g., a mitochondrial donor should not have parental rights). We should rather *recognise* when this is happening, and state those reasons directly.

Many of the theories that have been outlined here might be seen as attempts to produce descriptive, or at least stipulative, definitions of 'genetic parenthood'. However, attempts to provide a descriptive definition of 'genetic parenthood' (in the sense of trying to find a definition for the future that is also adequate and concords with our existing usage) are dead in the water. Attempting to describe what makes someone a genetic 'parent' will not work, because we are heading into uncharted waters now. What is there to *describe*? As I have explored in this chapter, and previously, technologies like IVG produce a wide variety of situations that completely disrupt attempts to produce such definitions.[70] Stipulative definitions are easier but suffer from the limitation of having use in only very specific settings. Bioethics endeavours to connect with the realities of people's lives, and thus very narrow stipulative definitions cannot do what we need here—such as helping us figure out who may have custody of a child, how to navigate social and family relationships, decide what to do with frozen embryos, or when assisted reproductive technologies might get funding from the state.

However, I would argue that much of the literature really aims to provide an explicative definition of the concept—that is, a definition that aims to generally capture how the term is used, but that is stipulative in other parts.[71] From Carnap's theory of explication, an explicative definition of a concept is one that is relatively similar to an imprecise concept, but is also, among other things, 'fruitful'—in that it allows for the concept to further our inquiry and make progress towards achieving our relevant theoretical goals.[72]

Seeking an explicative definition of something like what we understand as 'genetic parenthood' is not necessarily a mistake; we have already explored some reasons why as a concept it might be important. However, when we talk about the concept of 'genetic parenthood', what we are usually talking about is the various forms of a type of genetic relation—the direct inheritance of genetic material or information from a particular source, or the use of this genetic material or

[69] Beatrijs Haverkamp, Bernice Bovenkerk, and Marcel F. Verweij, 'A Practice-Oriented Review of Health Concepts', *The Journal of Medicine and Philosophy: A Forum for Bioethics and Philosophy of Medicine*, 43/4 (2018), 381–401.

[70] Bowman-Smart, 'Orphans by Design'.

[71] James Justus, 'Carnap on Concept Determination: Methodology for Philosophy of Science', *European Journal for Philosophy of Science*, 2/2 (2012), 161–179.

[72] Mark Pinder, 'What Ought a Fruitful Explicatum to Be?', *Erkenntnis*, 87/2 (2022), 913–932.

information to generate another individual. This source may or may not be an existing person (e.g., *in vitro* embryos), and critically, it may not correlate anywhere near at all with our understandings of how someone might assume moral parenthood, or the rights, responsibilities, and duties of a parent raising a child.

Therefore, while it may potentially be useful to seek an explicative definition *of some sort* for this type of genetic relation, it is not necessarily genetic *parenthood* at all. The 'fruitfulness' of such explicative definitions of 'genetic parenthood' are limited, because, as Strawson states in his critique of Carnap's theory of explication, 'typical philosophical problems about the concepts used in non-scientific discourse cannot be solved by laying down the rules of exact and fruitful concepts in science. To do this last is not to solve the typical philosophical problem, but to *change the subject*.'[73]

However, perhaps we *should*, as it were, change the subject. I would argue that all work focussing on the question of 'biological' or 'genetic parenthood', in the sense of some types of biological or genetic relations, should use language that avoids the paradigm of parenthood. As Cutas has stated, 'Family language is normatively loaded and provocative.'[74] The term 'parent', and the parenthood paradigm more generally, should be reserved for discussion of moral, legal, or social parenthood. Instead, when thinking about genetic relations, we must use more neutral framing around the source or origin of genetic material—of which there may be multiple, and of different types. Otherwise, it becomes challenging to distinguish normative disagreement from conceptual disputes. Words like 'parent' and 'parenthood' are too sticky with moral and social norms, and it is unlikely these can ever be unstuck.

The descriptive and explicative endeavours to define 'genetic parenthood' represent a fundamental mistake about the utility and normative value of doing so, and why we want to understand what it is. People want to define 'genetic parenthood' because they believe or feel that it has some normative or conceptual value in our understandings of parenthood more generally. However, this is a general privileging or enforcement (deliberate or otherwise) of the bionormative paradigm of parenthood. Bionormativity positions biological and genetic forms of relation over others when understanding what it means to be a parent and be a family in the social and legal senses. This situates other forms of parenthood (e.g., adoptive) as less valuable or authentic and can generate significant material, legal and social harms.[75] We should therefore be wary of it.

[73] Justus, 'Carnap on Concept Determination'; Peter F. Strawson, 'Carnap's Views on Conceptual Systems Versus Natural Languages in Analytic Philosophy', in *The Philosophy of Rudolf Carnap*, ed. Paul Arthur Schilpp (1963), 503–518.

[74] Daniela Cutas, 'Nothing If Not Family? Genetic Ties beyond the Parent/Child Dyad', *Bioethics*, 37/8 (2023), 763–770.

[75] Maung, 'Parenthood and the Concept of the Biological Tie'; Baker, 'Bionormativity and the Construction of Parenthood'.

Using terminology like that described in Table 9.1 is not accompanied by discretely bounded definitions precisely because it should not be. We can take something like what Ausín describes as a fuzzy logic approach to bioethics—'facts, states of affairs, and categories in Bioethics are fuzzy and, therefore, should be treated in a fuzzy way'.[76] This should not be understood as vagueness or imprecision, but rather an acknowledgment of a gradualist approach that allows us to step away from strict dichotomies, and reckon more readily with malleability and flexibility.[77]

This sort of approach is well established elsewhere in bioethics and philosophy of medicine—for example, in understanding the nature of disease, or the concept of the therapeutic misconception.[78] Sadegh-Zadeh, in his work on the concept of disease, drew on Wittgenstein's idea of 'family resemblance' and Eleanor Rosch's 'prototype theory' to distinguish between classical and non-classical concepts. A classical concept denotes a category where all members share a common property (e.g., a square). A non-classical concept—which nearly all real-world categories fall under—denotes a category where all members have a relation of similarity to a central example or prototype.[79] Genetic progeneration, much like disease, is a non-classical concept covering various types of genetic relations.

The use of terms like progeneration, progenitor (person), and precursor (non-person) allow us to discuss these types of genetic relation without falling into the trap of assuming not only a particular type of normative value, but also particular types of temporal[80] and social relations between individual persons or things or some other defining common feature. The parenthood paradigm makes it easier to fall into these traps. When discussing one's source or origin, usage or references can be qualified and specified where necessary. For example, consider the following examples:

[76] Txetxu Ausín, 'A Fuzzy-Logic Approach to Bioethics', in *Fuzziness and Medicine: Philosophical Reflections and Application Systems in Health Care: A Companion Volume to Sadegh-Zadeh's Handbook of Analytical Philosophy of Medicine*, ed. Rudolf Seising and Marco Elio Tabacchi (Berlin: Springer, 2013), 133–143, https://doi.org/10.1007/978-3-642-36527-0_9.

[77] Ausín, 'A Fuzzy-Logic Approach to Bioethics'.

[78] Lucien Karhausen, 'Medicine', in *The Fragility of Philosophy of Medicine*, Philosophy and Medicine 147 (Cham: Springer, 2023), 15–19, https://doi.org/10.1007/978-3-031-41633-0; Neil Pickering, 'Extending Disorder: Essentialism, Family Resemblance and Secondary Sense', *Medicine, Health Care and Philosophy*, 16/2 (2013), 185–195; Daniel S. Goldberg, 'Eschewing Definitions of the Therapeutic Misconception: A Family Resemblance Analysis', *The Journal of Medicine and Philosophy: A Forum for Bioethics and Philosophy of Medicine* 36/3 (2011), 296–320.

[79] Kazem Sadegh-Zadeh, 'The Prototype Resemblance Theory of Disease', *The Journal of Medicine and Philosophy*, 33/2 (2008), 106–139.

[80] Although acknowledging that the prefix '*pro-*' can mean 'before' in the temporal sense, here I mean to point out that while some sort of 'starting material' must come before (by definition)—the temporal relations between *persons* may be quite disrupted (e.g., as in the example of someone choosing to gestate the embryos that were their precursors, thus giving birth to their 'genetic parents').

'I only have one immediate progenitor, as I was born through cloning'.

'I was born using mitochondrial replacement, so I have three progenitors—two nuclear, and one who provided the mitochondria. One of the progenitors is my father, who you met the other day'.

'My precursors were in vitro embryos, and the progenitors of those four embryos were my four parents'.

'I'm not very close to my father's family, but I am connected to another family through a progenitor of mine, though I don't talk to him [the progenitor]'.

'I was born using gamete donation—I don't know who the progenitor is personally, but I was told that she developed Alzheimer's when she was older'.

The use and definition of something like 'progeneration' or 'progenitor' does not need to be strictly defined in a way that future-proofs for all technologies and reproductive scenarios. This is because its utility lies precisely in its ability to capture a general grouping of a type of genetic relations, while avoiding the invocation of the paradigm of 'parenthood'. The nature of who is a 'progenitor' can be a matter of degree or kind, rather than a binary situation of whether one *is* or is *not*.

Conclusion

The concept of 'genetic parenthood' comes under pressure from two directions. The first is the questioning of its normative value or utility when put into service for ethical analysis and decision-making. The impetus to define 'genetic parenthood' so precisely is largely in part because of the weight it is given when we discuss how to respond to technologies like IVG, cloning, mitochondrial replacement therapy, or heritable human genome editing. We draw on 'parenthood' as a familiar framework for using biological or genetic relations to assign moral, social, and legal parenthood, or at least connecting them with these ideas. However, while the 'parenthood' paradigm is familiar, technologies like IVG are demonstrating its decreasing usefulness for debates in reproductive ethics.

The second direction is a more abstract, conceptual one: these technologies produce such a range of complicated reproductive scenarios that our attempts to *describe* the concept of 'genetic parenthood' end up just turning us around in circles. We seek to create definitions by drawing on our intuitions, but the real potential of these technologies is how they might fundamentally transform social dynamics and our understanding of our own relationships. The future landscape of reproduction will likely be quite alien to our intuitions, and so they cannot guide us here, even if we wanted them to.

IVG is only the beginning. Using the parenthood paradigm to try and describe where we *come from*, or where we *started from*, may be completely inapplicable for some future scenarios. I have explored these previously elsewhere[81]—these could include synthetic embryos, synthetic genomes, or the use of 'information' as 'starting material' rather than actual cells or biological matter. In these cases, we would straightforwardly say that these resulting individuals do not have *anything* or *anyone* resembling a genetic or biological 'parent'. We will not be able to defer to 'genetic parenthood'. Therefore, we must reckon with the use of the 'parenthood' paradigm in discussing genetic relations, and the genetic information or material that someone comes from.

As reproductive technologies like IVG develop, we must accept the necessity of this approach, rather than attempting to define 'genetic parenthood' in such a way that allows us to retain some utility for assigning moral, legal or social parenthood. IVG, as a technology, undermines the reason for its existence. People seek it because they want to be a 'genetic parent' to their child—but IVG calls into question the very nature and value of that genetic relation.

Bibliography

Adashi, E. Y., Hayashi, K., and Cohen, I. G., 'Ethical and Legal Challenges in Assisted Same-Sex Conception through In Vitro Gametogenesis', *Nature Medicine*, 30/2, 2024, 322–323, https://doi.org/10.1038/s41591-023-02689-7.

Ausín, T., 'A Fuzzy-Logic Approach to Bioethics', in R. Seising and M. E. Tabacchi, eds., *Fuzziness and Medicine: Philosophical Reflections and Application Systems in Health Care* (Berlin: Springer, 2013), 133–143, https://doi.org/10.1007/978-3-642-36527-0_9.

Baker, K., 'Bionormativity and the Construction of Parenthood', *Georgia Law Review*, 42 (2008), 649.

Bowman-Smart, H., 'Orphans by Design: The Future of Genetic Parenthood', *Bioethics*, 35/1 (2021), 23–30, https://doi.org/10.1111/bioe.12828.

Bowman-Smart, H., and Taylor-Sands, M., 'Fetal Information as Shared Information: Using NIPT to Test for Adult-Onset Conditions', *Monash Bioethics Review*, 31 December 2021, https://doi.org/10.1007/s40592-021-00142-4.

Buckleton, J. S., Taylor, D., and Bright, J., 'Parentage Testing', in *Forensic DNA Evidence Interpretation*, 2nd ed. (Boca Raton: CRC Press, 2018), 411–430.

Cohen, I. G., and Chen, D. L., 'Trading-Off Reproductive Technology and Adoption: Does Subsidising IVF Decrease Adoption Rates and Should It Matter?', *Minnesota Law Review*, 95/2 (2010), 485–577.

Cordeiro, J. J., 'How Ectogestation Can Impact the Gestational versus Moral Parenthood Debate', *Journal of Medical Ethics*, 51 (2025), 98–99, https://doi.org/10.1136/jme-2024-110416.

Cutas, D., 'Nothing If Not Family? Genetic Ties beyond the Parent/Child Dyad', *Bioethics*, 37/8 (2023), 763–770, https://doi.org/10.1111/bioe.13213.

Cutas, D., and Smajdor, A., '"I Am Your Mother and Your Father!' In Vitro Derived Gametes and the Ethics of Solo Reproduction', *Health Care Analysis*, 25/4 (2017), 354–369, https://doi.org/10.1007/s10728-016-0321-7.

[81] Bowman-Smart, 'Orphans by Design'.

Dempsey, D., 'Conceiving and Negotiating Reproductive Relationships: Lesbians and Gay Men Forming Families with Children', *Sociology*, 44/6 (2010), 1145–1162, https://doi.org/10.1177/0038038510381607.

Douglas, T., and Devolder, K., 'A Conception of Genetic Parenthood', *Bioethics*, 33/1 (2019), 54–59, https://doi.org/10.1111/bioe.12493.

von Estorff, F., Mochtar, M. H., Lehmann, V., and van Wely, M., 'Driving Factors in Treatment Decision-Making of Patients Seeking Medical Assistance for Infertility: A Systematic Review', *Human Reproduction Update*, 30/3 (2024), 341–354, https://doi.org/10.1093/humupd/dmae001.

Ferracioli, L., *Parenting and the Goods of Childhood* (Oxford: Oxford University Press, 2023).

'*Friends with Kids*', *Wikipedia*, 21 December 2024, https://en.wikipedia.org/w/index.php?title=Friends_with_Kids&oldid=1264405061.

Gheaus, A., 'Biological Parenthood: Gestational, Not Genetic', *Australasian Journal of Philosophy*, 96/2 (2018), 225–240, https://doi.org/10.1080/00048402.2017.1354389.

Gillett, G., 'Intention, Autonomy, and Brain Events', *Bioethics*, 23/6 (2009), 330–339, https://doi.org/10.1111/j.1467-8519.2009.01726.x.

Le Goff, A. Jeffries Hein, R., Hart, A. N., Roberson, I., and Landecker, H. L., 'Anticipating In Vitro Gametogenesis: Hopes and Concerns for IVG among Diverse Stakeholders', *Stem Cell Reports*, 19/7 (2024), 933–945, https://doi.org/10.1016/j.stemcr.2024.05.002.

Goldberg, D. S., 'Eschewing Definitions of the Therapeutic Misconception: A Family Resemblance Analysis', *Journal of Medicine & Philosophy*, 36/3 (2011), 296–320, https://doi.org/10.1093/jmp/jhr014.

Griesemer, J., 'The Units of Evolutionary Transition', *Selection*, 1/1–3 (2005), 67–80, https://doi.org/10.1556/select.1.2000.1-3.7.

Griesemer, J. R., 'Reproduction and the Reduction of Genetics', in H.-J. Rheinberger, P. J. Beurton and R. Falk, eds., *The Concept of the Gene in Development and Evolution* (Cambridge: Cambridge University Press, 2000), 240–285, https://doi.org/10.1017/CBO9780511527296.013.

Hall, G. A., 'Reconceiving Reproduction: Removing "Rearing" from the Definition—and What This Means for ART', *Journal of Bioethical Inquiry*, 21/1 (2023), 117–129, https://doi.org/10.1007/s11673-023-10281-4.

Hanna, J., 'Causal Parenthood and the Ethics of Gamete Donation', *Bioethics*, 33/2 (2019), 267–273, https://doi.org/10.1111/bioe.12537.

Haverkamp, B., Bovenkerk, B., and Verweij, M. F., 'A Practice-Oriented Review of Health Concepts', *The Journal of Medicine and Philosophy: A Forum for Bioethics and Philosophy of Medicine*, 43/4 (2018), 381–401, https://doi.org/10.1093/jmp/jhy011.

Hendriks, S., van Wely, M., D'Hooghe, T. M., Meissner, A., Mol, F., Peeraer, K., et al., 'The Relative Importance of Genetic Parenthood', *Reproductive Biomedicine Online*, 39/1 (2019), 103–110, https://doi.org/10.1016/j.rbmo.2019.02.008.

Hertz, R., and Nelson, M. K., *Random Families: Genetic Strangers, Sperm Donor Siblings, and the Creation of New Kin* (Oxford: Oxford University Press, 2018).

Hoppe, A., Fritsche, I., and Koranyi, N., 'Self-transcendence as a Psychological Parenthood Motive: When Mortality Salience Increases the Desire for Non-biological Children', *European Journal of Social Psychology*, 47/4 (2017), 488–500, https://doi.org/10.1002/ejsp.2218.

Horn, C., 'Gender, Gestation and Ectogenesis: Self-determination for Pregnant People Ahead of Artificial Wombs', *Journal of Medical Ethics*, 46/11 (2020), 787–788, https://doi.org/10.1136/medethics-2020-106156.

Indekeu, A., Bolt, S. H., and Maas, A. J. B. M., 'Meeting Multiple Same-Donor Offspring: Psychosocial Challenges', *Human Fertility*, 25/4 (2022), 677–687, https://doi.org/10.1080/14647273.2021.1872804.

Isaksson, S., Sydsjö, G., Skoog Svanberg, A., and Lampic, C., 'Managing Absence and Presence of Child–Parent Resemblance: A Challenge for Heterosexual Couples Following Sperm

Donation', *Reproductive Biomedicine & Society Online*, 8 (2019), 38–46, https://doi.org/10.1016/j.rbms.2019.07.001.

Justus, J., 'Carnap on Concept Determination: Methodology for Philosophy of Science', *European Journal for Philosophy of Science*, 2/2 (2012), 161–179, https://doi.org/10.1007/s13194-011-0027-5.

Karhausen, L., 'Medicine', in *The Fragility of Philosophy of Medicine* (Cham: Springer, 2023), 15–19, https://doi.org/10.1007/978-3-031-41633-0.

Kim, J., and Chung-Sherman, M., 'Critiques of Adoption', *Encyclopedia of Social Work* (2024), https://doi.org/10.1093/acrefore/9780199975839.013.1678.

Kolers, A., and Bayne, T., '"Are You My Mommy?' On the Genetic Basis of Parenthood', *Journal of Applied Philosophy*, 18/3 (2001), 73–285.

Komen, H., and Thorgaard, G. H., 'Androgenesis, Gynogenesis and the Production of Clones in Fishes: A Review', *Aquaculture*, 269/1 (2007), 150–173, https://doi.org/10.1016/j.aquaculture.2007.05.009.

Lawton, B. L., Pyott, L. C., Rubinstein Deyerin, K., and Foeman, A. K., 'Experiences of Misattributed Parentage Communities: Impacts of Discovering New Familial Kinships', *Journal of Family History*, 49/1 (2024), 75–97, https://doi.org/10.1177/03631990231156176.

Martin, E., 'The Egg and the Sperm: How Science Has Constructed a Romance Based on Stereotypical Male–Female Roles', *Signs*, 16/3 (1991), 485–501, https://doi.org/10.1086/494680.

Maung, H., 'Parenthood and the Concept of the Biological Tie', *DiGeSt: Journal of Diversity and Gender Studies*, 7/2 (2021), https://doi.org/10.21825/digest.v7i2.15950.

Mertes, H., 'Gamete Derivation from Stem Cells: Revisiting the Concept of Genetic Parenthood', *Journal of Medical Ethics* 40/11 (2014), 744–747, https://doi.org/10.1136/medethics-2013-101830.

Mertes, H., and Pennings, G., 'Embryonic Stem Cell-Derived Gametes and Genetic Parenthood: A Problematic Relationship', *Cambridge Quarterly of Healthcare Ethics*, 17/1 (2008), 7–14, https://doi.org/10.1017/S096318010808002X.

Mills, C., 'Nuclear Families: Mitochondrial Replacement Techniques and the Regulation of Parenthood', *Science, Technology, & Human Values*, 46/3 (2021), 507–527, https://doi.org/10.1177/0162243920934542.

Millum, J., 'How Do We Acquire Parental Rights?', *Social Theory & Practice*, 36/1 (2010), 112–132, https://doi.org/10.5840/soctheorpract20103615.

Millum, J., *The Moral Foundations of Parenthood* (Oxford: Oxford University Press, 2017).

National Academies of Sciences, Engineering, and Medicine, 'Potential Clinical Implications of IVG', in *In Vitro-Derived Human Gametes as a Reproductive Technology: Scientific, Ethical, and Regulatory Implications—Proceedings of a Workshop* (Washington, DC: National Academies Press, 2023), https://www.ncbi.nlm.nih.gov/books/NBK599667/.

National Academies of Sciences, Engineering, and Medicine, 'Social, Ethical, and Legal Considerations Raised by IVG', in *In Vitro-Derived Human Gametes as a Reproductive Technology: Scientific, Ethical, and Regulatory Implications – Proceedings of a Workshop* (Washington, DC: National Academies Press, 2023), https://www.ncbi.nlm.nih.gov/books/NBK599672/.

NeJaime, D., 'The Nature of Parenthood', *Yale Law Journal*, 126/8 (2017), 2260–2381.

Notini, L., Gyngell, C., and Savulescu, J., 'Drawing the Line on In Vitro Gametogenesis', *Bioethics*, 34/1 (2020), 123–134, https://doi.org/10.1111/bioe.12679.

Palacios-González, C., 'Genetic Parenthood and Causation: An Objection to Douglas and Devolder's Modified Direct Proportionate Genetic Descent Account', *Bioethics*, 33/9 (2019), 1085–1090, https://doi.org/10.1111/bioe.12639.

Palacios-González, C., 'Human Dignity and the Creation of Human–Nonhuman Chimeras', *Medicine, Health Care & Philosophy*, 18/4 (2015), 487–499, https://doi.org/10.1007/s11019-015-9644-7.

Palacios-González, C., 'What Is to Reproduce? On the Overlap, Development, and Persistence Account of Reproduction', *The Philosophical Quarterly*, 74/4 (2024), 1376–1385, https://doi.org/10.1093/pq/pqae071.

Palacios-González, C., Harris, J., and Testa, G., 'Multiplex Parenting: IVG and the Generations to Come', *Journal of Medical Ethics*, 40/11 (2014), 752–758, https://doi.org/10.1136/medethics-2013-101810.

Pennings, G., 'Why We Need Stem-Cell Derived Gametes', *Reproductive Biomedicine Online*, 47/5 (2023), 103322, https://doi.org/10.1016/j.rbmo.2023.103322.

'The People Whose DNA Held a Secret', *BBC Future*, 4 October 2023, https://www.bbc.com/future/article/20231004-these-people-took-dna-tests-the-results-changed-their-lives.

Pettit, P., 'Preference, Deliberation and Satisfaction', *Royal Institute of Philosophy Supplement*, 59 (2006), 131–153, https://doi.org/10.1017/S1358246100009498.

Pickering, N., 'Extending Disorder: Essentialism, Family Resemblance and Secondary Sense', *Medicine, Health Care & Philosophy*, 16/2 (2013), 185–195, https://doi.org/10.1007/s11019-011-9372-6.

Pinder, M., 'What Ought a Fruitful Explicatum to Be?', *Erkenntnis*, 87/2 (2022), 913–932, https://doi.org/10.1007/s10670-020-00223-6.

Piotrowska, M., 'Is "Assisted Reproduction" Reproduction?', *The Philosophical Quarterly*, 68/270 (2018), 138–157, https://doi.org/10.1093/pq/pqx032.

Prince, S., McGee, A., Bowman-Smart, H., and Savulescu, J., 'In vitro Fertilisation Mix-ups and Contested Parenthood', *Journal of Medical Ethics* (2026), https://doi.org/10.1136/jme-2025-111285.

Romanis, E. C., 'Parenthood', in *Biotechnology, Gestation, and the Law* (Oxford: Oxford University Press, 2025), https://doi.org/10.1093/9780191986574.003.0006.

Romanis, E. C., and Brown, A., 'Legal Parenthood, Novel Reproductive Practices, and the Disruption of Reproductive Biosex', *The Modern Law Review*, 88/1 (2025), 155–191, https://doi.org/10.1111/1468-2230.12914.

Romualdez-Tan, M. V., 'Modelling In Vitro Gametogenesis using Induced Pluripotent Stem Cells: A Review', *Cell Regeneration*, 12/1 (2023), 33, https://doi.org/10.1186/s13619-023-00176-5.

Roth, W. D., and Ivemark, B., 'Genetic Options: The Impact of Genetic Ancestry Testing on Consumers' Racial and Ethnic Identities', *American Journal of Sociology*, 124/1 (2018), 150–184, https://doi.org/10.1086/697487.

Rulli, T., 'Preferring a Genetically-Related Child', *Journal of Moral Philosophy*, 13/6 (2016), 669–698, https://doi.org/10.1163/17455243-4681062.

Sadegh-Zadeh, K., 'The Prototype Resemblance Theory of Disease', *Journal of Medicine & Philosophy*, 33/2 (2008), 106–139, https://doi.org/10.1093/jmp/jhn004.

Saitou, M., and Hayashi, K., 'Mammalian In Vitro Gametogenesis', *Science*, 374/6563 (2021), eaaz6830, https://doi.org/10.1126/science.aaz6830.

Segers, S., 'The IVG "Relatedness Paradox": Researchers Should Mind Speculation', *Trends in Biotechnology*, 41/10 (2023), 1220–1222, https://doi.org/10.1016/j.tibtech.2023.04.008.

Smajdor, A., and Cutas, D., 'Artificial Gametes and the Ethics of Unwitting Parenthood', *Journal of Medical Ethics*, 40/11 (2014), 748–751, https://doi.org/10.1136/medethics-2013-101824.

Sparrow, R., 'Cloning, Parenthood, and Genetic Relatedness', *Bioethics*, 20/6 (2006), 308–318, https://doi.org/10.1111/j.1467-8519.2006.00509.x.

Sparrow, R., 'In Vitro Eugenics', *Journal of Medical Ethics*, 40/11 (2014), 725–731, https://doi.org/10.1136/medethics-2012-101200.

Strawson, P. F., 'Carnap's Views on Conceptual Systems versus Natural Languages in Analytic Philosophy', in P. A. Schilpp, ed., *The Philosophy of Rudolf Carnap* (La Salle: Open Court, 1963), 503–518.

Tober, D., Garibaldi, C., Blair, A., and Baltzell, K., 'Alignment between Expectations and Experiences of Egg Donors: What Does It Mean to Be Informed?', *Reproductive Biomedicine & Society Online*, 12 (2021), 1–13, https://doi.org/10.1016/j.rbms.2020.08.003.

Wesevich, V. G., Arkfeld, C., and Seifer, D. B., 'In Vitro Gametogenesis in Oncofertility: A Review of Its Potential Use and Present-Day Challenges in Moving toward Fertility Preservation and Restoration', *Journal of Clinical Medicine*, 12/9 (2023), 3305, https://doi.org/10.3390/jcm12093305.

Zarrugh, A., and Romero, L., 'Genetic Racialisation: Ancestry Tests and the Reification of Race', *Social Problems* (2023), spad056, https://doi.org/10.1093/socpro/spad056.

10

Artificial Gametes and the Deconstruction of Reproduction

Anna Smajdor

The Early Days of Artificial Gametes

When people started to publish papers detailing steps towards the development of artificial gametes (henceforth AGs) in the early 2000s, AGs were already being discussed as a potential remedy for infertility. In this chapter, I will highlight the ways in which new biotechnological developments, including AGs, actually call into question the ways that we define gametes and reproduction, and hence, infertility. I will argue that biological facts alone cannot give us answers as to what constitutes reproduction. The trajectory of research and innovation has tended to reinforce the idea that reproduction must necessarily involve genes. However, it is becoming increasingly clear that this genetic focus is unsatisfactory from a biological, philosophical, and ethical perspective. Nevertheless, while genetic essentialism is being rejected by bioethicists, it continues to be reinforced by the fertility industry as well as by the ongoing trajectory of research into artificial gametes for the treatment of infertility. In turn, this feeds into policymaking and eligibility criteria in ways that continue to reproduce and reinforce existing inequalities.

How to Define a Gamete

The first reports of the development of gamete-type cells from other types of cells simply described the novel creations as 'gametes'. For example, a headline in 2003 in *Science* announced that oocytes had been derived from mouse embryonic stem cells, without further qualifications as to what manner of oocytes these were.[1] In the bioethics literature, however, a number of different terms

[1] Karl Hubner et al., 'Derivation of Oocytes from Mouse Embryonic Stem Cells', *Science*, 300/5623 (2003), 1251–1256.

Anna Smajdor, *Artificial Gametes and the Deconstruction of Reproduction* In: *Revolutionizing Reproduction*.
Edited by: Christopher Gyngell, Julian Savulescu, and Tsutomu Sawai, Oxford University Press.
 DOI: 10.1093/9780197810415.003.0010

were used, including 'artificial gametes',[2] 'ES-derived gametes',[3] and 'synthetic gametes'.[4,5]

When the *Journal of Medical Ethics* published a call for a special issue on this topic in 2014, the term it used was 'stem cell derived gametes', while the submissions themselves used a variety of expressions.[6] Ultimately, the term '*in vitro* derived gametes' has come to dominate in the literature. This accurately represents the fact that biotechnological intervention is involved in the production of these cells, while avoiding the negative associated with terms such as 'artificial' or 'synthetic'. Nevertheless, in this chapter, I choose to use the older term 'artificial gametes' (AGs) both for brevity and because the name raises questions that I regard as being important for bioethicists.

How should one define a gamete? What makes something artificial? Is the artificial necessarily bad in some sense that the 'natural' or 'real' is not? Given that AGs have never yet been used in human reproduction at the time of writing, what enables us to call them gametes at all? This question has not been widely discussed in the bioethics literature, but it is a fundamental issue when we think about how to understand the ethics of creating, using, and regulating such cells. In order to think about the ethical questions, we need to be clear about exactly *what* we are discussing.

It is possible that one could define a gamete as a cell that has the capability of creating an embryo when united with a complementary gamete. However, this definition would suggest that eggs and sperm that lack this capacity (and there are many such egg and sperm cells) could no longer be termed gametes. Another alternative would be to define gametes in genetic terms. That is, gametes contain twenty-three chromosomes, unlike most other somatic cells that contain forty-six. However, having twenty-three chromosomes in itself guarantees nothing in terms of producing a child. Moreover, if it becomes possible to remove twenty-three chromosomes from a somatic cell, it would not necessarily seem satisfactory to conclude that that cell is now a gamete. Another possible approach is to say that a gamete is a cell that emerges organically from a certain set of biological processes—this would put emphasis on the derivation rather than the ultimate composition of the cell. On this view, anything that emerges from this series of events could be defined as a gamete, however many chromosomes it contains, and regardless of its morphology or capacity to generate offspring. Thus, if a cell

[2] Ainsley Newson and Anna Smajdor, 'Artificial Gametes: New Paths to Parenthood?', *Journal of Medical Ethics* 31/3 (2005), 184–186.

[3] Giuseppe Testa and John Harris, 'Ethical Aspects of ES Cell-Derived Gametes', *Science*, 305/5691 (2004), 1719.

[4] Giuseppe Testa and John Harris, 'Ethics and Synthetic Gametes', *Bioethics* 19/2 (2005), 146–166.

[5] Testa and Harris, 'Ethical Aspects of ES Cell-Derived Gametes'.

[6] Anna Smajdor and Daniela Cutas, 'Artificial Gametes', background paper, Nuffield Council of Bioethics, 2016.

is ejaculated by a man, after being matured in the testes and passing through the various processes of spermatogenesis and meiosis, we would not necessarily deem that it is not a sperm at all, even if it should turn out to lack the genetic, morphological, and other features that usually make up a normal sperm cell.

A further possibility is that morphological features are what enable us to define some cells as 'gametes'. In this case, anything purporting to be an egg would need to be extremely large in comparison to other cells. Sperm would perhaps have to exhibit the familiar tadpole-like structure. Still another possibility in relation to sperm is the question of *action*. Typically, eggs are regarded as passive, even if this has come under some criticism over the past decades.[7] However, it is not contentious to suggest that one of the things that differentiates sperm from other cells, including eggs, is their ability to *swim*. Yet given that we often categorize non-swimming sperm as sperm in the context of ICSI, it seems that swimming cannot be a necessary or sufficient condition to be classified as sperm.

In reproductive terms, it is the functionality of a gamete that is important for fertility treatment. Eggs are eggs because they have the capacity to fuse with a sperm cell, and give rise to an embryo. Their size, or morphology, or biological trajectory might seem irrelevant in this respect. With respect to sperm, as noted above, immotile sperm cells are routinely used in ICSI procedures, so it would seem that the most basic requirement for a sperm cell in terms of reproductive success is simply that, on being injected into an appropriate egg cell, it can give rise to an embryo. It is important to note, however, that this commonsense functional view does not correlate perfectly with current practice, which tends rather to follow the 'biological processes' account above. That is, if a cell emerges from the testes or ovaries in the expected way, it is typically regarded as an egg or sperm, even though it might be morphologically unusual, genetically anomalous, or incapable of giving rise to an embryo.

While these considerations may seem merely semantic, they are of particular relevance for bioethicists. New developments in biotechnology force us to ask questions that would not have been necessary in the past. With the creation of gamete-type cells in the laboratory, we have to decide what features a 'gamete' must have in order to be treated as such. These questions also have a bearing on how the law relating to fertility treatments and embryo research is interpreted and applied. In these situations we cannot turn to biology to give us answers. There are many different *biological* options when we are thinking about defining gametes. To choose one over another is not a matter of biological fact, but a

[7] Martin, Emily. 'The Egg and the Sperm: How Science Has Constructed a Romance Based on Stereotypical Male-Female Roles', in Suzanne Forman Ragone, Louise Lamphere, and Micaela di Leonardo, eds., *Situated Lives: Gender and Culture in Everyday Life* (New York; London: Routledge, 2014), 85–98.

question of human judgment and evaluation, as Sally Haslanger has noted.[8] Such choices can have far-reaching impacts. Our decisions in this context may undermine or reinforce the idea that reproduction is primarily a matter of transmitting genes, for example—a question I will return to later in this chapter.

In what follows, I will consider the question of whether artificial gametes—however defined—facilitate reproduction, or offer a remedy for infertility. I will show that, just as we now have to re-define what we mean by gametes, we need to reconsider what is meant by reproduction and infertility, in the context of the new biotechnological opportunities offered by artificial gametes.

Reproduction and Genes

'Children are a parent's most direct route to genetic immortality'.[9] This portentous statement comes from the *Handbook of Parenting*, published in 2002, just when artificial gametes were starting to be discussed as a potential remedy for infertility. Few people would express the value of parenthood as succinctly as this, and yet the discourse of genes, immortality, and reproduction remains a common trope. In a culture that prioritizes an understanding of reproduction as essentially genetic, artificial gametes are clearly of immense significance.

Following the early news announcements about the development of artificial gametes, a flurry of comments and discussions emerged. While the actual feasibility of using AGs in reproduction remained uncertain, there was a palpable air of excitement and new possibilities. An article in *New Scientist* eagerly proclaimed, 'Artificial gametes will end infertility'.[10] The reasoning behind this pronouncement was that artificial gametes can in principle be provided for any patient regardless of their own biological capacity for making gametes. Of course, infertility is not usually understood simply as a failure to produce gametes. If it were, prepubescent children would be deemed infertile, as would post-menopausal women.

A paper published in 2005 in the journal *Reproductive Biomedicine Online* declared that artificial gametes held the promise of addressing the 'final frontier' of fertility treatment. That is to say, also in the words of the relevant

[8] Sally Haslanger, 'The Sex/Gender Distinction and the Social Construction of Reality', in Ann Garry, Serene Khader, and Alison Stone, eds., *The Routledge Companion to Feminist Philosophy* (Taylor and Francis, 2017), 157–167.

[9] David Bjorklund et al., 'The Evolution of Parenting and Evolutionary Approaches to Childrearing', in Marc Bornstein, ed., *Handbook of Parenting*, vol. 2 (Mahwah: Lawrence Erlbaum Associates, 2002), 169.

[10] Shaoni Bhattacharya, 'Stem Cells Can End Infertility, Say IVF Pioneers', *New Scientist.com News Service*, 24 July 2004.

paper: 'patients with absent gametes or gonads'.[11] It is interesting to note that the authors located the frontier specifically in the gonad region. Absent gonads were at that time very far from the main reason why many people could not achieve their reproductive goals. During the years following Louise Brown's birth, it was becoming increasingly clear that what makes people unable to reproduce is in many cases the fact of being disbarred from access to treatment. Many prospective patients were and still are denied access to fertility treatment because of funding or eligibility issues. I will discuss this further in a later section.

Finally, in a paper published in 2005 in the journal *Bioethics*, Testa and Harris announced that artificial gametes would 'democratize reproduction'. Again, it seems that Testa and Harris, along with many of the other authors cited here, took reproduction to be essentially a *genetic* business. Further, where people's reproductive wishes were thwarted, the assumption was that a *biomedical* innovation would offer the solution.[12] They, and others, do not engage with the fact that society, not biology, dictates whether people have access to a new biomedical innovation. And as long as infertility was, and continues to be, interpreted by medical professionals and society more broadly in normative terms, reproduction would not be democratized. That is to say, biotechnological innovation does not in and of itself change entrenched social norms (though it may challenge them).

How Could AGs Cure Infertility?

The underlying assumption highlighted above is that infertility is essentially a genetic problem. Accordingly, reproduction can be understood as the transmission of genes. Some people are able to do this without medical help simply as a result of following their sexual inclinations. Others are not. In cases where they are not, there is often very little that can be done to facilitate genetic transmission in people who cannot produce gametes. IVF may work for people with blocked fallopian tubes, or varicocelectomy for blockages of the spermatic cord. But if gametes are not being produced *at all*, something more than surgical unblocking is needed. It is in this context that AGs have been regarded as being the potential to democratize reproduction. But what do we really mean by reproduction here?

The question of how, if at all, particular treatments remedy infertility is not a new one. Writing in 1998, on the subject of artificial insemination, Simone Bateman asks: '[w]hat reasoning led physicians to assume that they were providing treatment for infertility, when the technical act they were proposing

[11] Zsolt Peter Nagy and Ching-Chien Chang, 'Current Advances in Artificial Gametes', *Reproductive BioMedicine Online*, 11/3 (2005), 332–339.

[12] Giuseppe Testa and John Harris, 'Ethics and Synthetic Gametes', *Bioethics*, 19/2 (2005), 165.

may be understood as an alternative mode of conception to heterosexual intercourse?'.[13] Bateman's point is important. Fertility treatments may take many forms, but their relationship with infertility and reproduction is not always clear. For example, suppose a heterosexual couple of 'normal' reproductive age—let's call them Anne and Brad—want to have a child. Brad cannot produce sperm. They use donated sperm from Colin, and Anne gives birth to a baby, Delia. But how does this treat Brad's infertility? Here, Bateman's point seems persuasive. It is not at all obvious that Brad's infertility has been treated. He remains unable to produce sperm. Delia's birth has arisen from interventions involving people other than Brad. It would seem odd, on the other hand, to suggest that *Anne*'s infertility has been treated, since she already had the capacity to produce eggs, and has not been 'treated' in this respect, even though she has undergone an intervention—it was not an intervention that somehow remedied her inability to have a genetically related child. She never had such an inability. Rather, she—and Brad—had an inability to produce a child related to Brad.

Thus, the donation of Colin's sperm, and the use of that sperm in a medical intervention performed on Anne, seem hard to locate in any medical relationship with respect to the couple's infertility, or with their claim to have reproduced following the birth of Delia. Anne may indeed have reproduced, assuming that she produced the egg that gives rise to Delia, and also gestated and gave birth to the child. But it is hard to claim that Brad has reproduced. Rather, as Bateman suggests, it seems that Anne and *Colin* have been enabled to produce a child together. And they have achieved this without having to have intercourse with each other. Thus, on Bateman's view the 'treatment' is an alternative to intercourse rather than a remedy for infertility per se.

However, Bateman's question relates specifically to donated sperm. The point about AGs is precisely that they *do* enable a person in Brad's position to reproduce. In this respect perhaps they can be regarded as a bona fide treatment for infertility. Yet as I will show, things are not as straightforward as they seem.

Let us return to the scenario involving Anne and Brad, in a context where AGs are available. The couple long for a baby, but Brad has no sperm. Hearing of a new treatment option involving artificial gametes, the couple refuse Colin's offer of donated sperm, and instead they use AG technology to create sperm from Brad's skin cells. In this scenario it is perhaps easier to view AGs as a treatment that offers a remedy for Brad's infertility in a way that the use of donor sperm does not. If we explore this further, it appears that genes are the key to the question.

[13] Simone Bateman Novaes, 'The Medical Management of Donor Insemination', in Katrina Daniels and Erica Haimes, eds., *Donor Insemination: International Social Science Perspectives* (Cambridge University Press, 1998), 105–130.

When donor sperm is obtained from Colin, there is no genetic connection between Brad and Delia. Instead, there is a genetic relationship between Delia and Colin. However, when Brad uses his somatic cells to produce sperm, there *is* a genetic connection between him and Delia. It is because Delia 'contains' his genes that we can understand Brad as having reproduced, and in turn, this implies that his infertility has been successfully treated. Brad has transmitted his genes. Yet it does not seem as if genes can really be the whole answer here. If we return to the donor-sperm scenario, we can envisage a context in which the potential donor is in fact Brad's identical twin, Boris. In this case, if Brad uses Boris' donated sperm, there *is* a genetic connection between Boris and Delia. In fact—genetically—Delia is indistinguishable from any child Brad would have conceived with his own gametes. If we accept the claim that it is genetic connections that are central here, it seems that we are committed to saying that when *Boris* donates sperm, this effectively treats Brad's infertility, but using sperm from the unrelated donor—Colin—would not. Again, there is something odd here in that neither of these interventions, whether it involves Colin or Boris, has any connection with Brad's body at all.

A further counterintuitive implication of the idea of reproduction as essentially genetic is that Brad need not really bother having Delia at all, since if Boris already has his own child—let us call her Emily—this child will *also* have as much genetic resemblance to Brad as she does to Boris. Thus, if infertility really consists in the inability to have genetically related children, it seems that Brad was never infertile in the first place! In a sense, Boris functions as Brad's proxy, ensuring that the world is populated with Brad's genetic offspring, even though Brad himself has played no active part in bringing this about.

Intuitively, however, it is not easy to reconcile the idea that Emily's existence somehow renders Brad fertile, or can be taken as evidence that he has reproduced. If so, this suggests that genes are not the sole answer to what constitutes either infertility or reproduction. Yet the whole point of AGs is that they enable people to transmit genes. Their entire claim to democratize reproduction, end infertility, and push past the final frontier lies in their *genetic* power. If genes are, after all, not the answer to what constitutes reproduction or infertility, it seems that the pursuit of AGs for use in reproductive technologies may be fundamentally misguided. But can it really be the case that genes do not hold the answers to reproduction, infertility, and hence parenthood?

Biology, Genes, and Reproduction

The role of genes in reproduction and infertility is not entirely clear. In recent years, some philosophers and bioethicists have begun to cast doubt on the idea that genetic connections are the best way of understanding what matters in infertility treatment.[14] So how did genes come to play such a central role in our understandings of reproduction? It seems that there has been a subtle transition from the discourse of *biology* per se to genes specifically. Arguably, it is biotechnology itself that drives this genetic focus. The years that followed the birth of the first IVF baby were dominated by news of developments in genetic science. The human genome project was underway, and governments and research funders hastily diverted money towards the vital project of ensuring that societies would be ready to deal with the implications of the genetic revolution. Richard Dawkins published his famous book *The Selfish Gene*, which effectively captured the public imagination with its claim that we are essentially machine programmed by our genes. According to the selfish gene theory, genes have one single imperative, which is to perpetuate themselves.[15,16]

In the context of this popular genetic frenzy, perhaps it is not surprising that genes began to take centre stage in our understandings of reproduction, and to form the basis of new arguments relating to infertility and access to new reproductive technologies. For example, John Harris, writing on the ethics of reproductive cloning, asserted that such a procedure could be construed as an aspect of reproductive autonomy.[17] It seems that in Harris' analysis, it is the *genetic* connection that cloning facilitates that enables him to treat cloning as being reproductive. (The World Health Organization, however, took a very different view, stating that cloning replicates individuals, and is 'contrary to human dignity and integrity'.)[18]

John Robertson's conception of reproductive autonomy is rooted in its connection with biology. According to him, there is a 'basic biologic . . . drive to have a biologically related family'.[19] Biology in Robertson's sense has been largely interpreted to mean simply genes, even if this was not necessarily what Robertson himself envisaged. Buchanan et al. also emphasize the link between genes and biology in reproduction where they assert that reproductive autonomy

[14] Heidi Mertes, 'Gamete Derivation from Stem Cells: Revisiting the Concept of Genetic Parenthood', *Journal of Medical Ethics* 40/11 (2014), 744–747.

[15] Richard Dawkins, *The Selfish Gene* (Oxford: Oxford University Press, 2006).

[16] I should mention here that Dawkins himself is at pains to claim that he does not place any moral weight on his description of the role of genes in our lives. He is merely, he says, describing how things are, not how things ought to be.

[17] John Harris, 'Goodbye Dolly? The Ethics of Human Cloning', *Journal of Medical Ethics*, 23 (1997), 353–360.

[18] World Health Organization, *Ethical, Scientific and Social Implications of Cloning in Human Health*, WHA41.10. 16 May 1998, https://apps.who.int/gb/ebwha/pdf_files/EB101/pdfangl/angr25.pdf.

[19] John Robertson, *Children of Choice* (Princeton University Press, 1994), 39.

encompasses the choice to have biologically related children—including cloning.[20]

Derek Parfit, whose work has been profoundly significant in the context of bioethics and reproductive ethics, has also been taken to endorse a view of reproduction as being fundamentally genetic. Parfit's famous 'non-identity problem' posits that a child born from gametes produced today must necessarily be different from a child born in a few months—from different gametes. This has been interpreted by many to indicate that it is the genes that form the basis of the child's identity.[21] However, others have argued that Parfit's work can be understood to place emphasis on the gametes themselves rather than on the genes they transmit per se, or alternatively that it is the passage of time itself that makes a difference to a child's identity when its conception is delayed by a few months.[22]

Either way, it is undeniable that Parfit's account lends itself readily to a genetic essentialist account of reproduction, even if he may not have intended this himself. Indeed, at the time Parfit was writing, it would have made little sense to suppose that gametes could be separated from genes, or that gametes could be derived from skin cells, or that a single sperm or egg cell could be isolated for use now, or frozen for use at any point in future. Now, however, it does make sense to ask whether gametes have any function or significance aside from their status as carriers of genes. This is especially the case in the context of artificial gametes, as I will discuss later.

The Warnock Report, too, endorsed this message that the longing to pass on genes is what matters in biological reproduction: 'people experience a powerful urge to perpetuate their genes through a new generation. This desire cannot be assuaged by adoption'.[23] And a joint report published by the Human Genetics Advisory Committee and the HFEA states: 'The wish for *genetic* offspring is a natural human aspiration'[24] (my emphasis).

Yet with the mapping of the human genome, it appeared that genes could not really be the answer to everything after all. Humans have relatively few genes—tomatoes and grapes have more than we do. With only between 20,000 and 30,000 genes in total, it does not seem plausible that there is 'a gene' for everything.[25]

[20] Alan Buchanan et al., *From Chance to Choice: Genetics and Justice* (Cambridge University Press, 2000).

[21] Anthony Wrigley, 'Harm to Future Persons: Non-identity Problems and Counterpart Solutions', *Ethical Theory and Moral Practice*, 15 (April 2012), 175–190.

[22] Tim Lewens, 'The Fragility of Origin Essentialism: Where Mitochondrial "Replacement" Meets the Non-identity Problem', *Bioethics*, 35/7 (2021), 615–622.

[23] Mary Warnock, *A Question of Life* (Blackwell, 1985), 8.

[24] Human Genetics Advisory Commission and the Human Fertilisation and Embryology Authority, 'Cloning Issues in Reproduction, Science and Medicine', January 1998, https://wellcomecollection.org/works/qghye2vw.

[25] Adam Rutherford, 'Eugenics and the Misuse of Mendel', *The American Journal of Human Genetics*, 111/7 (2024), 1254–1257.

Two additional developments have cast further doubt on the degree to which we can think of genes as being the central or only significant aspect of reproduction. Firstly, it has become increasingly clear that simply having a gene is not enough to guarantee that it will have a noticeable impact. Genes may be active or inactive, or somewhere in between. If inactive, they will not produce the proteins that characterize their specific function. Thus, even if I transmit 'my' genes to my children, it does not necessarily follow that they will exhibit the same traits that I do. This process of turning off and on—'epigenetics'—further complicates the genetic landscape. Moreover, it calls into question the idea that reproduction is genetic per se. If I wish my offspring to resemble me, it is not enough to expect that they should have my genes. Epigenetic effects are hard to predict or control. To a significant degree, they are brought about by environmental conditions. A striking example of this occurred when a wealthy Chinese man had his beloved pet cat cloned. The cloning procedure was successful, but the cloned kitten had different markings from the original, despite being genetically identical. Epigenetics influences in utero had affected the expression of the genes that govern the colour and pattern of the fur.

An additional challenge to geneticized accounts of reproduction lies in the possibility of replacing mitochondrial DNA. Some people have mutations in the mitochondria that can then be transmitted to offspring. Mitochondrial disease can be devastating. Rather than risk transmitting the disease, some people who know they are at risk may seek to use mitochondrial DNA from donor eggs (sperm do not carry mitochondrial DNA; it is inherited solely from the mother). To return to the original characters in the scenario, let us suppose that Anne discovers she is at risk of transmitting a mitochondrial disease to her offspring. She therefore asks a friend, Fiona, to donate mitochondrial DNA. Since mitochondrial donation involves the invasive process of egg retrieval from two women—Anne and Fiona—one might ask whether simply using Fiona's eggs would be a simpler and less invasive solution, rather than attempting to join Fiona's mitochondrial DNA with Anne's egg. The process in this case would be far less technically complex and would spare Anne at least one surgical intervention. But to approach it in this way would not enable Anne to transmit her own genes to the offspring.

Here, something odd seems to be going on. In choosing to use donor mitochondrial DNA, Anne *already accepts* the idea that Fiona's genes will be transmitted to the offspring. The issue seems to be a question of whether these genes really 'count'. In the United Kingdom, legislation has been passed to assert that they do *not* count. Accordingly, the mitochondrial donor is not a mother and has not reproduced. The mitochondrial DNA is described as being of negligible significance, a mere battery that powers the cell.

Here we have an indication of what happens when we want to make distinctions and differentiations between genes in ways that preserve the intuition that a child has only two genetic parents. At this stage, it is no longer the case that reproduction is 'genetic' per se, or that passing on genes is what counts. It is a question of *which* genes. Chromosomal genes are those that are reproductively important while mitochondrial genes are not. In this way, parents who use donated mitochondrial DNA are able to reassure themselves that their offspring are their own and no one else's.

All of these challenges and questions tend to undermine the idea that reproduction is a simple matter of genetic transmission, and that infertility is remedied by ensuring that offspring display a sufficient genetic resemblance to their parents. Offspring can display genetic similarities to individuals in some instances (the identical twin scenario, and the mitochondrial donation scenario) where we would not necessarily agree that every person who shares a genetic resemblance has reproduced. The question here is not *just* about the degree of resemblance either. After all, in the mitochondrial case, the resemblance is not simply resemblance (assuming the procedure has been maximally successful). Rather, the mitochondrial DNA that Delia inherits from Fiona has *exactly the same genes* as those in Fiona's own mitochondrial DNA. In contrast, the DNA that Delia inherits from her 'parents', Anne and Brad, are reshuffled *and* mixed with each other. In terms of pure genetic resemblance, therefore, Fiona is more closely related to Delia than either of Delia's real' parents.

This example seems to show that in fact it is not simply the closeness of the genetic resemblance that counts. Perhaps it matters *which* genes resemble those of the prospective parent. Technology has pushed us towards discriminating between chromosomal genes and mitochondrial genes. Perhaps further technological innovations will lead to further schisms between different types of genes in ways that we are not yet able to predict. If so, then we may see new developments in terms of both how fertility treatments are construed, and how genetic relationships correspond with reproduction and parenthood. If we accept that this is the case, it seems to imply that there is no simple biological answer to the question of how genes relate to reproduction and infertility. The ways in which genes are interpreted and valued may differ over time, in response to both social values and biotechnological possibilities.

Conclusion

If we are not entirely certain what reproduction is, or how infertility is cured, or what the role of genes is in these processes, can we really be sure that AGs are likely to help in remedying infertility at all? Insofar as reproduction is deemed to be essentially genetic, perhaps AGs will indeed enable people to ensure that offspring share a genetic resemblance to them. But whether this resemblance is sufficient to imply a parent-child relationship is another question. This depends not on genes at all, but on social and legal structures. It is not biology, but the law, that decrees a mitochondrial donor is not a mother. At the same time, it is the social practice of medicine that determines whether mitochondrial replacement therapy is available for people who want a mitochondrial link with 'their' child.[26] Some wishes are treated as medical needs, and others are not, even though they may be very similar.

What we *can* see for now, though, is that the scurry towards creating AGs is in itself an endorsement and reinforcement of the idea that we should prioritize a genetic understanding of infertility and work towards biotechnological solutions. It is already at the time of writing over twenty years since AGs started to generate excitement as a potential cure for infertility. Yet they have not yet made an appearance in the clinic. Even if we do endorse a genetic account of infertility, AGs have not lived up to their promise of a cure.

Bibliography

Bhattacharya, Shaoni, 'Stem Cells Can End Infertility, Say IVF Pioneers', *New Scientist.com News Service*, 24 July 2004.

Bjorklund, David F., Yunger, Jennifer L., and Pellegrini, Anthony D., 'The Evolution of Parenting and Evolutionary Approaches to Childrearing', in M. Bornstein, ed., *Handbook of Parenting*, Volume 2 (Mahwah: Lawrence Erlbaum Associates, 2002), 3–29.

Buchanan, Allen, Brock, Dan W., Daniels, Norman, and Wikler, Daniel, *From Chance to Choice: Genetics and Justice* (Cambridge University Press, 2000).

Cavaliere, Giulia, and Palacios-González, César, 'Lesbian Motherhood and Mitochondrial Replacement Techniques: Reproductive Freedom and Genetic Kinship', *Journal of Medical Ethics*, 44/12 (December 2018), 835–842.

Dawkins, Richard, *The Selfish Gene* (Oxford: Oxford University Press, 2006).

Harris, John, 'Goodbye Dolly? The Ethics of Human Cloning', *Journal of Medical Ethics*, 23 (1997), 353–360.

Haslanger, Sally, 'The Sex/Gender Distinction and the Social Construction of Reality', in Ann Garry, Serene Khader, and Alison Stone, eds, *The Routledge Companion to Feminist Philosophy* (Taylor and Francis, 2017), 157–167.

Hübner, Karin, Fuhrmann, Guy, Christenson, Lane K., Kehler, James, Reinbold, Rolland, De La Fuente, Rabindranath, et al., 'Derivation of Oocytes from Mouse Embryonic Stem Cells', *Science*, 300/5623 (2003), 1251–1256.

Human Genetics Advisory Commission and the Human Fertilisation and Embryology Authority, 'Cloning Issues in Reproduction, Science and Medicine', January 1998, https://wellcomecollection.org/works/qghye2vw.

[26] Giulia Cavaliere and Cesar Palacios-González, 'Lesbian Motherhood and Mitochondrial Replacement Techniques: Reproductive Freedom and Genetic Kinship', *Journal of Medical Ethics*, 44/12 (2018), 835–842.

Lewens, Tim, 'The Fragility of Origin Essentialism: Where Mitochondrial "Replacement" Meets the Non-identity Problem', *Bioethics*, 35/7 (September 2021), 615–622.

Martin, Emily, 'The Egg and the Sperm: How Science Has Constructed a Romance Based on Stereotypical Male-Female Roles', in *Situated Lives* (Routledge, 2014), 85–98.

Mertes, Heidi, 'Gamete Derivation from Stem Cells: Revisiting the Concept of Genetic Parenthood', *Journal of Medical Ethics*, 40/11 (2014), 744–747.

Nagy, Zsolt P., and Chang, Ching-Chien, 'Current Advances in Artificial Gametes', *Reproductive BioMedicine Online*, 11/3 (2005), 332–339.

Newson, Ainsley J., and Smajdor, Anna C., 'Artificial Gametes: New Paths to Parenthood?', *Journal of Medical Ethics*, 31/3 (2005), 184–186.

Novaes, Simone B., 'The Medical Management of Donor Insemination', in K. Danies and E. Haimes, eds., *Donor Insemination: International Social Science Perspectives* (Cambridge University Press, 1998), 105–130.

Robertson, John A., *Children of Choice* (Princeton University Press, 1994).

Rutherford, Adam, 'Eugenics and the Misuse of Mendel', *The American Journal of Human Genetics*, 111/7 (2024), 1254–1257.

Smajdor, Anna, and Cutas, Daniela, 'Artificial Gametes', background paper, Nuffield Council of Bioethics, 2016.

Testa, Giuseppe, and Harris, John, 'Ethical Aspects of ES Cell-Derived Gametes', *Science*, 305/5691 (2004), 1719.

Testa, Giuseppe, and Harris, John, 'Ethics and Synthetic Gametes', *Bioethics*, 19/2 (2005), 146–166.

Warnock, Mary A., *A Question of Life* (Blackwell, 1985).

World Health Organization, 'Ethical, Scientific and Social Implications of Cloning in Human Health', WHA41.10, 16 May 1998, https://apps.who.int/gb/ebwha/pdf_files/EB101/pdfangl/angr25.pdf.

Wrigley, Anthony, 'Harm to Future Persons: Non-identity Problems and Counterpart Solutions', *Ethical Theory and Moral Practice* 15 (April 2012), 175–190.

11

Gendered and Ageist Norms of Parenthood Manifested in Public Attitudes Towards IVG

Heidi Mertes

Some IVF Patients Are More Equal Than Others

In vitro gametogenesis (IVG), or 'growing eggs and sperm in the lab', is only one of a long list of innovations in the field of reproductive medicine. While many of these innovations are not free from risks for both the intended parents and the resulting offspring, and although some groups of people object to IVF in general for various reasons, acceptance of these innovations is overall quite high in the general population. For example, in a Belgian survey only 9.2% of respondents disagreed with the statement that 'progress in medically assisted reproduction is a positive development', while 68.5% agreed and 22.3% was neutral.[1] This can most likely be explained by the fact that many people desire to have children, causing a high level of empathy and sympathy with those who need medical assistance to achieve their dream of parenthood. However, this empathy is not equally directed at all groups of people. This chapter particularly zooms in on attitudes towards women who are no longer able to conceive spontaneously due to their age. This chapter builds on the results of previously published surveys on people's attitudes towards IVG in the Netherlands,[2] Japan,[3] and Belgium,[4] on interesting findings of the citizens' forum that was conducted as a follow-up to the Belgian survey as a public engagement initiative by a Belgian non-profit organization, De Maakbare Mens,

[1] H. Mertes et al., 'Enthusiasm, Concern and Ambivalence in the Belgian Public's Attitude towards In-Vitro Gametogenesis', *Reproductive Biomedicine & Society Online*, 14 (2022), 160.

[2] S. Hendriks et al., 'The Acceptability of Stem Cell-Based Fertility Treatments for Different Indications', *MHR: Basic Science of Reproductive Medicine*, 23/12 (2017), 855–863.

[3] K. Akatsuka et al., 'Public Attitudes in Japan toward the Reproductive Use of Gametes Derived from Human-Induced Pluripotent Stem Cells', *Future Science OA*, 7/10 (2021), FSO754.

[4] Mertes et al., 'Enthusiasm, Concern and Ambivalence in the Belgian Public's Attitude towards In-Vitro Gametogenesis', 156–168.

Heidi Mertes, *Gendered and Ageist Norms of Parenthood Manifested in Public Attitudes Towards IVG*
In: *Revolutionizing Reproduction*. Edited by: Christopher Gyngell, Julian Savulescu, and Tsutomu Sawai,
Oxford University Press. © Oxford University Press 2026. DOI: 10.1093/9780197810415.003.0011

but will also link these results to a Belgian survey on mitochondrial replacement therapy,[5] and to previous work in the context of oocyte cryopreservation for age-related fertility decline.[6] What we see in all these contexts is that the general attitude towards the use of reproductive technologies by women in their forties and beyond is a lot less positive than for younger women. It appears that there are strong societal norms not only about whether people should desire to have children, but also about when people ought to pursue this desire and when they ought not to pursue this desire. While these norms exist for both men and women, they are more pronounced for women than for men.

Attitudes towards IVG to Counter Age-Related Infertility

Several surveys have explored the attitudes of the general public towards IVG, and at least three of them included a question measuring the levels of acceptance of the use of IVG in the specific setting of infertility due to reproductive aging. A first survey, conducted by Hendriks et al. in the Netherlands, presented the following case: 'Sarah and John want to have a child together. However, Sarah is 50 years old and has already entered menopause. Sarah cannot become a genetic parent unless she uses stem cell-based fertility treatments. An egg cell is made from Sarah's stem cells, which is fertilized by John's sperm. Sarah and John are the genetic parents of the conceived child'.[7] Participants could either indicate 'acceptable under certain conditions', 'not acceptable', or 'no opinion'. Only 17% of respondents who expressed an opinion judged this scenario to be acceptable, contrasting with 94% who accepted the scenario involving IVG for a young heterosexual couple faced with female infertility.

A second survey, conducted by Akatsuka et al. in Japan, asked participants, 'To what degree do you agree that [heterosexual married couples who cannot have children due to menopause] use hiPSC-derived sperm/ova to produce children?'.[8] Possible responses were 'I strongly agree', 'I agree', 'I do not agree', 'I do not agree at all', or 'I don't know'. In this survey, there was a much greater acceptance of IVG for post-menopausal women compared to the Dutch study, with 40.5% of respondents (strongly) agreeing, although it still contrasts with 70.8%

[5] G. Pennings et al., 'Attitude of Belgian Women towards Enucleated Egg Donation for Treatment of Mitochondrial Diseases and Infertility', *Reproductive Biomedicine Online*, 49/3 (2024), 104101.

[6] H. Mertes, 'The Portrayal of Healthy Women Requesting Oocyte Cryopreservation', *Facts, Views & Vision in ObGyn*, 5/2 (2013), 141–146. H. Mertes, 'Ethical Aspects of AGE Banking', in *Preventing Age-Related Fertility Loss* (Cham: Springer International Publishing, 2018), 157–168.

[7] Hendriks et al., 'The Acceptability of Stem Cell-Based Fertility Treatments for Different Indications', 857.

[8] Akatsuka et al., 'Public Attitudes in Japan toward the Reproductive Use of Gametes Derived from Human-Induced Pluripotent Stem Cells', 3.

of participants (strongly) agreeing with IVG for other heterosexual married couples who cannot rely on standard infertility treatments.[9]

Finally, in a Belgian survey,[10] respondents' attitudes were gathered on different potential scenarios in which IVG might be used, including 'for older women couples to conceive a genetically related child after menopause'. While for all other scenarios presented (IVG for heterosexual couples, same-sex couples, and solo reproduction), the number of positive responses (acceptable or fully acceptable) outnumbered the negative ones (unacceptable or fully unacceptable) convincingly (by 21% to 69%), the single scenario for which the reverse was true was the one referring to post-menopausal motherhood, with 62.7% of respondents who expressed an opinion judging that use of IVG for 'older' women was (fully) unacceptable, while only 18.2% thought it was (fully) acceptable.

While it is perhaps not surprising that the acceptance of 'older' motherhood is lower than the acceptance of younger motherhood, a different point of reference is more remarkable, notably the comparison with the acceptance rates for the scenario in which egg and sperm from the same person (one natural, one generated through IVG) would be used to reproduce, which is considered highly undesirable on medical grounds[11] as this kind of solo reproduction comes with serious health risks for the resulting offspring due to a high possibility of autosomal recessive genetic conditions (the resulting child's DNA would on average be 50% homozygous, which is the same rate as one would expect in the case of incest between full siblings[12]). Moreover, reproductive cloning, which is also a form of solo reproduction, but without this risk of increased homozygosity, generally has very low acceptance rates in the general population. We therefore would expect people to object more to solo reproduction than to the arguably less alienating scenario of post-menopausal women reproducing. When we compare the acceptance rates of IVG for solo-reproduction versus post-menopausal reproduction, however, we see that while acceptance of post-menopausal reproduction is larger in the Japanese study (18.7% vs 40.5%), it is smaller in both the Dutch survey (27% vs 17%) and the Belgian survey (51% vs 18.2%). Mayeur et al.[13] have previously argued that rather than dismissing the

[9] Akatsuka et al., 'Public Attitudes in Japan toward the Reproductive Use of Gametes Derived from Human-Induced Pluripotent Stem Cells'.

[10] Mertes et al., 'Enthusiasm, Concern and Ambivalence in the Belgian Public's Attitude towards In-Vitro Gametogenesis'.

[11] See, e.g., L. Notini, C. Gyngell, and J. Savulescu, 'Drawing the Line on In Vitro Gametogenesis', *Bioethics*, 34/1 (2020), 123–134.

[12] For an elaboration on this issue, see D. Cutas and A. Smajdor, '"I Am Your Mother and Your Father!" In Vitro Derived Gametes and the Ethics of Solo Reproduction', *Health Care Analysis*, 25 (2017), 354–369.

[13] C. Mayeur, H. Mertes, and W. Van Hoof, 'How to Deal with Uninformed and Poorly Informed Opinions of Citizens? A Critical Approach to Online Public Engagement', *Citizen Science: Theory and Practice*, 9/1 (2024), 1–10.

participants' unexpected responses based on a perceived flaw in their insight or knowledge, it can be enriching for the ethical debate to uncover the reasons behind these opinions, as these can (1) uncover implicit values, fears, or needs, and (2) highlight misconceptions for which more targeted education is necessary. This chapter might add that it can also uncover societal norms. It is therefore interesting to further explore the arguments presented by the participants against the use of IVG with the aim of enabling late motherhood.

Justifications for Negative Attitudes towards IVG for Post-menopausal Women

Previous academic work by Cutas and Smajdor engaged with the most important concerns linked to post-menopausal motherhood, namely arguments from nature, risks and efficacy, reduced energy of the mother due to age, maternal life expectancy, overvaluing genetic relationships, and the reinforcement of a pro-reproductive culture, concluding that '[m]any of the arguments formulated against older others appear to be based on concern for the offspring, but on closer inspection they reveal distaste or prejudice against older mothers'.[14]

While the surveys mentioned above do not offer an insight into the underlying motivations of the respondents, we can learn something from a citizens' forum that was organized by De Maakbare Mens ('The Makeable Human'), a nonprofit organization encouraging public dialogue on the societal impact of new medical technologies, as a follow-up of the Belgian survey.[15] All participants in the survey were asked whether they would be willing to participate in the citizens' forum and had the option of sharing their contact information (which was not linked to their responses to the survey questions) with a representative of De Maakbare Mens. The goal of this forum was to explore some of the responses more in depth, in order to elicit the reasons behind the responses. Twenty people with diverse demographic profiles were selected, of which sixteen participated in the forum. During the citizens' forum, one of the discussion tables dealt with taboos, more specifically: use of IVG by post-menopausal women, single reproduction, and the role of government in facilitating or hindering research. The participants consented to the public sharing of the results of the citizens' forum by De Maakbare Mens, for example through their website (https://www.demaakbaremens.org/burgerforumivg/). The arguments presented below were obtained in an anonymized format by personal communication from De Maak

[14] D. Cutas and A. Smajdor, 'Postmenopausal Motherhood Reloaded: Advanced Age and In Vitro Derived Gametes', *Hypatia*, 30/2 (2015), 393.

[15] Information and output of this citizens' forum can be found at https://www.demaakbaremens.org/burgerforumivg/ (accessed 7 October 2024) or requested at De Maakbare Mens, vzw.

bare Mens, and verified by the author, who was the moderator for the discussion table at which the arguments were presented.[16] While many of the citizens' arguments presented overlap with the academic concerns previously outlined by Cutas and Smajdor,[17] there are also some additional ones that appear to confirm Cutas and Smajdor's remark that prejudice may be at play.

The arguments presented for not accepting the use of IVG for postmenopausal women were related to nature, the well-being of the offspring, and risks for the woman, but also to the character of the woman, overpopulation, entitlement, and alternative options.

Type or argument	Quote (own translation from Dutch to English)
Unnatural	*After menopause, natural reproduction is no longer possible. When nature says 'no', is it really necessary to do this [IVG]? There is a reason why there is a menopause.*
	*The body of an older woman is not fit for pregnancy and childbirth.**
Negative impact on well-being of offspring	*There will be a large generation gap between the parent and child, your mother will look like your grandmother, and that is already an extra burden that you put on that child.*
	The caring role can turn around quickly, when a twenty-year-old child needs to take care of its seventy-year-old parent.
	Will the parent live long enough to raise the child?
	The risk of abnormalities in the child increases.
Risks and burdens for the woman	*At an older age you are less flexible and resilient to raise a child.**
	The risk of a difficult pregnancy is larger.
Character of the woman	*This is selfish.*
	An older woman has had her chance; she should have thought about her desire for a child earlier.
Overpopulation	*There is overpopulation now already, should we be doing this then?*
Entitlement	*The interests of the child need to outweigh the desire of the parents.**
	The right to a child does not exist.
Alternative options	*There is an alternative: adoption.*

* These quotes also refer to the well-being of the child.

[16] Given that these arguments were not gathered in a properly designed qualitative empirical study, but in a public engagement initiative separate from the scientific research, they should be interpreted as anecdotal information that formed the starting point for the reflections in this chapter, but are not research findings in themselves.

[17] Cutas and Smajdor, 'Postmenopausal Motherhood Reloaded: Advanced Age and In Vitro Derived Gametes'.

Counterarguments that were presented, by either the moderator or participants, were that same-sex couples cannot reproduce naturally either (while only two participants objected to IVG for same-sex couples); that the health and resilience of the parents are verified in the trajectory leading up to fertility treatment, as these will differ between different patients of the same age; that the second parent might be younger; that older fathers seem to be more socially accepted although many of the presented arguments would also apply to them; that adoption also has an age limit and is not easily accessible; that a woman might not have been able to have a child earlier; and that the desire for parenthood is always self-interested, also for younger women.

During the debates, despite being confronted with counterarguments and inconsistencies, opinions did not shift towards greater acceptance of IVG for infertility due to aging. On the contrary, attitudes shifted to being less accepting, rather than more.

Late Motherhood through IVG, Mitochondrial Replacement Therapy, or Egg Freezing

The finding that many people do not have a positive attitude towards women becoming mothers after menopause through IVG is in line with attitudes in several other contexts. Reluctance to support the use of new reproductive technologies to achieve late motherhood has for example also been noticed in the contexts of mitochondrial replacement therapy and egg freezing, as will be outlined below.

There are several different techniques under the umbrella of the term 'mitochondrial replacement therapy', but what they all aim at achieving is an embryo with cells that contain the nuclear DNA from the intended parents, but the cytoplasm (including mitochondria) of an egg cell donor.[18] There are, broadly speaking, two potential areas of application for this technology: first, to significantly reduce the risk that women who are carriers of pathogenic mitochondrial DNA mutations would pass on their defective mitochondria to their offspring,[19] and second, to enable women with repeated fertilization failure due to oocyte aging to 'rejuvenate' their oocytes and increase their chances of conceiving.[20] Before licensing the technology for the United Kingdom, the HFEA (Human Fertilization and Embryology Authority, the UK regulator for fertility treatments) conducted a large-scale public consultation and decided to

[18] D. P. Wolf, N. Mitalipov, and S. Mitalipov, 'Mitochondrial Replacement Therapy in Reproductive Medicine', *Trends in Molecular Medicine*, 21/2 (2015), 68–76.

[19] E. Kang et al., 'Mitochondrial Replacement in Human Oocytes Carrying Pathogenic Mitochondrial DNA Mutations', *Nature*, 540/7632 (2016), 270–275.

[20] E. Labarta et al., 'Mitochondria as a tool for oocyte rejuvenation', *Fertility and Sterility*, 111/2 (2019), 219–226.

restrict access to the technology to women at high risk of passing on severe mitochondrial diseases.[21] The rationale for this limitation has been questioned,[22] and the argument has been made that as long as the technique can only reduce the risk of passing on faulty mitochondria but not eliminate it (due to mitochondrial carryover and the reversion phenomenon[23]), it might actually be wiser to limit current clinical trials to those who might benefit from oocyte rejuvenation, rather than to women with mitochondrial disorders.[24] The framing of one application aiming to prevent a serious disorder and the second as 'merely' curing infertility is misleading in the sense that both start from a desire to have a (healthy) genetically related child and that childlessness or donor conception are possible alternatives for both. The only difference is that the group of people faced with a mitochondrial disorder have—as it were—the additional options of natural conception (with or without prenatal testing) or embryo selection after IVF and pre-implantation genetic testing to select embryos with a low mutation load (except when one is homozygous for the faulty mutation). Here again, it appears that the difference in empathy towards young people with mitochondrial disorders, on the one hand, and older women with age-related infertility, on the other, might be weighing on this debate. In a recent survey in Belgium, women were asked how acceptable they rated three scenarios: use of the technique for women who are at high risk of having a child with a mitochondrial disease, for women with unexplained infertility, and for women between forty and fifty years old to rejuvenate their eggs. Whereas the first two scenarios were rated as acceptable by 76.1% and 72% of respondents, the latter scenario was rated as acceptable by only 26.7% of respondents.[25] What we see here is that even when the age is capped at fifty years (which was not the case in the IVG survey), respondents are still reluctant to approve of late motherhood. This is consistent with previous research indicating that in European countries, the average age at which people judge women to be too old to bear an(other) child is 41.8 years. For Belgium—where this survey and the IVG survey with the citizens' forum took place—this is even lower, at 40.7 years.[26] An interesting observation to make here

[21] HFEA, *Mitochondrial donation treatment*, 2017, https://www.hfea.gov.uk/treatments/embryo-testing-and-treatments-for-disease/mitochondrial-donation-treatment/.

[22] E. Braun, 'Mitochondrial Replacement Techniques for Treating Infertility', *Journal of Medical Ethics*, 2024, https://doi.org/jme-2023-109660.

[23] N. Costa-Borges et al., 'First Pilot Study of Maternal Spindle Transfer for the Treatment of Repeated In Vitro Fertilization Failures in Couples with Idiopathic Infertility', *Fertility and Sterility*, 119/6 (2023), 964–973.

[24] G. Pennings, 'Enucleated Oocyte Donation: First for Infertility Treatment, then for Mitochondrial Diseases', *Journal of Assisted Reproduction and Genetics*, 39/3 (2022), 605–608.

[25] G. Pennings, 'Attitude of Belgian Women towards Enucleated Egg Donation for Treatment of Mitochondrial Diseases and Infertility'.

[26] V. Paksi and I. Szalma, 'Age Norms of Childbearing: Early, Ideal and Late Childbearing in European Countries', *Review of Sociology*, 2 (2009), 57–80.

is that the response may have been quite different if men were concerned, as the same study shows that the average age at which men are deemed too old to become a parent is 47.4 years (or 5.6 years later than for women).[27]

A similar story can be told about egg freezing. Here, too, there are broadly speaking two groups of people who might benefit: those who fear losing their ovarian reserve due to medical reasons (either a disease itself or the treatment for a disease) and those who fear losing their ovarian reserve—and egg quality—due to reproductive aging. And also here, the initial reflex of clinicians and regulators was to make the technology only available to the former category, but not the latter. The arguments for making this distinction were quite similar to the arguments that were mentioned above in the context of IVG for 'older' women: the argument from nature, undue medicalization of reproduction, and physical and psychological risks for mother and child.[28] These arguments were supplemented with stereotyping that framed the women interested in egg freezing for age-related fertility decline either as selfish, career-oriented women who brought infertility upon themselves (in line with the focus on women's character in the citizens' forum) or as victims of a male-oriented society who are now also being pushed to freeze their eggs in order to cope with structural barriers to motherhood.[29] As ethicists have argued against the legitimacy and consistency of several arguments, and as empirical research has painted a picture of the women who request egg freezing that is very different from the original portrayals, the original strict divide between 'medical egg freezing' (= good) and 'social egg freezing' (= bad) has fortunately become more nuanced in recent years, and the debate about this technology has become more constructive. Concerns about the technology's potential to support late motherhood are still present, but the rhetoric has become less stigmatizing, and the best interest of these women has come centre stage (e.g., De Clercq et al.[30]).

An important remark I would like to make at this point is that I am not advocating for a widespread implementation, let alone promotion, of new reproductive technologies for late motherhood, nor that I am blind to the challenges that come with it. As I have previously argued in the context of fertility preservation, IVG, and uterus transplantation, by developing ever more costly,

[27] Paksi and Szalma, 'Age Norms of Childbearing: Early, Ideal and Late Childbearing in European Countries'.

[28] W. J. Dondorp and G. M. de Wert, 'Fertility Preservation for Healthy Women: Ethical Aspects', *Human Reproduction*, 24/8 (2009), 1779–1785; S. Bernstein and C. Wiesemann, 'Should Postponing Motherhood via "Social Freezing" Be Legally Banned? An Ethical Analysis', *Laws*, 3/2 (2014), 282–300; H. Mertes, 'Ethical Aspects of AGE banking', in *Preventing Age-Related Fertility Loss*, 157–168.

[29] H. Mertes, 'The Portrayal of Healthy Women Requesting Oocyte Cryopreservation', *Facts, Views & Vision in ObGyn*, 5/2 (2013), 141–146.

[30] E. De Clercq et al., 'Rethinking Advanced Motherhood: A New Ethical Narrative', *Medicine, Health Care and Philosophy*, 26/4 (2023), 591–603.

burdensome, and risky technologies to enable infertile people to reproduce, implicit messages about the importance of (genetic and/or gestational) parenthood are sent along with these technologies,[31] which in turn reinforces deep-seated socio-cultural discourses on motherhood.[32] In that sense the offer of a new technology is never neutral, but might be taken to signal that women ought to want to become mothers and that this desire ought to be so strong that great sacrifices are justified. That this norm is very strong is, for example, very tangible when the attitudes towards women requesting medical help to conceive are compared to the attitudes towards women requesting medical help to prevent conceiving[33] and to women who are childfree by choice in general (see below). Rather than throwing ever more reproductive technologies at women faced with age-related fertility decline, we should therefore also make an effort to change the narrative that parenthood is the (only) road to happiness so that fewer childfree people are being stigmatized and so that more involuntarily childless people can find peace with the fact that parenthood is not in the cards for them,[34] an objective that should also be adopted by fertility clinics.[35]

Biased Arguments against Late Motherhood?

Returning now to the findings from the citizens' forum in IVG, it is clear that a number of legitimate concerns linked to the possibility of producing oocytes *in vitro* for women past their reproductive lifespan were presented, most notably related to health risks and well-being for both the women and their potential offspring. The aim of this chapter, however, is not to make an ethical cost-benefit analysis.[36] Instead, the focus here will be on a bias in the arguments. Two observations can be made: first, that age is merely a proxy for the legitimate issues at stake and that beneficial aspects of motherhood at a

[31] H. Mertes, 'Stem Cell Derived Gametes and Uterus Transplants: Hurray for the End of Third Party Reproduction! Or Not?', in *Personalised Medicine, Individual Choice and the Common Good* (Cambridge: Cambridge University Press, 2018), 37–51.; H. Mertes and G. Pennings, 'Ethical Considerations of Fertility Preservation', in *Female and Male Fertility Preservation* (Cham: Springer International Publishing, 2022), 627–640.

[32] De Clercq et al., 'Rethinking Advanced Motherhood: A New Ethical Narrative'.

[33] See, e.g., H. Mertes, 'The Role of Anticipated Decision Regret and the Patient's Best Interest in Sterilisation and Medically Assisted Reproduction', *Journal of Medical Ethics*, 43/5 (2017), 314–318.

[34] M. Mertens and H. Mertes, 'Deconstructing Self-fulfilling Outcome Measures in Infertility Treatment', *Bioethics*, 38/7 (2024), 616–623.

[35] S. Gameiro, D. Leone, and H. Mertes, 'Fertility Clinics Have a Duty of Care towards Patients Who Do Not Have Children with Treatment', *Human Reproduction* 39/8 (2024), deae128.

[36] But for such an analysis for IVG in general, see: S. Segers et al., 'Balancing Ethical Pros and Cons of Stem Cell Derived Gametes', *Annals of Biomedical Engineering*, 45 (2017), 1620–1632; and for an analysis about IVG for late motherhood specifically, see D. Cutas and A. Smajdor, 'Postmenopausal Motherhood Reloaded: Advanced Age and In Vitro Derived Gametes'.

later age are oftentimes absent in the debate, and second, that many arguments appear to be presented rather ad hoc, to justify a disapproving attitude, rather than genuinely being the ground for disapproval. I will elaborate on both remarks below.

Regarding the first, while there is a correlation between the age of a woman and possible pregnancy and labour complications like hypertension, diabetes, premature birth, and caesarean sections,[37] it is arguably not the case that beyond a certain age—say forty-five or fifty—these are so substantial for *all* women that a pregnancy becomes immoral. In fact, as noticed earlier by Bernstein and Wiesemann in the context of negative attitudes towards egg freezing for late motherhood, 'underlying pre-existing conditions as well as general physical health, possibly influenced by excessive nicotine use or obesity, seem to be more significant predictors of complications in pregnancy than chronological age'.[38] An individual health assessment would therefore be more appropriate to determine this risk than an age check. The same applies to concerns regarding the psychological well-being of the child. If one is truly concerned about these risks for the child, the more logical consequence would be to install safeguards aimed at assessing those risks directly, rather than focussing on a proxy that is not always reliable, especially if we are using that proxy to deny access to a medical intervention.[39] Such safeguards would also take into account the *positive* impact of late motherhood on the offspring's well-being, linked to more financial and relational stability of the parents, lower levels of parental stress, and higher rates of parental involvement in raising the children.[40] Regarding the physical health-related risks for the offspring, we could make a distinction between risks linked to pregnancy and labour complications (addressed above) and (epi)genetic risks. It is well established that the risk of chromosomal abnormalities increases with maternal age[41] and that the risk of harmful genetic mutations increases

[37] B. Luke and M. B. Brown, 'Elevated Risks of Pregnancy Complications and Adverse Outcomes with Increasing Maternal Age', *Human Reproduction*, 22/5 (2007), 1264–1272; J. Luo et al., 'Pregnancy Complications among Nulliparous and Multiparous Women with Advanced Maternal Age: A Community-Based Prospective Cohort Study in China', *BMC Pregnancy and Childbirth*, 20 (2020), 1–9.

[38] Bernstein and Wiesemann, 'Should Postponing Motherhood via "Social Freezing" Be Legally Banned? An Ethical Analysis', 287.

[39] S. R. Piek, G. Pennings, and V. Provoost, 'Age-Based Restrictions on Reproductive Care: Discerning the Arbitrary from the Necessary', *Theoretical Medicine and Bioethics*, 45/1 (2024), 41–56.

[40] Bernstein and Wiesemann, 'Should Postponing Motherhood via "Social Freezing" Be Legally Banned? An Ethical Analysis'; I. Goold and J. Savulescu, 'In Favour of Freezing Eggs for Non-medical Reasons', *Bioethics*, 23/1 (2009), 47–58.

[41] S. Munné et al., 'Embryo Morphology, Developmental Rates, and Maternal Age Are Correlated with Chromosome Abnormalities', *Fertility and Sterility*, 64/2 (1995), 382–391; L. Elmerdahl Frederiksen et al., 'Maternal Age and the Risk of Fetal Aneuploidy: A Nationwide Cohort Study of More than 500,000 Singleton Pregnancies in Denmark from 2008 to 2017', *Acta Obstetricia et Gynecologica Scandinavica*, 103/2 (2024), 351–359.

with increased maternal and paternal age.[42] This effect can be circumvented by conceiving with donor gametes from young donors. The process of IVG will remain suboptimal compared to young donor gametes in the sense that it will not be able to erase genetic mutations and completely 'reset' aged somatic cells to prevent the effects of aging on the resulting offspring. Also the reconstitution of epigenetic reprogramming has been an important hurdle towards clinical implementation, although recently, significant advances have been reported.[43] Epigenetic reprogramming is important not only to reduce the effect of aging, but also to undo epigenetic changes caused by the manipulations inherent in IVG itself.[44] The question that concerns us here, however, as we are looking for ageist and gender biases in public attitudes, is not so much whether there is a health risk, but rather whether this risk is significantly higher than for other scenarios that were found to be more acceptable. The two most interesting scenarios to compare with are IVG for same-sex reproduction and solo reproduction. An important preliminary note to be made here is that there is not a heightened risk for the psychological well-being of children born to singles or same-sex couples, an often-heard concern that has been convincingly invalidated by empirical research,[45] and which contributes to the discrimination that these children suffer from.[46] However, there are physical safety concerns linked to IVG for solo reproduction and same-sex reproduction that are—particularly in the first scenario but arguably also in the second—more concerning than the risks involved in IVG for post-menopausal women. As outlined above, solo reproduction entails a heightened chance of the child suffering from autosomal recessive disorders. For same-sex reproduction, the danger resides in the derivation of sperm from female somatic cells and oocytes from male somatic cells, which would be necessary in order to reach a complement of one female and one male gamete, but is fraught with difficulties on the level of imprinting. Returning to my argument, the fact that concerns over physical health risks carry less weight in the context of public opinions on IVG for same-sex or solo reproduction than in the context

[42] A. N. Yatsenko and P. J. Turek, 'Reproductive Genetics and the Aging Male', *Journal of Assisted Reproduction and Genetics*, 35 (2018), 933–941; Z. Gao et al., 'Overlooked Roles of DNA Damage and Maternal Age in Generating Human Germline Mutations', *Proceedings of the National Academy of Sciences*, 116/19 (2019), 9491–9500.

[43] Y. Murase et al., 'In Vitro Reconstitution of Epigenetic Reprogramming in the Human Germ Line', *Nature*, 631/8019 (2024), 170–178.

[44] N. El Hajj and T. Haaf, 'Epigenetic Disturbances in In Vitro Cultured Gametes and Embryos: Implications for Human Assisted Reproduction', *Fertility and Sterility*, 99/3 (2013), 632–641.

[45] E.g., S. Golombok and S. Badger, 'Children Raised in Mother-Headed Families from Infancy: A Follow-Up of Children of Lesbian and Single Heterosexual Mothers, at Early Adulthood', *Human Reproduction*, 25/1 (2010), 150–157.

[46] G. Pennings, 'Evaluating the Welfare of the Child in Same-Sex Families', *Human Reproduction*, 26/7 (2011), 1609–1615; F. Oberklaid et al., 'The Kids Are OK: It Is Discrimination, Not Same-Sex Parents, that Harms Children', *The Medical Journal of Australia*, 207/9 (2017), 1–2.e1.

of IVG for post-menopausal women, although the former potential applications are arguably more risky in terms of possible epigenetic and genetic disorders in the offspring, hints at the fact that safety may not have been the driving underlying concern.

This brings us to the second point, which also presents the crux of this chapter. Many of the arguments why IVG for post-menopausal women was labelled as unacceptable also apply to other scenarios, but only come up—or are significantly expanded—in the discussion regarding post-menopausal women. For example, the argument that having children past menopause is unnatural and should therefore not be pursued is quite ironic in a context where the overall opinion about IVG—which is by definition unnatural—was positive. Perhaps most striking was the participants' attitudes towards the desire to have a child. In the general discussion on IVG in the Belgian survey and in the citizens' panel, the desire for parenthood was framed in very positive terms as a widely shared and deeply rooted desire, and there was widespread empathy with people confronted with infertility.[47] However, in the context of making assisted reproduction available after menopause, it was labelled as self-centred and as opposed to the best interest of the resulting children. It must be remarked here that also motherhood in (very) young women is frequently criticized and stigmatized in a similar way.[48] At the same time, and seemingly contradictory, voluntarily childfree women *also* face reproaches of selfishness and stigmatization.[49] In the latter case, the reproach cannot be based on the idea that they act contrary to the well-being of their offspring—as there are none—but is rather explained by backlash theory:[50] when people violate a cultural norm (in this case the norm that people should want to become a parent) their behaviour is likely to be disapproved of, leading to various kinds of penalization.[51] Empirical research by Ashburn-Nardo confirms not only that adults who have chosen not to become parents are viewed as less psychologically fulfilled and more poorly adjusted than parents, but also that they elicited significantly greater moral outrage, in

[47] Mertes et al., 'Enthusiasm, Concern and Ambivalence in the Belgian Public's Attitude towards In-Vitro Gametogenesis'.

[48] See, e.g., B. Hanna, 'Negotiating Motherhood: The Struggles of Teenage Mothers', *Journal of Advanced Nursing*, 34/4 (2001), 456–464; E. Yardley, 'Teenage Mothers' Experiences of Stigma', *Journal of Youth Studies*, 11/6 (2008), 671–684; R. Whitley and L. J. Kirmayer, 'Perceived Stigmatisation of Young Mothers: An Exploratory Study of Psychological and Social Experience', *Social Science & Medicine*, 66/2 (2008), 339–348.

[49] See, e.g., R. Harrington, 'Childfree by Choice', *Studies in Gender and Sexuality*, 20/1 (2019), 22–35; L. Ashburn-Nardo, 'Parenthood as a Moral Imperative? Moral Outrage and the Stigmatization of Voluntarily Childfree Women and Men', *Sex Roles*, 76/5 (2017), 393–401.

[50] L. A. Rudman and K. Fairchild, 'Reactions to Counterstereotypic Behavior: The Role of Backlash in Cultural Stereotype Maintenance', *Journal of Personality and Social Psychology*, 87/2 (2004), 157–168.

[51] C. Verniers, 'Behind the Maternal Wall: The Hidden Backlash toward Childfree Working Women', *Journal of Theoretical Social Psychology*, 4/3 (2020), 107–124.

the form of anger, disgust, and disapproval.[52] Unfortunately, this moral outrage towards voluntarily childfree women and the devaluing rhetoric that comes with it were also illustrated during the 2024 presidential campaign in the United States.[53] Ekelund and Ask, building on Ashburn-Nardo's work, also suggest that childfree people are seen as morally inferior by those who do intend to have children.[54] We appear to be confronted with a paradox here: women are stigmatized when they choose to remain childfree, but they are equally stigmatized when they choose to become mothers too early or too late. The paradox is, however, easily resolved when looking at it through the lens of norm conformity, whereby a descriptive (gender) stereotype (in this case 'women have children between the ages of twenty-five and forty') becomes a prescriptive (gender) stereotype ('women should have children between the ages of twenty-five and forty'). If the prescriptive stereotype is framed in such a way (rather than merely 'women should have children'), then both women within the designated age bracket who do not want to become parents are deviating from the norm and those outside this age bracket who do want to become parents are deviating from the norm. As mentioned earlier by Cutas and Smajdor: 'Women, therefore, have to do anything it takes to become mothers, but not after a certain age'.[55]

Conclusion and Moving Forward

Even when citizens are generally supportive of reproductive technologies, opinions regarding late motherhood in the context of IVG, mitochondrial replacement therapy, and egg freezing indicate that oftentimes there are strong societal norms regarding the appropriate time during one's life trajectory when a desire for parenthood should or should not be present, let alone acted on.

Several evolutions might occur in the future. As the average age of new parents is increasing and as more people remain childless, either by choice or due to circumstances, also norms regarding whether and when people should become parents may adapt and conform to new social realities. Importantly, however, while policymakers should be aware of existing norms and concerns in society, they should also remain vigilant that not all norms and the public

52 Ashburn-Nardo, 'Parenthood as a Moral Imperative? Moral Outrage and the Stigmatization of Voluntarily Childfree Women and Men'.

53 G. Allen, 'Election 2024: The Economic Reality of Vance's "Childless Cat Ladies"', *Forbes*, 4 August 2024, https://www.forbes.com/sites/gemmaallen/2024/08/03/race-2024-vances-childless-cat-ladies-the-economic-realities/.

54 M. Ekelund and K. Ask, 'Stigmatization of Voluntarily Childfree Women and Men in the UK', *Social Psychology* 52/5 (2021), 275–286.

55 D. Cutas and A. Smajdor, 'Postmenopausal Motherhood Reloaded: Advanced Age and In Vitro Derived Gametes', 397.

attitudes following from them are rooted in sound arguments. When deciding who should have access to new reproductive technologies such as IVG and who should not, it is important to look beyond the current societal norms and to critically challenge them, in order to distinguish between valid concerns and prejudices. When late motherhood is concerned, both valid concerns and prejudices will be found whereby the prejudices may explain why some valid concerns are amplified and some reassuring facts are downplayed. The aim of this chapter is to illustrate this phenomenon of how ageist norms, which are oftentimes also gendered, may interfere with an objective balancing of arguments for and against making IVG available to women beyond their natural reproductive lifespan.

Bibliography

Akatsuka, K., Hatta, T., Sawai, T., and Fujita, M., 'Public Attitudes in Japan toward the Reproductive Use of Gametes Derived from Human-Induced Pluripotent Stem Cells', *Future Science OA*, 7/10 (2021), FSO754.

Allen, G., 'Election 2024: The Economic Reality of Vance's "Childless Cat Ladies"', *Forbes*, 4 August 2024, https://www.forbes.com/sites/gemmaallen/2024/08/03/race-2024-vances-childless-cat-ladies-the-economic-realities/.

Ashburn-Nardo, L., 'Parenthood as a Moral Imperative? Moral Outrage and the Stigmatization of Voluntarily Childfree Women and Men', *Sex Roles*, 76/5 (2017), 393–401.

Bernstein, S., and Wiesemann, C., 'Should Postponing Motherhood via "Social Freezing" Be Legally Banned? An Ethical Analysis', *Laws*, 3/2 (2014), 282–300.

Brand, G., Morrison, P., and Down, B., '"You Don't Know Half the Story": Deepening the Dialogue with Young Mothers in Australia', *Journal of Research in Nursing*, 20/5 (2015), 353–369.

Braun, E., 'Mitochondrial Replacement Techniques for Treating Infertility', *Journal of Medical Ethics* (2024), https://doi.org/jme-2023-109660.

Costa-Borges, N., Nikitos, E., Späth, K., Miguel-Escalada, I., Ma, H., Rink, K. et al., 'First Pilot Study of Maternal Spindle Transfer for the Treatment of Repeated In Vitro Fertilization Failures in Couples with Idiopathic Infertility', *Fertility and Sterility*, 119/6 (2023), 964–973.

Cutas, D., and Smajdor, A., '"I Am Your Mother and Your Father!' In Vitro Derived Gametes and the Ethics of Solo Reproduction', *Health Care Analysis*, 25 (2017), 354–369.

Cutas, D., and Smajdor, A., 'Postmenopausal Motherhood Reloaded: Advanced Age and In Vitro Derived Gametes', *Hypatia*, 30/2 (2015), 386–402.

De Clercq, E., Martani, A., Vulliemoz, N., Elger, B. S., and Wangmo, T., 'Rethinking Advanced Motherhood: A New Ethical Narrative', *Medicine, Health Care and Philosophy*, 26/4 (2023), 591–603.

Dondorp, W. J., and de Wert, G. M., 'Fertility Preservation for Healthy Women: Ethical Aspects', *Human Reproduction*, 24/8 (2009), 1779–1785.

Ekelund, M., and Ask, K., 'Stigmatization of Voluntarily Childfree Women and Men in the UK', *Social Psychology* 52/5 (2021), 275–286.

El Hajj, N., and Haaf, T., 'Epigenetic Disturbances in In Vitro Cultured Gametes and Embryos: Implications for Human Assisted Reproduction', *Fertility and Sterility*, 99/3 (2013), 632–641.

Elmerdahl Frederiksen, L., Ølgaard, S. M., Roos, L., Petersen, O. B., Rode, L., Hartwig, T. et al., 'Maternal Age and the Risk of Fetal Aneuploidy: A Nationwide Cohort Study of More than 500,000 Singleton Pregnancies in Denmark from 2008 to 2017', *Acta Obstetricia et Gynecologica Scandinavica*, 103/2 (2024), 351–359.

Gameiro, S., Leone, D., and Mertes, H., 'Fertility Clinics Have a Duty of Care towards Patients Who Do Not Have Children with Treatment', *Human Reproduction*, 39/8 (2024), deae128.

Gao, Z., Moorjani, P., Sasani, T. A., Pedersen, B. S., Quinlan, A. R., Jorde et al., 'Overlooked Roles of DNA Damage and Maternal Age in Generating Human Germline Mutations', *Proceedings of the National Academy of Sciences*, 116/19 (2019), 9491–9500.

Golombok, S., and Badger, S., 'Children Raised in Mother-Headed Families from Infancy: A Follow-Up of Children of Lesbian and Single Heterosexual Mothers, at Early Adulthood', *Human Reproduction*, 25/1 (2010), 150–157.

Goold, I., and Savulescu, J., 'In Favour of Freezing Eggs for Non-medical Reasons', *Bioethics*, 23/1 (2009), 47–58.

Hanna, B., 'Negotiating Motherhood: The Struggles of Teenage Mothers', *Journal of Advanced Nursing*, 34/4 (2001), 456–464.

Harrington, R., 'Childfree By Choice', *Studies in Gender and Sexuality*, 20/1 (2019), 22–35.

Hendriks, S., Dancet, E. A. F., Vliegenthart, R., and Repping, S., 'The Acceptability of Stem Cell-Based Fertility Treatments for Different Indications', *MHR: Basic Science of Reproductive Medicine*, 23/12 (2017), 855–863.

HFEA, *Mitochondrial Donation Treatment*, 2017, https://www.hfea.gov.uk/treatments/embryo-testing-and-treatments-for-disease/mitochondrial-donation-treatment/.

Kang, E., Wu, J., Gutierrez, N. M., Koski, A., Tippner-Hedges, R., Agaronyan, K. et al., 'Mitochondrial Replacement in Human Oocytes Carrying Pathogenic Mitochondrial DNA Mutations', *Nature*, 540/7632 (2016), 270–275.

Labarta, E., de Los Santos, M. J., Escribá, M. J., Pellicer, A., and Herraiz, S., 'Mitochondria as a Tool for Oocyte Rejuvenation', *Fertility and Sterility*, 111/2 (2019), 219–226.

Luke, B., and Brown, M. B., 'Elevated Risks of Pregnancy Complications and Adverse Outcomes with Increasing Maternal Age', *Human Reproduction*, 22/5 (2007), 1264–1272.

Luo, J., Fan, C., Luo, M., Fang, J., Zhou, S., and Zhang, F., 'Pregnancy Complications among Nulliparous and Multiparous Women with Advanced Maternal Age: A Community-Based Prospective Cohort Study in China', *BMC Pregnancy and Childbirth*, 20 (2020), 1–9.

Mayeur, C., Mertes, H., and Van Hoof, W., 'How to Deal with Uninformed and Poorly Informed Opinions of Citizens? A Critical Approach to Online Public Engagement', *Citizen Science: Theory and Practice*, 9/1 (2024), 4.

Mertes, H., 'Ethical Aspects of AGE Banking', in *Preventing Age Related Fertility Loss* (Cham: Springer International Publishing, 2018), 157–168.

Mertes, H., 'The Portrayal of Healthy Women Requesting Oocyte Cryopreservation', *Facts, Views & Vision in ObGyn*, 5/2 (2013), 141.

Mertes, H., 'The Role of Anticipated Decision Regret and the Patient's Best Interest in Sterilisation and Medically Assisted Reproduction', *Journal of Medical Ethics*, 43/5 (2017), 314–318.

Mertes, H., 'Stem Cell Derived Gametes and Uterus Transplants: Hurray for the End of Third Party Reproduction! Or Not?', in *Personalised Medicine, Individual Choice and the Common Good* (Cambridge: Cambridge University Press, 2018), 37–51.

Mertes, H., Goethals, T., Segers, S., Huysentruyt, M., Pennings, G., and Provoost, V., 'Enthusiasm, Concern and Ambivalence in the Belgian Public's Attitude towards In-Vitro Gametogenesis', *Reproductive Biomedicine & Society Online*, 14 (2022), 156–168.

Mertes, H., and Pennings, G., 'Ethical Considerations of Fertility Preservation', in *Female and Male Fertility Preservation* (Cham: Springer International Publishing, 2022), 627–640.

Mertens, M., and Mertes, H., 'Deconstructing Self-fulfilling Outcome Measures in Infertility Treatment', *Bioethics*, 38/7 (2024), 616–623.

Munné, S., Alikani, M., Tomkin, G., Grifo, J., and Cohen, J., 'Embryo Morphology, Developmental Rates, and Maternal Age Are Correlated with Chromosome Abnormalities', *Fertility and Sterility*, 64/2 (1995), 382–391.

Murase, Y., Yokogawa, R., Yabuta, Y., Nagano, M., Katou, Y., Mizuyama, M. et al., 'In Vitro Reconstitution of Epigenetic Reprogramming in the Human Germ Line', *Nature*, 631/8019 (2024), 170–178.

Notini, L., Gyngell, C., and Savulescu, J., 'Drawing the Line on In Vitro Gametogenesis', *Bioethics*, 34/1 (2020), 123–134.

Oberklaid, F., Little, M. H., Patton, G. C., Sawyer, S. M., Skinner, S. R., Telfer, M. M. et al., 'The Kids Are OK: It Is Discrimination, Not Same-Sex Parents, that Harms Children', *The Medical Journal of* Australia, 207/9 (2017), 1–2.e1.

Paksi, V., and Szalma, I., 'Age Norms of Childbearing: Early, Ideal and Late Childbearing in European Countries', *Review of Sociology*, 2 (2009), 57–80.

Pennings, G., 'Enucleated Oocyte Donation: First for Infertility Treatment, then for Mitochondrial Diseases', *Journal of Assisted Reproduction and Genetics*, 39/3 (2022), 605–608.

Pennings, G., 'Evaluating the Welfare of the Child in Same-Sex Families', *Human Reproduction*, 26/7 (2011), 1609–1615.

Pennings, G., Heindryckx, B., Stoop, D., and Mertes, H., 'Attitude of Belgian Women towards Enucleated Egg Donation for Treatment of Mitochondrial Diseases and Infertility', *Reproductive Biomedicine Online*, 49/3 (2024), 104101.

12

Japanese Perspectives on *In Vitro* Gametogenesis

Insights from a Questionnaire Survey

Kyoko Akatsuka, Tsutomu Sawai, Taichi Hatta, and Misao Fujita

Background and Purpose

Since 2010, the Japanese research guidelines 'Guidelines on the Utilization of Human Embryonic (ES) Stem Cells' and the 'Guidelines for Research on the Creation of Germ Cells from Human Induced Pluripotent Stem (iPS) Cells or Human Tissue Stem Cells' have allowed the creation of stem cell-derived gametes (SCDG) through IVG, but have prohibited the creation of embryos using SCDG.[1,2] In 2015, the Expert Panel on Bioethics of the Cabinet Office considered whether to allow the creation of embryos using SCDG, but decided that it was too early to allow the creation of embryos, given that the creation of SCDG had not been successfully achieved. Thus, to date, there has been no relaxation of the regulations.[3] In Japan, although the use of human embryos for research is prohibited, except in very limited cases, the creation of human embryos by the combination of adult-derived sperm and eggs is permitted for research purposes under certain conditions.[4] Furthermore, from a scientific perspective, it is important to confirm whether SCDG is functionally equivalent to actual gametes, regardless of whether IVG will be used for reproductive

[1] Ministry of Education, Culture, Sports, Science, and Technology (MEXT), 'Guidelines on the Utilization of Human Embryonic Stem Cells' (in Japanese), 2010, https://warp.ndl.go.jp/info:ndljp/pid/1255146/www.lifescience.mext.go.jp/files/pdf/n592_S01.pdf.

[2] Ministry of Education Culture, Sports, Science, and Technology (MEXT), 'Guidelines on the Research on Producing Germ Cells from Human Induced Pluripotent Stem Cells or Human Tissue Stem Cells', 2010, https://www.mext.go.jp/lifescience/bioethics/files/pdf/n1567_02r2.pdf.

[3] Cabinet Office, 'The Interim Report of Issues Regarding Fertilization of Human Germ Cells from Pluripotent Stem Cells' (in Japanese), 2015, https://www8.cao.go.jp/cstp/tyousakai/life/chukanmatome_150909.pdf.

[4] Council for Science, Technology, and Policy (CSTP), 'Basic Ideas on the Handling of Human Embryos' (in Japanese), 2004, https://www8.cao.go.jp/cstp/tyousakai/life/haihu39/siryo5-1-1.pdf.

Kyoko Akatsuka, Tsutomu Sawai, Taichi Hatta, and Misao Fujita, *Japanese Perspectives on* In Vitro *Gametogenesis*
In: *Revolutionizing Reproduction.* Edited by: Christopher Gyngell, Julian Savulescu, and Tsutomu Sawai,
Oxford University Press. © Oxford University Press 2026. DOI: 10.1093/9780197810415.003.0012

purposes in the future.[5,6] Thus, the creation of embryos using SCDG is not permanently prohibited, and regulations may be relaxed in the future as IVG research progresses.

However, if it is possible to produce a large number of SCDGs through IVG, it would also theoretically be possible to produce a large number of embryos, which raises ethical issues, such as the mass production of embryos intended solely for research purposes and their mass disposal.[7,8] This type of human embryo handling would not be acceptable in countries and regions that recognize the moral value of embryos. In addition, if it becomes possible to use SCDG to create something functionally equivalent to actual embryos, there may be moves to use these embryos not only for basic research but also for clinical applications.

Because IVG can have a significant impact on society and individuals, depending on how they are used, it is essential to discuss in society (including stakeholders and experts) which technology usage should be allowed.[9,10] In 2017, we identified a few studies that had conducted surveys on the attitudes of the public and experts towards the use of IVG,[11,12] but no survey was conducted on the Japanese public. Thus, there was a lack of data on how Japanese society regards IVG research. To prepare for future discussions, we conducted a questionnaire survey in May 2017 and collected responses from 3,096 members of the public in Japan to clarify public attitudes towards IVG. From this survey, we published two papers in 2021 on *Future Science OA*.[13,14] Below, we summarize the main results of the survey published in these papers.

[5] The Hinxton Group (THG), 'Consensus Statement: Science, Ethics, and Policy Challenges of Pluripotent Stem Cell-Derived Gametes', 2008, http://www.hinxtongroup.org/Consensus_HG08_FINAL.pdf.

[6] Tetsuya Ishii, Renee A.Reijo Pera, and Henry T. Greely, 'Ethical and Legal Issues Arising in Research on Inducing Human Germ Cells from Pluripotent Stem Cells', *Cell Stem Cell*, 13/2 (2013), 145–148.

[7] Annelien L. Bredenoord and Insoo Hyun, 'Ethics of Stem Cell-Derived Gametes Made in a Dish: Fertility for Everyone?', *EMBO Molecular Medicine*, 9/4 (2017), 396–398.

[8] I. Glenn Cohen, George Q. Daley, and Eli Y. Adashi, 'Disruptive Reproductive Technologies', *Science Translational Medicine*, 372/9 (2017), 2959.

[9] Eli Y. Adashi et al., 'Stem Cell-Derived Human Gametes: The Public Engagement Imperative', *Trends in Molecular Medicine*, 25/3 (2019), 165–167.

[10] Amander T. Clark et al., 'Human Embryo Research, Stem Cell-Derived Embryo Models and in Vitro Gametogenesis: Considerations Leading to the Revised ISSCR Guidelines', *Stem Cell Reports*, 16/6 (2021), 1416–1424.

[11] Saskia Hendriks et al., 'Couples with Non-obstructive Azoospermia Are Interested in Future Treatments with Artificial Gametes', *Human Reproduction*, 31/8 (2016), 1738–1748.

[12] Saskia Hendriks et al., 'The Acceptability of Stem Cell-Based Fertility Treatments for Different Indications', *Molecular Human Reproduction*, 23/12 (2017), 855–863.

[13] Kyoko Akatsuka et al., 'Public Attitudes in Japan toward the Reproductive Use of Gametes Derived from Human-Induced Pluripotent Stem Cells', *Future Science OA*, 7/10 (2021), FSO754.

[14] Tsutomu Sawai et al., 'Public Attitudes in Japan toward the Creation and Use of Gametes Derived from Human-Induced Pluripotent Stem Cells', *Future Science OA*, 7/10 (2021), FSO755.

In our survey, to explore the extent to which the public would accept the creation and use of induced pluripotent stem cell-derived gametes (iPSCDG) via IVG (hereafter, the degree of acceptance by stage), we divided the process from the creation of iPSCDG to reproductive use into three stages. We asked the respondents up to which stage they would accept. Based on this delineation, 26.8% of the respondents accepted up to the first stage (the creation of iPSCDG through IVG and its use for research), 25.8% accepted up to the second stage (the creation of embryos using iPSCDG and its use in research), and 25.9% accepted up to the third stage (the reproductive use of embryos created using iPSCDG). Meanwhile, 21.4% of the respondents did not accept any of the stages, including the first stage. For this question, it was assumed that respondents who accepted up to the second stage accepted the first stage, while those who accepted up to the third stage accepted both the first and second stages. It was found that 78.6% of all respondents accepted the creation of iPSCDG, and 51.7% accepted the creation of embryos using iPSCDG and its research use.[15]

Furthermore, we asked who should be allowed to use IVG for reproduction, considering the possibility that IVG may be used for reproduction in the future. To this question, 70.8% of all respondents favoured the use of IVG by couples who were not expected to have children through infertility treatment, and 63.8% favoured the use of IVG by couples who were undergoing infertility treatment. Meanwhile, 49.2% of respondents opposed the reproductive use of IVG by same-sex couples who wanted to have children, and 62.7% opposed its use by single individuals; fewer than 30% of respondents supported such uses of IVG. Furthermore, the results for the use of IVG by post-menopausal women and their husbands were split (40.5% in favour of and 40.6% against).[16] The results suggest that the public is more accepting of IVG use in harmony with the traditional family and reproductive system and current medical treatment goals, rather than for uses inconsistent with them.

As IVG technology develops, the advantages and disadvantages of creating embryos using iPSCDG and using them for research will become important points of discussion in Japan's policy debate. What values underlie the respondents' opposition to creating embryos using iPSCDG and using them for research? In this study, we considered what the public thinks about the creation of embryos using iPSCDG and its research use, based on unpublished data from the 2017 questionnaire survey.

[15] Sawai et al., 'Public Attitudes in Japan toward the Creation and Use of Gametes Derived from Human-Induced Pluripotent Stem Cells'.

[16] Akatsuka et al., 'Public Attitudes in Japan toward the Reproductive Use of Gametes Derived from Human-Induced Pluripotent Stem Cells'.

Methods

Participants

In May 2017, we conducted an online survey of the Japanese public. We created a questionnaire in advance and commissioned a private research company to collect demographic data from the Japanese population, which were sex- and age-matched. The survey company sent an email to all its members aged between twenty and seventy-nine years, and the email contained a URL to a webpage where the individuals could provide informed consent for this survey. This webpage clearly stated the purpose of the survey, the estimated time required to complete it (ten to fifteen minutes in total), the rewards for participating (exchangeable points issued by the company), the policy on handling personal information and anonymized responses, and information about the institutions conducting the survey. At the bottom of the webpage, there were 'Next' and 'Cancel' buttons. By clicking on the 'Next' button, participants were considered to have agreed to take part in the survey, whereas if they clicked on 'Cancel', they were deemed not to have agreed to participate in the study. When the number of participants exceeded 3,000, the company closed the URL and ended participant recruitment and response collection. For details on the sampling strategy, refer to the published papers.[17,18]

Question Items

In this survey, we set up a question to determine the extent to which the public favours the research use of surplus embryos that were created for infertility treatment and are no longer necessary, embryos created solely for research purposes, and embryos created using iPSCDG, and whether the degree of acceptance varies depending on the type of embryo (hereafter, the acceptability of different types of embryos) (Figure 12.1). As it was assumed that many respondents would not be familiar with such topics as stem cell research involving iPS cells and IVG, we presented respondents with five options—'strongly disagree', 'disagree', 'strongly agree', 'agree', and 'do not know'—and asked them to choose one option for each type of embryo. Of the five types of embryos, (1) are embryos created for infertility treatment, and (2) to (5) are newly created embryos for research purposes. Embryos (3) to (5) were created using iPSCDG.

[17] Akatsuka et al., 'Public Attitudes in Japan toward the Reproductive Use of Gametes Derived from Human-Induced Pluripotent Stem Cells'.

[18] Sawai et al., 'Public Attitudes in Japan toward the Creation and Use of Gametes Derived from Human-Induced Pluripotent Stem Cells'.

Question: So far, embryos used for research have been surplus embryos that are no longer used in infertility treatment (these embryos are frozen). Do you agree with the following five ways of creating and using human embryos for scientifically significant research?

① Using human embryos that were created for infertility treatment but are no longer used for that purpose in research.

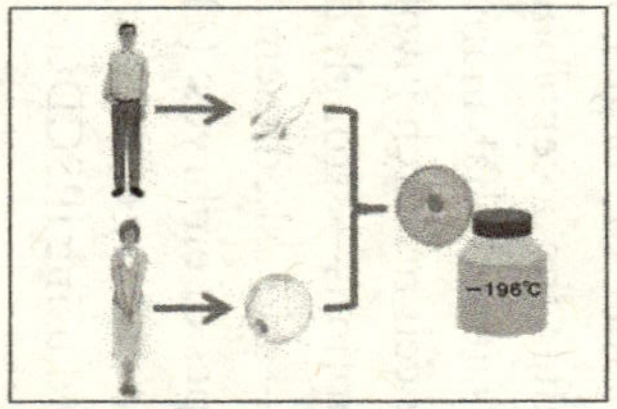

Answer:
- ◯ I do not agree at all
- ◯ I do not agree
- ◯ I agree
- ◯ I strongly agree
- ◯ I don't know

② Creating human embryos from sperm and eggs donated by adults and using them for research.

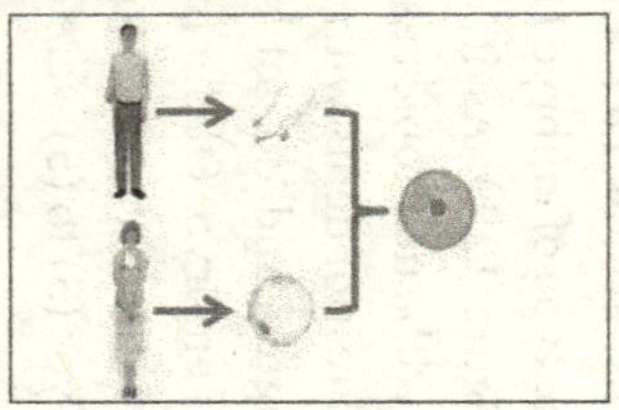

Answer:
- ◯ I do not agree at all
- ◯ I do not agree
- ◯ I agree
- ◯ I strongly agree
- ◯ I don't know

③ Creating embryos from sperm donated by adults and eggs derived from human iPS cells, and using them in research.

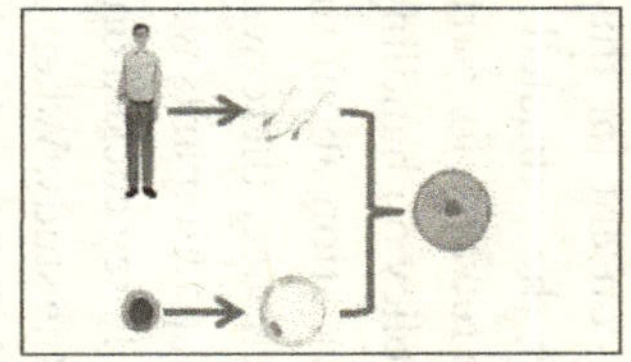

Answer:
- ◯ I do not agree at all
- ◯ I do not agree
- ◯ I agree
- ◯ I strongly agree
- ◯ I don't know

④ Creating embryos from sperm derived from human iPS cells and eggs donated by adults, and using them in research.

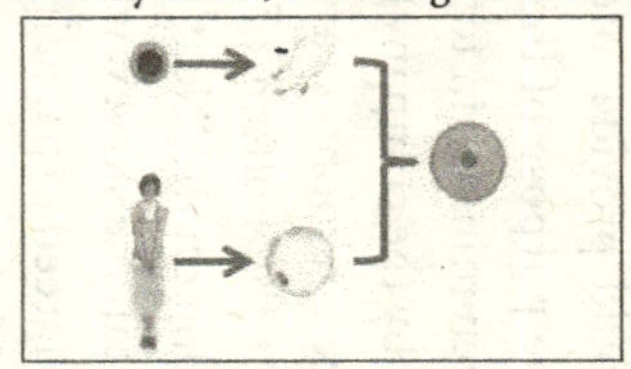

Answer:
- ◯ I do not agree at all
- ◯ I do not agree
- ◯ I agree
- ◯ I strongly agree
- ◯ I don't know

⑤ Creating embryos from human iPS cell-derived sperm and eggs for research.

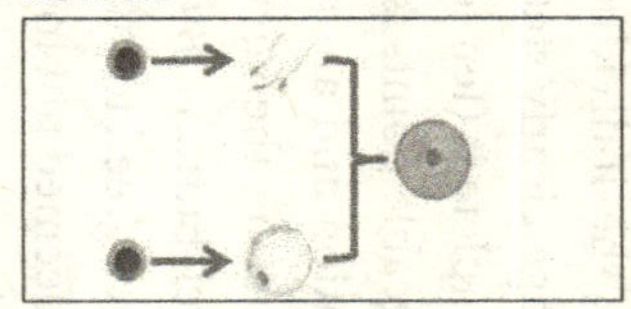

Answer:
- ◯ I do not agree at all
- ◯ I do not agree
- ◯ I agree
- ◯ I strongly agree
- ◯ I don't know

Figure 12.1 Question on the acceptability of different types of embryos.

(1) Embryos were created for infertility treatment and were not used for reproduction (surplus embryos).
(2) Embryos were created using sperm and eggs derived from adults (research embryos).
(3) Embryos were created using sperm derived from adults and iPSCD-eggs.
(4) Embryos were created using iPSCD-sperm and eggs derived from adults.
(5) Embryos were created using iPSCD sperm and eggs (iPSCDG embryos).

We presented respondents with twelve items regarding their expectations for the creation and use of iPSCDG and asked them to respond using a Likert scale (1 = not expected at all, 2 = not expected, 3 = expected, 4 = strongly expected). Similarly, sixteen items were presented to respondents regarding concerns about the creation and use of iPSCDG, and respondents were asked to respond using a Likert scale (1 = not concerned at all, 2 = not concerned, 3 = concerned, 4 = strongly concerned). For details on these items, see Tables 12.4 and 12.5 or refer to Sawai et al.[19]

Statistical Analysis

We conducted an exploratory description to clarify the public's attitudes towards IVG. After illustrating the distribution of responses for the five acceptance levels for the five types of embryos, we converted these levels into three to examine the degree of consistency of the respondents' acceptance levels ('strongly disagree' and 'disagree' were combined into 'disagree', 'strongly agree' and 'agree' were combined into 'agree', and 'do not know' was left as it was). The kappa coefficient was calculated to indicate the consistency of the five embryo acceptance levels after conversion to three levels.

In addition, we conducted a binary logistic regression analysis to explore the expectations and concerns of respondents who consistently opposed the creation of embryos using iPSCDG under conditions (3) to (5). The response variable was the acceptance of (3) to (5), with the group that consistently opposed (selecting 'strongly disagree' or 'disagree' for all three items) given a value of 1, and the other groups (including at least one 'strongly agree', 'agree', or 'do not know' for all three items) given a value of 0. The twelve items of expectations regarding the use of iPSCDG were used as explanatory variables and were converted to a binary scale (with 'not expected at all' and 'not expected' converted to 0, and 'expected' and 'strongly expected' converted to 1). The sixteen items of

[19] Sawai et al., 'Public Attitudes in Japan toward the Creation and Use of Gametes Derived from Human-Induced Pluripotent Stem Cells'.

concern regarding the use of iPSCDG were converted in the same manner and used as explanatory variables. In the multivariate logistic regression analysis, the explanatory variables were entered using the forced entry method, and to avoid multicollinearity, we confirmed that the variance inflation factor for each explanatory variable was less than 10.[20]

Ethical Considerations

This survey was approved by the Joint Medical Ethics Committee of the Institute for Frontier Life and Medical Sciences, and the Centre for Cell Research and Application, Kyoto University (approval number: 81; date of approval: 11 May 2017).

Results

Demographic Characteristics of the Respondents

After conducting the online survey, we collected responses from 3,096 individuals. The demographic characteristics of the respondents are shown in Table 12.1, which was extracted from papers we published in 2021.[21,22]

Acceptability of Different Types of Embryos

Regarding the use of surplus embryos (1) for research, 60.5% of respondents selected 'strongly agree' or 'agree', whereas 17.6% selected 'strongly disagree' or 'disagree'. Of the respondents, 21.8% selected 'do not know'. Meanwhile, for the four types of embryos other than surplus embryos, 46.3% to 50.9% of respondents selected 'strongly agree' or 'agree', 25.8% to 28.9% selected 'strongly disagree' or 'disagree', and 23.3% to 24.9% selected 'do not know' (Figure 12.2).

According to the results of the analysis of the response agreement rate for each type of embryo, the responses for (3) embryos created using sperm derived from adults and iPSCD-eggs, and (4) embryos created using iPSCD-sperm

[20] Robert M. O'Brien, 'A Caution Regarding Rules of Thumb for Variance Inflation Factors', *Quality and Quantity*, 41/5 (2007), 673–690.

[21] Akatsuka et al., 'Public Attitudes in Japan toward the Reproductive Use of Gametes Derived from Human-Induced Pluripotent Stem Cells'.

[22] Sawai et al., 'Public Attitudes in Japan toward the Creation and Use of Gametes Derived from Human-Induced Pluripotent Stem Cells'.

Table 12.1 Demographic characteristics of the respondents (n = 3,096).

	n	%
Sex		
Male	1533	49.5
Female	1563	50.5
Age (years)		
20–29	419	13.5
30–39	525	17
40–49	593	19.2
50–59	507	16.4
60–69	597	19.3
70–79	455	14.7
Marital status		
Married	2066	66.7
Unmarried	1030	33.3
Presence or absence of children		
Yes	1871	60.4
No	1225	39.6
Experience of infertility treatment		
No	2772	89.5
Yes	261	8.4
I do not want to answer	63	2
Educational background		
Elementary school	4	0.1
Junior high school	75	2.4
High school	1041	33.6
Technical college	351	11.3
Two-year college	327	10.6
Four-year college	1159	37.4
Postgraduate studies (master's degree)	105	3.4
Postgraduate studies (doctorate)	23	0.7
Other	11	0.4

(continued)

Table 12.1 Continued

	n	%
Personal income (yen/year)		
Less than two million	1175	38
Two to four million	698	22.5
Four to six million	410	13.2
Six to eight million	185	6
Eight to ten million	66	2.1
Ten to twelve million	33	1.1
Twelve to fifty million	7	0.2
Fifty to twenty million	6	0.2
Over twenty million	7	0.2
I do not know	195	6.3
I do not want to answer	314	10.1
Religious affiliation		
No, I do not have a religious affiliation	2646	85.5
Yes, I have a religious affiliation	313	10.1
I do not want to answer	137	4.4
How much did you know about iPSCs?		
I knew enough about them to be able to provide an explanation to a certain extent	607	19.6
I had only heard a bit about them	2267	73.2
I was not aware of them	222	7.2
Are you interested in iPSCs?		
Yes, I am interested	2220	71.7
No, I am not interested	876	28.3
How much did you know about the 'research for creating sperm/ova from hiPSCs?'		
I knew enough about it to be able to provide an explanation to a certain extent	212	6.8
I had only heard a bit about it	1322	42.7
I was not aware of it	1562	50.5

Table 12.1 Continued

	n	%
How much did you understand the explanation on 'the creation and use of hiPSC-derived sperm/ova'?		
I fully understood it	128	4.1
I mostly understood it	2101	67.9
I could not understand it very well	743	24
I could not understand it at all	124	4

hiPSC, human-induced pluripotent stem cell; iPSC, induced pluripotent stem cell.

and eggs derived from adults, were almost perfect (κ =.861). In addition, the responses to (3) and (4) were substantially consistent with the responses to (5) iPSCDG embryos created using iPSCD-sperm and iPSCD-eggs (κ = .736;.780) (Table 12.2).

(1) Surplus embryos and (5) iPSCDG embryos had the lowest agreement rates (κ =.525). To show divergence, Table 12.3 presents a cross-tabulation stratified by these three levels of agreement. Of the respondents, 12.5% opposed the use of both surplus embryos and iPSCDG embryos for research, whereas 42.3% favoured their use. In addition, 12.8% of the respondents favoured the use of (1) surplus embryos but opposed the creation and research use of (5) iPSCDG embryos. Of the respondents, 4.6% opposed the use of (1) surplus embryos but favoured the creation and research use of (5) iPSCDG embryos. Meanwhile, 18.7% of respondents chose 'I do not know' for the use of both types of embryos in research.

(Figures in parentheses are percentages)

Correlation with Expectations and Concerns Related to IVG

A total of 662 respondents (21.4%) consistently opposed the three types of embryo creation using iPSCDG. We explored the relationship between the acceptability of the three types of embryo creation and expectations regarding the use of IVG by conducting a binary logistic regression analysis (Table 12.4). The results of the univariate regression analysis showed that all expectations were statistically significant and had an odds ratio of <1, indicating that respondents who consistently opposed the creation of embryos using iPSCDG tended not

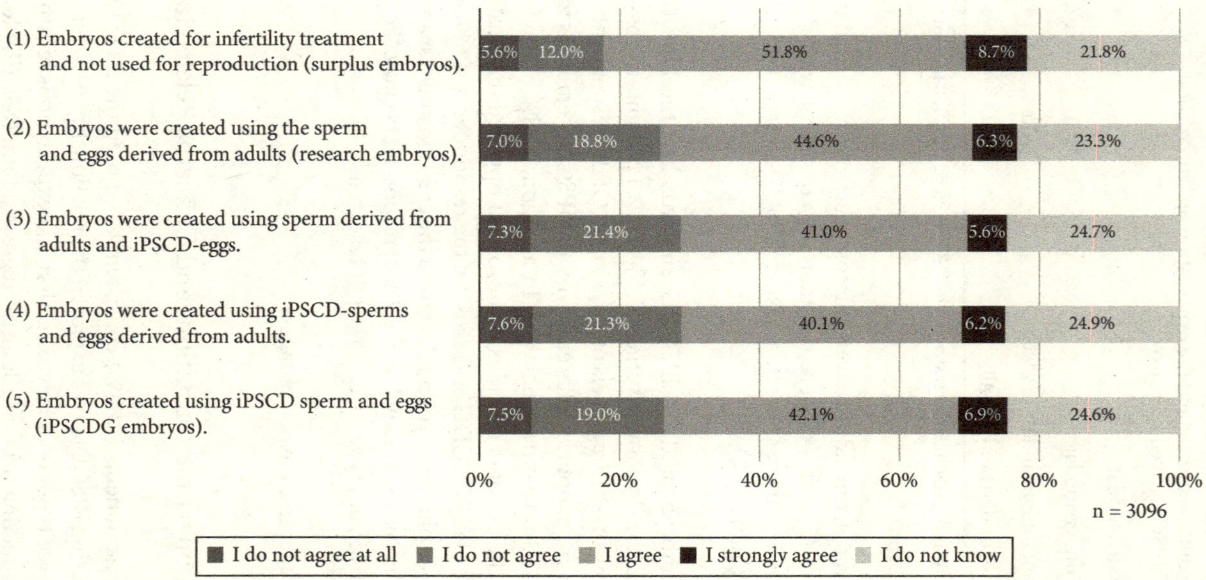

Figure 12.2 Acceptability of different types of embryos.

Table 12.2 Concordance rate of responses for the acceptability of different types of embryos.

	1. Surplus embryos	2. Sperm & eggs	3. Sperm & iPSCD-eggs	4. iPSCD-sperm & eggs
1. Surplus embryos				
2. Sperm & eggs	0.650			
3. Sperm & iPSCD-eggs	0.573	0.754		
4. iPSCD-sperm & eggs	0.556	0.716	0.861	
5. iPSCD-sperm& iPSCD-eggs	0.525	0.619	0.736	0.780

iPSCD-eggs, induced pluripotent stem cell-derived eggs; iPSCD-sperm, induced pluripotent stem cell-derived sperm.

Table 12.3 Distribution of responses for surplus embryos and iPSCDG embryos.

		5. iPSCDG embryos		
		Disagree	**Agree**	**Do not know**
1. Surplus embryos	Disagree	386 (12.5%)	143 (4.6%)	18 (0.6%)
	Agree	397 (12.8%)	1311 (42.3%)	165 (5.3%)
	Do not know	35 (1.1%)	63 (2.0%)	578 (18.7%)

iPSCDG embryos, embryos created using induced pluripotent stem cell-derived sperm and eggs.

to have expectations regarding the use of IVG. The results of multivariate regression analysis showed that there was statistical significance for the following items: 'Anyone who wants a child can have a genetically related child' (OR = 0.74, 95% CI: 0.57–0.94), 'By using a large number of sperm and eggs with certain characteristics, it is possible to reduce genetic diseases in society' (OR = 0.75, 95% CI: 0.59–0.94), and 'It is possible to make designer babies' (OR = 0.55, 95% CI: 0.42–0.72). An odds ratio of less than 1 indicates that respondents who do not have such expectations tend to oppose the creation of embryos using iPSCDG.

Similarly, we explored the relationship between the acceptability of creating and using embryos involving iPSCDG and concerns about using IVG by conducting a binary logistic regression analysis (Table 12.5). The results of the univariate regression analysis showed that all concerns were statistically significant and exceeded an odds ratio of 1, indicating that respondents who consistently opposed the creation of embryos using iPSCDG tended to be concerned

Table 12.4 Expectations of respondents who do not accept the creation of the three types of embryos using iPSCDG and their research use.

Expectations	Univariate binomial logistic regression analysis								Multivariate binomial logistic regression analysis[a]							
	B	*SE B*	*Wald*	*OR*	*95% CI*		*df*	*P*	*B*	*SE B*	*Wald*	*OR*	*95% CI*		*df*	*P*
					LL	*UL*							*LL*	*UL*		
Elucidation of the mechanism of embryonic development	−0.82	0.10	63.64	0.44	0.36	0.54	1	0.00	−0.15	0.13	1.29	0.86	0.66	1.12	1	0.26
Elucidation of the aetiology of infertility	−0.95	0.11	75.18	0.39	0.31	0.48	1	0.00	0.17	0.20	0.73	1.19	0.80	1.76	1	0.39
Development of therapeutic modalities and drugs for infertility	−1.05	0.11	92.47	0.35	0.28	0.43	1	0.00	−0.32	0.20	2.68	0.73	0.50	1.07	1	0.10
Elucidation of the aetiology of congenital diseases	−0.90	0.13	51.62	0.41	0.32	0.52	1	0.00	−0.13	0.22	0.36	0.88	0.57	1.35	1	0.55
Development of therapeutic modalities and drugs for congenital diseases	−0.88	0.13	50.00	0.41	0.32	0.53	1	0.00	0.17	0.22	0.58	1.18	0.77	1.83	1	0.45
Availability of gametes for research	−1.03	0.09	132.75	0.36	0.30	0.42	1	0.00	0.21	0.22	0.88	1.23	0.80	1.89	1	0.35
Availability of embryos for research	−1.11	0.09	152.16	0.33	0.28	0.39	1	0.00	−0.37	0.22	2.74	0.69	0.45	1.07	1	0.10
Birth of genetically related children	−1.18	0.09	170.58	0.31	0.26	0.37	1	0.00	−0.31	0.13	5.95	0.74	0.57	0.94	1	0.02
Availability of gametes for reproduction	−1.32	0.09	208.69	0.27	0.22	0.32	1	0.00	−0.34	0.22	2.35	0.71	0.47	1.10	1	0.13
Availability of embryos for reproduction	−1.32	0.09	207.62	0.27	0.22	0.32	1	0.00	−0.33	0.22	2.32	0.72	0.47	1.10	1	0.13
Reduction in genetic diseases	−1.09	0.09	143.17	0.34	0.28	0.40	1	0.00	−0.29	0.12	6.32	0.75	0.59	0.94	1	0.01
Designer babies	−1.16	0.12	86.93	0.31	0.25	0.40	1	0.00	−0.59	0.14	18.86	0.55	0.42	0.72	1	0.00

B, partial regression coefficient; OR, odds ratio; CI, confidence interval; LL, lower limit; UL, upper limit. a) The coefficients of determination, Cox–Snell $R^2 = 0.09$, Nagelkerke $R^2 = 0.14$.

Table 12.5 Concerns of respondents who do not accept the creation of the three types of embryos using iPSCDG and their research use.

Concerns	Univariate binomial logistic regression analysis								Multivariate binomial logistic regression analysis[a]							
	B	*SE B*	*Wald*	*OR*	*95% CI*		*df*	*P*	*B*	*SE B*	*Wald*	*OR*	*95% CI*		*df*	*P*
					LL	*UL*							*LL*	*UL*		
Unnaturalness	1.30	0.13	103.23	3.67	2.86	4.72	1	0.00	0.30	0.17	3.09	1.34	0.97	1.87	1	0.08
Playing God	1.11	0.10	129.17	3.03	2.50	3.66	1	0.00	0.25	0.13	3.90	1.29	1.00	1.66	1	0.05
Infringement of human dignity	1.38	0.10	172.99	3.97	3.23	4.88	1	0.00	0.68	0.15	20.28	1.96	1.46	2.63	1	0.00
Loss of values	1.16	0.10	126.73	3.18	2.60	3.89	1	0.00	0.29	0.14	4.21	1.33	1.01	1.75	1	0.04
Social turmoil	0.71	0.10	52.06	2.03	1.67	2.45	1	0.00	−0.17	0.12	1.96	0.84	0.66	1.07	1	0.16
Commercialization of reproductive medicine	1.16	0.13	79.06	3.19	2.47	4.12	1	0.00	0.41	0.18	5.19	1.50	1.06	2.14	1	0.02
Loss of genetic diversity	1.04	0.13	68.04	2.83	2.21	3.63	1	0.00	−0.01	0.18	0.01	0.99	0.70	1.40	1	0.94
Unknown risk to children	0.88	0.14	41.73	2.40	1.84	3.13	1	0.00	−0.08	0.19	0.18	0.92	0.64	1.33	1	0.67
Infringement of identity	0.95	0.12	67.89	2.59	2.07	3.25	1	0.00	0.14	0.16	0.72	1.15	0.84	1.57	1	0.40
Unknown risk to future generations	0.96	0.13	58.55	2.61	2.04	3.33	1	0.00	−0.06	0.18	0.10	0.95	0.67	1.34	1	0.75
Use of gametes as tools	1.09	0.13	69.28	2.99	2.31	3.86	1	0.00	0.21	0.24	0.74	1.23	0.77	1.98	1	0.39
Use of embryos as tools	1.05	0.13	66.22	2.87	2.23	3.70	1	0.00	−0.25	0.25	1.06	0.78	0.48	1.26	1	0.30
Promotion of embryo destruction	1.08	0.11	99.37	2.95	2.38	3.65	1	0.00	0.38	0.15	6.54	1.46	1.09	1.95	1	0.01
Pressure on parents	0.74	0.09	63.12	2.10	1.75	2.53	1	0.00	0.15	0.12	1.62	1.16	0.92	1.47	1	0.20
Limited access to reproduction	0.49	0.10	23.68	1.63	1.34	1.99	1	0.00	−0.24	0.13	3.56	0.79	0.61	1.01	1	0.06
Designer babies	0.88	0.13	47.59	2.41	1.88	3.09	1	0.00	0.17	0.16	1.18	1.19	0.87	1.62	1	0.28

B, partial regression coefficient; OR, odds ratio; CI, confidence interval; LL, lower limit; UL, upper limit. a) The coefficients of determination, Cox–Snell $R^2 = 0.08$, Nagelkerke $R^2 = 0.13$.

about the use of IVG. The results of multivariate regression analysis showed that there was a significant association between the items 'Human dignity is undermined' (OR = 1.96, 95% CI: 1.46–2.63), 'Traditional values regarding family and reproduction are undermined' (OR = 1.33, 95% CI: 1.01–1.75)), 'It promotes the destruction of fertilized eggs (embryos)' (OR = 1.46, 95% CI: 1.09–1.95), and 'It accelerates the commercialization of assisted reproductive technology through the production of sperm and eggs' (OR = 1.50, 95% CI: 1.06–2.14). As these items had an odds ratio greater than 1, they indicated that respondents with these concerns tended to oppose the creation of embryos using iPSCDG.

Discussion

The Attitudes of the Public towards Human Embryos Do Not Change Depending on the Origin of the Gametes

According to the results on the acceptability of different types of embryos, 46.3% to 60.5% of respondents favoured using human embryos for crucial scientific research, although there were only minor differences depending on the origin of the embryos. Excluding those who answered, 'I do not know', 17.6% to 28.9% of the respondents explicitly opposed the use of human embryos for research (Figure 12.2). These results suggest that the Japanese public does not strongly oppose the use of human embryos in research.

Mertes et al.[23] asked the Belgian public about their attitudes towards the use of IVG in a survey conducted in 2018. It was found that 55.8% of the respondents supported the use of surplus embryos to investigate the potential risks of IVG, 23.8% opposed it, and 20.3% were neutral. This distribution of responses is similar to that in our survey. Meanwhile, 37.1% to 37.5% of respondents opposed the creation and use of research embryos to investigate the potential risks of IVG and induce gametes, while 38.1% to 41.2% favoured it.[24] Therefore, there is a common tendency for the Japanese people and Belgians to support the research use of surplus embryos over research embryos, but the Japanese public is more accepting of the creation and use of research embryos than the Belgian public.

In our survey, there was a slight tendency to support the use of surplus embryos over the use of other types of embryos; however, as the rate of agreement in the

[23] Heidi Mertes et al., 'Enthusiasm, Concern, and Ambivalence in the Belgian Public's Attitude towards In-Vitro Gametogenesis', *Reproductive Biomedicine and Society Online*, 14 (2022), 156–168.

[24] Mertes et al., 'Enthusiasm, Concern, and Ambivalence in the Belgian Public's Attitude towards In-Vitro Gametogenesis'.

responses shows, there was no significant difference in the degree of acceptance of the four types of embryos other than surplus embryos (Table 12.2). In addition, given that the response concordance rate was high for (3) and (4) embryos, it can be said that the difference in the gametes used for embryo creation did not affect the degree of acceptance. Under the current Japanese guidelines, eggs used for the creation of research embryos are defined as eggs collected for *in vitro* fertilization but are no longer required or eggs collected from ovaries that were removed during disease treatment and are not intended for reproductive purposes. Therefore, eggs are not typically obtained from women solely to create research embryos. Nevertheless, to avoid forcing women to donate eggs for research purposes, which imposes a physical and mental burden, it is necessary to obtain consent for donation in an environment that is more considerate than that for sperm donation.[25] Our respondents may not have recognized the burdens associated with egg donation.

Furthermore, given that the responses to (3) and (4) were substantially consistent with the responses to (5) (Table 12.2), it seems that the respondents' acceptance of the embryos was independent of whether the gametes used to create them were artificially created. In other words, the Japanese public accepts embryos from iPSCDG as they do embryos from gametes taken from adults.

Approximately 80% of the General Public Does Not Oppose the Creation of Embryos Using iPSCDG and Their Research Use

According to our previous paper on the degree of acceptance by stage, approximately half of the respondents (48.2%) did not allow the creation of embryos using iPSCDG.[26] Currently, Japanese guidelines prohibit the creation of embryos using SCDG,[27,28] which is the point that divides the public.

However, based on these results, it is too early to conclude that approximately half of the respondents did not accept the creation of embryos using iPSCDG. According to the results of the responses regarding the acceptability

[25] Ministry of Education Culture, Sports, Science, and Technology (MEXT), Ministry of Health Labour and Welfare (MHLW), and Children and Family Agency (CFA), 'Ethical Guidelines of the Production and Use of Human Fertilized Embryos for the Research' (in Japanese), 2024, https://www.mext.go.jp/lifescience/bioethics/files/pdf/n2409_01.pdf.

[26] Sawai et al., 'Public Attitudes in Japan toward the Creation and Use of Gametes Derived from Human-Induced Pluripotent Stem Cells'.

[27] Ministry of Education, Culture, Sports, Science, and Technology (MEXT), 'Guidelines on the Utilization of Human Embryonic Stem Cells' (in Japanese), 2010, https://warp.ndl.go.jp/info:ndljp/pid/1255146/www.lifescience.mext.go.jp/files/pdf/n592_S01.pdf.

[28] Ministry of Education Culture, Sports, Science, and Technology (MEXT), 'Guidelines on the Research on Producing Germ Cells from Human Induced Pluripotent Stem Cells or Human Tissue Stem Cells', 2010, https://www.mext.go.jp/lifescience/bioethics/files/pdf/n1567_02r2.pdf.

of different types of embryos (Figure 12.2), the number of respondents who selected 'strongly disagree' or 'disagree' with the creation and research use of embryos using iPSCDG was 26.5% to 28.9%, which is lower than the number of respondents (48.2%) who did not accept the creation of embryos using iPSCDG.[29] Although the respondents did not always answer questions with a consistent attitude, among those who did not accept the creation and research use of embryos using iPSCDG, some people may have answered 'I do not know' to the question on the acceptability of different types of embryos and suspended their judgment on the issue.

According to the results of the binary logistic regression analysis, 21.4% of respondents who consistently opposed the creation and research use of the three types of embryos using iPSCDG tended to have broad concerns about the use of IVG and no expectations (Tables 12.4 and 12.5). In addition, this respondent group was particularly concerned about the use of IVG in terms of the following: the violation of human dignity, the collapse of traditional values regarding family and reproduction, the promotion of the destruction of embryos, and the promotion of the commercialization of reproductive medicine. Most of these concerns relate to the reproductive use of IVG, but strictly speaking, it could be said that this group is strongly opposed to the transformation of their existing worldview, which is based on their self-evident views of humanity, life, and the nature of reproduction and the family. This indicates that they may consider the creation of iPSCDG unacceptable, even without the creation of embryos or their use in research. Their consistent opposition to the creation of embryos using iPSCDG and their use in research may be a sign of their concern that in a society in which such technology is used naturally, human life is treated as if it were a commodity that could be replaced. In this sense, this group of respondents is firmly resistant to IVG and will not change even if technical safety is guaranteed.

Meanwhile, the 78.6% of respondents who chose 'agree' or 'do not know' at least once for the creation and research use of any of the three types of embryos that involved iPSCDG are considered less resistant to the creation and research use of embryos using iPSCDG than those who consistently expressed resistance. In other words, even if the respondents were opposed to the creation and use of any of the three types of embryos using iPSCDG, they may have been concerned about the combination of iPSCDG with adult-derived sperm/eggs and not the creation of embryos from iPSCDG.

Furthermore, it is notable that the proportion of respondents in this group (78.6%) was the same as the proportion of those who accepted the creation of

[29] Sawai et al., 'Public Attitudes in Japan toward the Creation and Use of Gametes Derived from Human-Induced Pluripotent Stem Cells'.

iPSCDG (78.6%).[30] It is unlikely that these two groups are entirely consistent; however, based on the discussion so far, it is suggested that attitudes towards the creation and research use of embryos using iPSCDG may be related to attitudes towards the creation of iPSCDG. In other words, respondents who accepted the creation of iPSCDG did not strongly resist the creation and research use of embryos from iPSCDG. This means that respondents who accepted the creation of iPSCDG but did not accept the creation and research use of embryos using iPSCDG were limited.

In our survey, approximately 20% of respondents strongly opposed the creation and research use of embryos using iPSCDG. This suggests that the remaining 80% either accepted or did not explicitly oppose the practice. Whether to allow the creation of embryos using iPSCDG is a crucial point of contention in the Japanese regulatory debate, but this survey suggests that there is a high likelihood of public support for the relaxation of current regulations.

The Public May Accept the Creation and Research Use of Embryos Using iPSCDG, but May Not Accept the Mass Production of Embryos

The Japanese government has taken the fundamental stance of defining human embryos as 'the sprouts of human life' and banning their research use in principle.[31] However, there are exceptions, such as research that contributes to the improvement of assisted reproductive medicine, research that contributes to the treatment of hereditary or congenital diseases,[32] and the use of human embryos for the establishment of ES cells (only surplus embryos can be used).[33] Currently, research on the creation of SCDGs is being conducted to investigate the process by which gametes are created and to elucidate the pathology of infertility and congenital diseases. However, it is important to consider whether to allow the creation of embryos for research when it becomes possible to create gametes that are morphologically similar to normal gametes.

[30] Sawai et al., 'Public Attitudes in Japan toward the Creation and Use of Gametes Derived from Human-Induced Pluripotent Stem Cells.'

[31] Council for Science, Technology, and Policy (CSTP), 'Basic Ideas on the Handling of Human Embryos' (in Japanese), 2004, https://www8.cao.go.jp/cstp/tyousakai/life/haihu39/siryo5-1-1.pdf.

[32] Ministry of Education Culture, Sports, Science, and Technology (MEXT), Ministry of Health Labour and Welfare (MHLW), and Children and Family Agency (CFA), 'Ethical Guidelines of the Production and Use of Human Fertilized Embryos for the Research' (in Japanese), 2024, https://www.mext.go.jp/lifescience/bioethics/files/pdf/n2409_01.pdf.

[33] Ministry of Education Culture, Sports, Science, and Technology (MEXT) and Ministry of Health Labour and Welfare (MHLW), 'Guidelines on the Derivation of Human Embryonic Stem Cells' (in Japanese), 2022, https://www.mext.go.jp/lifescience/bioethics/files/pdf/n2318_02.pdf.

Based on the results shown in Table 12.3, we can speculate on the respondents' attitudes towards research regulations from the distribution of responses regarding the acceptability of the research use of surplus embryos and iPSCDG embryos. First, the 12.5% of respondents who opposed the use of both surplus embryos or iPSCDG embryos for research were likely to have an aversion to the creation and use of human embryos in research and may be resistant to deregulation. Next, the belief of the 12.8% of respondents who supported the research use of surplus embryos but opposed the creation and use of iPSCDG embryos could be linked to current regulations. This group may include people who distinguish between surplus and iPSCDG embryos, those who consider the ethical concerns involved in creating new embryos for research purposes alone, and those who judge surplus embryos to be more suitable for research use. Furthermore, 42.3% of the respondents who supported the use of both surplus and iPSCDG embryos would agree with the idea of relaxing regulations to allow the creation of embryos using iPSCDG in the future. This group includes those who have little resistance to the creation and use of human embryos and those with high expectations for the development of IVG research.

Focussing only on the results for iPSCDG embryos, the respondents who agreed with the creation and research use of embryos (48.9%) could be considered to support regulations that allow for the creation of embryos. However, when we asked about expectations and concerns regarding IVG, we found that 64.6% of all respondents were concerned about promoting embryo destruction.[34] This indicates that even among those who support the creation of iPSCDG embryos, a certain number of people consider the destruction of embryos ethically undesirable. At first glance, the attitude of accepting the creation and research use of iPSCDG embryos and the attitude of being concerned about embryo destruction appear contradictory. In fact, it is not possible to speculate from the results of this survey why respondents were concerned about embryo destruction. However, it is possible that even respondents who accepted the creation and use of embryos using iPSCDG for research purposes would not accept the mass production and mass disposal of embryos.

Conclusion

In this study, we analysed responses to a question about the acceptability of different types of embryos from a survey about public attitudes towards the use of

[34] Sawai et al., 'Public Attitudes in Japan toward the Creation and Use of Gametes Derived from Human-Induced Pluripotent Stem Cells'.

IVG conducted in 2017. We considered how the Japanese public would react to the creation and research use of embryos using iPSCDG. Some limitations exist in the interpretation of the results of the binary logistic regression analysis. First, the goodness of fit of the models (Nagelkerke R^2 and Cox–Snell R^2) was limited, and factors beyond the explanatory variables may have affected the respondents' attitudes. Second, it is unclear whether expectations and concerns directly affect tolerance because they do not demonstrate causality. Furthermore, bias due to the region or social background of the survey participants and social desirability bias among the respondents might have affected the results. However, this exploratory descriptive study discusses the attitudes of the general public in Japan towards IVG, taking these limitations into account.

The results showed that 21.4% of respondents had strong resistance to the creation and research use of embryos using iPSCDG, and they may have also had negative notions about the use of IVG in general, including the creation of iPSCDG. It was also clear that approximately 80% of the respondents either accepted the creation and research use of embryos using iPSCDG or did not strongly oppose it. By explaining the scientific rationale and social appropriateness of the research to the public, there is a good chance that the creation and research use of embryos using iPSCDG will be accepted. However, to gain public trust in IVG research, regardless of whether regulations are relaxed, the government and scientific community must disseminate accurate information about the purpose of the research, its impact on society, and the differences between basic and clinical research.

In recent years, ethical issues surrounding the handling of human embryos have become increasingly apparent in research on human genome editing and stem cell-derived embryo models. Even if we accept the creation of embryos using SCDG in the future, as long as we recognize a certain moral value in embryos, it is necessary to take appropriate measures to address the ethical concerns associated with the mass creation and disposal of embryos in Japan. In addition, it is desirable to proceed carefully with the creation of rules that are consistent with the regulations of related research fields, such as research purposes and periods that allow for embryo cultivation.

Bibliography

Adashi, Eli Y., Cohen, I. Glenn, Hanna, Jacob H., Surani, Azim M., and Hayashi, Katsuhiko, 'Stem Cell-Derived Human Gametes: The Public Engagement Imperative', *Trends in Molecular Medicine*, 25/3 (2019), 165–167, https://doi.org/10.1016/j.molmed.2019.01.005.

Akatsuka, Kyoko, Hatta, Taichi, Sawai, Tsutomu, and Fujita, Misao, 'Public Attitudes in Japan toward the Reproductive Use of Gametes Derived from Human-Induced Pluripotent Stem Cells', *Future Science OA*, 7/10 (2021), FSO754, https://doi.org/10.2144/fsoa-2021-0065.

Bredenoord, Annelian. L., and Hyun, Insoo, 'Ethics of Stem Cell-Derived Gametes Made in a Dish: Fertility for Everyone?', *EMBO Molecular Medicine*, 9/4 (2017), 396–398, https://doi.org/10.15252/emmm.201607291.

Cabinet Office, 'The Interim Report of Issues Regarding Fertilisation of Human Germ Cells from Pluripotent Stem Cells' (in Japanese), 2015, https://www8.cao.go.jp/cstp/tyousakai/life/chukanmatome_150909.pdf.

Clark, Amander T., Brivanlou, Ali, Fu, Jianping, Kato, Kazuto, Mathews, Debra, Niakan, Kathy K., et al., 'Human Embryo Research, Stem Cell-Derived Embryo Models and In Vitro Gametogenesis: Considerations Leading to the Revised ISSCR Guidelines', *Stem Cell Reports*, 16/6 (2021), 1416–1424, https://doi.org/10.1016/j.stemcr.2021.05.008.

Cohen, I. Glenn, Daley, George Q., and Adashi, Eli Y., 'Disruptive Reproductive Technologies', *Science Translational Medicine*, 9/372 (2017), eaag2959, https://doi.org/10.1126/scitranslmed.aag2959.

Council for Science, Technology, and Policy (CSTP), 'Basic Ideas on the Handling of Human Embryos' (in Japanese), 2004, https://www8.cao.go.jp/cstp/tyousakai/life/haihu39/siryo5-1-1.pdf.

Hendriks, Saskia, Dancet, Eline A. F., Vliegenthart, Rens, and Repping, Sjoerd, 'The Acceptability of Stem Cell-Based Fertility Treatments for Different Indications', *Molecular Human Reproduction*, 23/12 (2017), 855–863, https://doi.org/10.1093/molehr/gax027.

Hendriks, Saskia, Hessel, M., Mochtar, Monique, Meissner, Andreas, van der Veen, Fulco, Repping, Sjoerd, et al., 'Couples with Non-obstructive Azoospermia Are Interested in Future Treatments with Artificial Gametes', *Human Reproduction*, 31/8 (2016), 1738–1748, https://doi.org/10.1093/humrep/dew095.

The Hinxton Group, 'Consensus Statement: Science, Ethics and Policy Challenges of Pluripotent Stem Cell-Derived Gametes', 2008, http://www.hinxtongroup.org/Consensus_HG08_FINAL.pdf.

Ishii, Tetsuya, Pera, Renee A. Reijo, and Greely, Henry T., 'Ethical and Legal Issues Arising in Research on Inducing Human Germ Cells from Pluripotent Stem Cells', *Cell Stem Cell*, 13/2 (2013), 145–148, https://doi.org/10.1016/j.stem.2013.07.005.

Mertes, Heidi, Goethals, Tina, Segers, Seppe, Huysentruyt, Marie, Pennings, Guido, and Provoost, Veerle, 'Enthusiasm, Concern, and Ambivalence in the Belgian Public's Attitude towards In Vitro Gametogenesis', *Reproductive Biomedicine & Society Online*, 14 (2022), 156–168, https://doi.org/10.1016/j.rbms.2021.10.005.

Ministry of Education, Culture, Sports, Science, and Technology (MEXT), 'Guidelines on the Utilisation of Human Embryonic Stem Cells' (in Japanese), 2010, https://warp.ndl.go.jp/info:ndljp/pid/1255146/www.lifescience.mext.go.jp/files/pdf/n592_S01.pdf.

Ministry of Education, Culture, Sports, Science and Technology (MEXT), 'Guidelines on the Research on Producing Germ Cells from Human Induced Pluripotent Stem Cells or Human Tissue Stem Cells', 2010, https://www.mext.go.jp/lifescience/bioethics/files/pdf/n1567_02r2.pdf.

Ministry of Education, Culture, Sports, Science, and Technology (MEXT) and Ministry of Health, Labour and Welfare (MHLW), 'Guidelines on the Derivation of Human Embryonic Stem Cells' (in Japanese), 2022, https://www.mext.go.jp/lifescience/bioethics/files/pdf/n2318_02.pdf.

Ministry of Education, Culture, Sports, Science, and Technology (MEXT), Ministry of Health, Labour and Welfare (MHLW) and Children and Family Agency (CFA), 'Ethical Guidelines on the Production and Use of Human Fertilised Embryos for Research' (in Japanese), 2024, https://www.mext.go.jp/lifescience/bioethics/files/pdf/n2409_01.pdf.

O'Brien, Robert M., 'A Caution Regarding Rules of Thumb for Variance Inflation Factors', *Quality & Quantity*, 41/5 (2007), 673–690, https://doi.org/10.1007/s11135-006-9018-6.

Sawai, Ttsutomu, Hatta, Taichi, Akatsuka, Kyoko, and Fujita Misao, 'Public Attitudes in Japan toward the Creation and Use of Gametes Derived from Human-Induced Pluripotent Stem Cells', *Future Science OA*, 7/10 (2021), FSO755, https://doi.org/10.2144/fsoa-2021-0066.

13
Critical Bioethics

On False Hope, Normative Assumptions, and Re-negotiation in the Ethical Debate about IVG

Seppe Segers and Dorian Accoe

Bioethical Attention to IVG

It may be an overly broad question to ask about the purposes of *in vitro* gametogenesis (IVG) development, but in view of an ethical evaluation of this assumedly 'nascent technology', it may be relevant to know about its anticipated uses, benefits, and potential downsides. Answering this question might seem deceptively self-evident. One response may be that IVG is being developed as a technology catering to a desire to have genetically related offspring, especially for people for whom existing assisted reproductive technologies (ARTs) do not presently offer that possibility. Yet, attempts to mark this response as true or false will likely depend on who one asks (e.g., an investor or a scientist). The rationale for developing IVG may be as much about 'making revenue' or 'biological puzzle-solving' (to put it in Kuhnian terms) as it is about 'catering to reproductive desires'. Moreover, the formulation of this question in terms of IVG as a 'nascent technology' and its 'uses', 'benefits', and 'downsides' presupposes a great normative deal: when a certain procedure under consideration in the lab becomes a 'nascent' 'technology' is not merely an observational fact, and propositions about its anticipated (let alone 'appropriate') use may be received as benefits by some while offending the moral convictions of others.

Even if one accepts that a driving motive behind IVG research is future reproductive application, the earliest pay-off may rather come from research applications. While this was already put forward by Testa and Harris,[1] starting the ethical debate on stem cell-derived gametes (SCDGs), IVG for research purposes has caught less bioethical attention (with some exceptions[2]).

[1] Giuseppe Testa and John Harris, 'Ethical Aspects of ES Cell-Derived Gametes', *Science*, 305 (2004), 1719. Giuseppe Testa and John Harris, 'Ethics and Synthetic Gametes', *Bioethics*, 19 (2005), 146–166.

[2] Annelien L. Bredenoord and Insoo Hyun, 'Ethics of Stem Cell-Derived Gametes Made in a Dish: Fertility for Everyone?', *EMBO Molecular Medicine*, 9 (2017), 396–398. Ursula Eichenlaub-Ritter, 'Female Meiosis and Beyond: More Questions than Answers?', *Reproductive Biomedicine*

Seppe Segers and Dorian Accoe, *Critical Bioethics* In: *Revolutionizing Reproduction.* Edited by: Christopher Gyngell, Julian Savulescu, and Tsutomu Sawai, Oxford University Press.
DOI: 10.1093/9780197810415.003.0013

Reproductive use of IVG can be most attention-grabbing, but it should not be ignored that an earlier practical payoff of IVG may be to provide a pool of gametes to facilitate fundamental (embryo) research and to supply (junior) medical personnel with materials to train certain operations, like intracytoplasmic sperm injection (ICSI) or somatic cell nuclear transfer.[3]

Taking a further step back, one may contemplate why IVG and the motives for its development should be the topic of a *bioethical* volume in the first place. Categorizing IVG as a matter of ethics and not 'merely' as a matter of biology, fundamental knowledge, or medicine, or as a phenomenon that may 'just' necessitate some refinements to legal frameworks, already uncovers normative assumptions. Even more: the observation that today's research into IVG is taken seriously as a *scientific* endeavour already suggests that it concerns a problem assumed to have a solution, which is worth the energy, money, and time, and which should not be dismissed as esoteric or *science-fiction*.

The answer to the question why this scientific endeavour should concern bioethicists may again seem deceptively self-evident. One may point out that studying IVG involves human and non-human materials that have value in some (moral) sense to moral subjects, and that potential applications of this envisioned technology may affect the interests of those counted among members of the moral community. This is true, yet it also already presumes an understanding of what morality, bioethics, and its study are about.[4] Apart from accounts that present bioethics as 'either a legitimating device for biomedicine or a radical critique on behalf of patients and the public', there are those, like Duncan Wilson, who believe that bioethics can function 'as both a critique and a safeguard'.[5] What should be clear, in our view, is that the praxis of bioethics and the 'topics' that become part of that praxis should be located within concrete social, political, psychological, and ideological contexts from which normative thinking emerges and to which it feeds back. 'The world is a knot in motion', as Donna Haraway has it.[6] This calls for a critical bioethics as an empirically contextualized and normative discipline: it has to be empirically grounded, but

Online, 24 (2012), 589–590. Seppe Segers et al., 'Balancing Ethical Pros and Cons of Stem Cell Derived Gametes', *Annals of Biomedical Engineering*, 45 (2017), 1620–1632. Anna Smajdor and Daniela Cutas, 'Will Artificial Gametes End Infertility?', *Health Care Analysis*, 23 (2015), 134–147.

[3] Zubin Master, 'Embryonic Stem-Cell Gametes: The New Frontier in Human Reproduction', *Human Reproduction*, 21 (2006), 857–863. Segers et al., 'Balancing Ethical Pros and Cons of Stem Cell Derived Gametes', 1620–1632.

[4] Nikolas Rose, *The Politics of Life Itself: Biomedicine, Power, and Subjectivity in the Twenty-first Century* (Princeton: Princeton University Press, 2007). Duncan Wilson, *The Making of British Bioethics* (Manchester: Manchester University Press, 2014).

[5] Wilson, *The Making of British Bioethics*.

[6] Donna J. Haraway, *The Companion Species Manifesto: Dogs, People, and Significant Otherness* (Chicago: Prickly Paradigm Press, 2003).

what is accepted as 'empirical fact' also requires normative scrutiny. This means that, in our view, bioethics as a species of ethics will always have to consider itself, and the ideas to which it subscribes, as an object of critique, where the category of descriptive, empirical facts is tested for value-leadenness, and where the category of prescriptive, normative valorization is tested for empirical grounding.[7] In this chapter, we hope to illustrate this dynamic pertaining to the imaginations about and anticipated applications of IVG.

IVG, as a topic of bioethical discussion, requires critical questioning of and empirical insight into hopes, doubts, imaginations, research prioritization, legal and governmental manoeuvres, financial investments, taboos, and other expressions of what people care about in this context. From early on in the conversation, IVG has sparked imaginations of how it can facilitate reproduction for 'virtually everyone' (we return to this below). At the same time, scepticism arose about how realistic some of the imagined scenarios are. Alongside optimistic expectations, concerns about undue speculation, unrealistic expectations, and 'false hope' have risen. Shared and equal genetic parenthood for same-sex couples is one example (IVG solo reproduction could be another). Statements that in light of a given 'reality in the lab' an anticipated or imagined application of IVG is either deemed possible or impossible, and statements that a still nascent technology will address 'the needs' of (medically and socially) infertile people, are instances of what interest us here.

In this chapter, our research question spans two broad aspects. First, in the context of IVG, what do 'impossible', 'feasible', 'difficult', and neighbouring concepts like 'false hope' mean when projected unto considered applications of IVG—who decides on its normative content and its empirical benchmarks, and based on which criteria? Second, how can ethical questions about IVG expand, subside, metamorphose, or otherwise come into 'motion' when they are subject to empirical investigation? On the one hand, telling 'realism' from 'science fiction' based on empirical facts is not a value-neutral matter. On the other hand, grounding bioethical discussions in empirical insight may shed a different light on unquestioned assumptions about why IVG ought to be a topic of bioethical interest. For both aspects, the discussion on IVG as possible future reproductive options for LGBTQIA+ (or queer)[8] persons will be illustrative.

[7] Jaap Kruithof, 'De plaats van de moraalwetenschap in de universiteit van morgen', *Ethiek & Maatschappij*, 6 (2003), 145–188.

[8] We use 'LGBTQIA+' (or 'LGBTQ') and 'queer' interchangeably. 'Same-sex couples' is sometimes used, but because this mainly relates to cisgender (gay) couples, it is important to emphasize that IVG could also offer possibilities to other people in queer communities (as well as non-queer people; we return to this).

Catering to Reproductive Desires

In terms of technical feasibility, the production of viable person-specific SCDGs suitable for human reproduction is no reality.[9] Non-human animal research has leaped beyond what is currently possible in humans,[10] which—at least in theory—indicates that similar results may one day be possible in humans. Bracketing scepticism about whether human IVG will ever be ready for the clinic, experts in the field have stated that 'it is in principle achievable in humans', with some forecasters estimating that 'the field is about 10 years away from a baby conceived via IVG'.[11] In terms of guidelines, the International Society for Stem Cell Research presently categorizes human reproductive use of IVG as 'a currently prohibited research activity until safety and ethical issues are resolved', with 'potential for use in human reproduction once safety and efficacy is proven'.[12] In terms of market potential, biotech companies and investors have raised their interests in this new endeavour to cater to the desire to parent 'genetically related offspring'.[13]

While still in its infancy,[14] IVG has been described—like *in vitro* fertilization at the end of the 1970s—as a 'reproduction revolution'.[15] It is fair to say that most of the attention—in the technical scientific literature, ethical and legal literature, and popular media—has gone to this perspective of IVG as 'an essential cure to infertility',[16] understood as both 'functional' and 'situational' infertility. If functional human gametes could be obtained from stem cells originating from somatic cells, this would mean that individuals who do not produce viable gametes, and people who do not produce the complementary type of gamete required

[9] Meghan Robinson et al., 'Human In Vitro Spermatogenesis as a Regenerative Therapy—Where Do We Stand?', *Nature Reviews Urology*, 20 (2023), 461–479. Genie Min Ju Wu et al., 'Current Progress on In Vitro Differentiation of Ovarian Follicles from Pluripotent Stem Cells', *Frontiers in Cell and Developmental Biology*, 11 (2023), 1166351.

[10] Zhi-Kun Li et al., 'Generation of Bimaternal and Bipaternal Mice from Hypomethylated Haploid ESCs with Imprinting Region Deletions', *Cell Stem Cell*, 23 (2018), 665–676. Kenta Murakami et al., 'Generation of Functional Oocytes from Male Mice In Vitro', *Nature*, 615 (2023), 900–906.

[11] National Academies of Sciences, Engineering, and Medicine, *In Vitro–Derived Human Gametes as a Reproductive Technology: Scientific, Ethical, and Regulatory Implications: Proceedings of a Workshop* (Washington: The National Academies Press, 2023).

[12] Amander T. Clark et al., 'Human Embryo Research, Stem Cell-Derived Embryo Models and In Vitro Gametogenesis: Considerations Leading to the Revised ISSCR Guidelines', *Stem Cell Reports*, 16 (2021), 1416–1424.

[13] David Cyranoski, Jorge L. Contreras, and Victoria T. Carrington, 'Intellectual Property and Assisted Reproductive Technology', *Nature Biotechnology*, 41 (2023), 14–20.

[14] Robinson et al., 'Human In Vitro Spermatogenesis as a Regenerative Therapy', 461–479. Wu et al., 'Current Progress on In Vitro Differentiation of Ovarian Follicles from Pluripotent Stem Cells', 1166351.

[15] Vitaly A. Kushnir, Gary D. Smith, and Eli Y. Adashi, 'The Future of IVF: The New Normal in Human Reproduction', *Reproductive Sciences*, 29 (2022), 849–856.

[16] Jasin Taelman et al., 'Stem Cell-Derived Spermatozoa', in José A. Horcajadas and Jaime Gosálvez, eds., *Reproductomics* (London: Academic Press, 2018), 315–345.

to conceive, could reproduce without recourse to donor gametes. In line with this, it has been explicitly stated that researchers are developing IVG in response to people's desire to have genetically related offspring, especially for people for whom existing ART does not yield this desired outcome.[17]

This motivation comes with an assumption that people care about becoming a genetic parent and that this striving is not hindered by philosophical confusion about what 'genetic parenthood' means.[18] To be sure, there is conceptual confusion about the understanding of genetic parenthood and whether IVG can help achieve it,[19] but despite this, several researchers cite genetic parenthood as a reason to develop IVG.[20] As such, being in line with a longer history of the field of ART—where catering to a reproductive interest has been a common thread—does not in itself substantiate assumptions about people's interest in becoming a genetic parent and the role IVG can and should play in such strivings. Indeed, it requires empirical insight in people's caring about genetic parenthood and their respective attitudes towards IVG as a means to achieve this (in both cases avoiding overgeneralizations), while assessing what can be derived from this in a prescriptive sense.

There is relatively limited empirical research on attitudes towards the development of IVG.[21] One study by Saskia Hendriks and colleagues concluded that there is high acceptance among the Dutch public and gynaecologists towards using IVG as a pathway towards genetic parenthood.[22] A study with a sample of Belgian adults indicated that the opportunity to have a genetically related child was the most important potential benefit of IVG.[23] While there is—overall—a positive attitude towards IVG, mainly supported by the benefits of establishing a genetic link between parents and children, this study also indicated that there is—overall—a negative attitude towards accepting risks to future offspring that exceed those of other ART.

[17] Bredenoord and Hyun, 'Ethics of Stem Cell-Derived Gametes Made in a Dish, 396–398. Taelman et al., 'Stem Cell-Derived Spermatozoa', 315–345.

[18] Heidi Mertes, 'Gamete Derivation from Stem Cells: Revisiting the Concept of Genetic Parenthood', *Journal of Medical Ethics*, 40 (2014), 744–747.

[19] Segers et al., 'Balancing Ethical Pros and Cons of Stem Cell Derived Gametes', 1620–1632.

[20] Bredenoord and Hyun, 'Ethics of Stem Cell-Derived Gametes Made in a Dish', 396–398. Taelman et al., 'Stem Cell-Derived Spermatozoa', 315–345.

[21] Kyoko Akatsuka et al., 'Public Attitudes in Japan toward the Reproductive Use of Gametes Derived from Human-Induced Pluripotent Stem Cells', *Future Science OA*, 7 (2021). Saskia Hendriks et al., 'The Acceptability of Stem Cell-Based Fertility Treatments for Different Indications', *MHR: Basic Science of Reproductive Medicine*, 23 (2017), 855–863. Heidi Mertes et al., 'Enthusiasm, Concern and Ambivalence in the Belgian Public's Attitude towards In-Vitro Gametogenesis', *Reproductive Biomedicine & Society Online*, 14 (2022), 156–168.

[22] Hendriks et al., 'The Acceptability of Stem Cell-Based Fertility Treatments for Different Indications', 855–863.

[23] Mertes et al., 'Enthusiasm, Concern and Ambivalence in the Belgian Public's Attitude towards In-Vitro Gametogenesis', 156–168.

If one agrees that ethical analysis should be tethered to moral praxis and the psychological, social, and political dynamics inherent to that reality, such empirical input is important for bioethical discussions of IVG. Yet, it is not because people may find this prospect of genetic parenthood important that there is an ethical imperative to cater to it, for example by developing IVG. Still from an empirical perspective there is also growing insight that this discussion about IVG as a pathway towards genetic parenthood hinges on generalizations obscuring the underrepresentation of certain voices in matters of reproductive care and bioethics (we return to this below).[24]

Also from a normative perspective, much bioethical attention has been paid to the relationship between the proposition that many people find genetic parenthood important, and the prescriptive proposition that (technical) efforts must be made to make genetic parenthood possible.[25] These two propositions are intertwined: the number of people deemed sufficient to justify the (technical) effort depends not only on the costs involved, but also on the value placed on their desires. While there are good pro tanto reasons to supply IVG as a means to help satisfy a desire to become a genetic parent, it must be held against possible reasons that problematize this endeavour.[26]

It is relevant for our purposes to repeat that from early on in the discussion on IVG, there has been the imagination that SCDGs will lead to 'fertility for everyone' as IVG 'might enable anyone to produce gametes regardless of whether they ever had "natural" gametes, and irrespective of their age, sex, relationship status, or sexuality'.[27] Yet, IVG could also offer possibilities to those who can already reproduce through existing technologies, but find them suboptimal for some reason, such as avoiding ovulation stimulation and egg retrieval related to *in vitro* fertilization.[28] Determining which reasons are valid (and which are not) is part of the normative puzzle of IVG. Also, the idea that IVG will enable virtually everyone to have genetically related offspring is contestable. It belongs to the sphere of possibilities that IVG could allow reproduction for the medically infertile, same-sex couples, transgender people, (post)menopausal people, prepubescent children, deceased people, singles, and groups of people together,[29] but that does not imply that these

[24] Anne Le Goff, Robbin Jeffries Hein, Ariel N. Hart, Isaias Roberson, and Hannah L. Landecker, 'Anticipating In Vitro Gametogenesis: Hopes and Concerns for IVG among Diverse Stakeholders', *Stem Cell Reports*, 19 (2024), 933–945.

[25] Segers et al., 'Balancing Ethical Pros and Cons of Stem Cell Derived Gametes', 1620–1632. Seppe Segers, Guido Pennings, and Heidi Mertes, 'Getting What You Desire: The Normative Significance of Genetic Relatedness in Parent–Child Relationships', *Medicine, Health Care and Philosophy*, 22 (2019), 487–495.

[26] Segers, Pennings, and Mertes, 'Getting What You Desire', 487–495.

[27] Smajdor and Cutas, 'Will Artificial Gametes End Infertility?', 134–147.

[28] Guido Pennings, 'Why We Need Stem-Cell Derived Gametes', *Reproductive Biomedicine Online*, 47 (2023), 103322.

[29] Segers et al., 'Balancing Ethical Pros and Cons of Stem Cell Derived Gametes', 1620–1632.

possibilities will materialize. Such an actualization depends on not only technical abilities, but also voluntaristic conditions and societal forces.

In the next sections, we describe how technical aspects have been presented as a reason to conclude that shared and equal genetic parenthood for same-sex couples through IVG will be very difficult if not impossible, how this comes with normative interpretations, and how that weighs in on judgments that some reproductive hopes are 'false hopes'. The view centring on the 'impossibility' of IVG to make shared and equal genetic parenthood possible for same-sex couples may be less dominant today, in a sense indicating the margin of interpretation of scientific reality. Yet, it seems no less relevant to observe that even though earlier categorical claims about the difficulty or impossibility of IVG-pathways to queer-parenthood seem to sound less loud now, possibly as an effect of science communication about findings in animal research (that altogether have not been translated to human models[30]), this attunement does not seem to be an effect of significant scientific advancements in this direction. What still deserves more attention (in our view) is how socio-political background conditions co-determine what becomes possible for whom in terms of healthcare services, parenthood options, access to ART, and other life plan choices.

Interpreting 'Potential Use' from Scientific Findings

When discussing the potential users and clinical indications for IVG, the academic and policy debate quickly turns to SCDGs for same-sex reproduction.[31] From the earliest reflections on IVG, it has been lauded as a strategy to 'democratize reproduction'.[32] 'Democratizing reproduction' is catchy as a phrase, though the best way to interpret it is to be determined. If it is to approximate the vague and general 'power (*kratos*) to the people (*demos*)', it may be relevant to consider what 'the people', in all its diversity and pluralism of value and preferences, may actually hope and expect to do with that supposed nascent repro-technological power (see below). The assumption seems to be that IVG will have 'an emancipating and liberating effect for other people who are unable to have biological children altogether, such as same-sex couples',[33] which, however, is

[30] Li et al., 'Generation of Bimaternal and Bipaternal Mice from Hypomethylated Haploid ESCs with Imprinting Region Deletions', 665–676. Murakami et al., 'Generation of Functional Oocytes from Male Mice In Vitro', 900–906.

[31] Eli Y. Adashi, Katsuhiko Hayashi, and I. Glenn Cohen, 'Ethical and Legal Challenges in Assisted Same-Sex Conception through In Vitro Gametogenesis', *Nature Medicine*, 30 (2024), 322–323. National Academies of Sciences, Engineering, and Medicine, *In Vitro–Derived Human Gametes as a Reproductive Technology*.

[32] Testa and Harris, 'Ethics and Synthetic Gametes', *Bioethics*, 19 (2005), 146–166.

[33] Bredenoord and Hyun, 'Ethics of Stem Cell-Derived Gametes Made in a Dish', 396–398.

premised on the unspoken assumption that this may only be 'emancipating and liberating' for those who have such a corresponding desire and for those who can afford it, or for whom access is not otherwise blocked. In this sense, Bredenoord and Hyun interpret IVG's potential to 'democratize reproduction' as the potential to 'result in "fertility for everyone"', thus reiterating Smajdor and Cutas' wording.[34] In the present-day discussion, this also transpires in the careful enthusiasm that this 'still nascent [technology] opens up for the first time the possibility that same-sex couples might achieve something that most opposite-sex couples readily pursue'.[35]

As described in Chapter 1 of this volume, several pathways are being explored to obtain SCDGs, but current findings in non-human animal models are preliminary. Earlier on in the debate, this resulted in a particular focal scepticism about same-sex reproduction through IVG.[36] The Hinxton Group concluded that it is 'not likely in the future of the science' that SCDGs will be used in same-sex reproduction due to 'significant if not insurmountable scientific barriers'.[37] They stated that it will be very difficult to derive usable eggs from men and that due to biological and technical reasons, it may even be 'impossible, to derive sperm that could be used for reproduction from XX (chromosomally female) cells'.[38] Moreno and colleagues professed that 'the most challenged goal in the artificial gamete generation seems to be the production of sperm and eggs from the opposite sex'.[39] It is notable that such categorical reservations now seem to have less traction in the current discussions of IVG.

On the one hand, marking one currently hypothetical application as less realistic than another currently hypothetical application might be protested because it twice concerns hypotheticals that are presently not part of reality. Stating that one is less realistic than the other might look arbitrary. It can be retorted that this need not be arbitrary, since such distinctive judgment can be based on the current scientific insights and state of the art. This is true, yet it also raises the question to what extent the current scientific state of the art is what it is because of prioritizing certain research questions, 'puzzles', and objectives over others. The belief that current research practice is closer to one scenario (say, person-specific SCDGs for opposite-sex couples) than to the other (person-specific SCDGs

[34] Bredenoord and Hyun, 'Ethics of Stem Cell-Derived Gametes Made in a Dish', 396–398. Smajdor and Cutas, 'Will Artificial Gametes End Infertility?', 134–147.

[35] Adashi, Hayashi, and Cohen, 'Ethical and Legal Challenges in Assisted Same-Sex Conception through In Vitro Gametogenesis', 322–323.

[36] Hinxton Group, 'Consensus Statement: Science, Ethics and Policy Challenges of Pluripotent Stem Cell-Derived Gametes' (2008). Debra J. H. Mathews et al., 'Pluripotent Stem Cell-Derived Gametes: Truth and (Potential) Consequences', *Cell Stem Cell*, 5 (2009), 11–14.

[37] Mathews et al., 'Pluripotent Stem Cell-Derived Gametes', 11–14.

[38] Hinxton Group, 'Consensus Statement'.

[39] Inmaculada Moreno, Jose Manuel Míguez-Forjan, and Carlos Simón, 'Artificial Gametes from Stem Cells', *Clinical and Experimental Reproductive Medicine*, 42 (2015), 33–44.

for same-sex couples) sparks the normative question why this should be so.[40] Shrugging off a hypothetical scenario as 'impossible' may crush into a conservative status quo, even discriminatory tendency if one posits the 'more likely' scenario based on heteronormative parenthood ideals. Who decides, and on which (normative, though often implicit) assumptions one scenario is framed as 'more likely' than another, should be part of bioethical critique.[41]

This is an important instance of how fact and value in bioethics feed into each other: the contours of bioethical considerations should preferably have some connection with practical–ethical assessments of relevant factors in the present and immediate future. Reality in the research laboratory is one of those factors, and given their respective expertise and responsibilities, scientists and ethicists have an obligation to introduce this to the conversation in and outside the laboratory.[42] Yet, science is not value neutral: what is considered a relevant research question and a suitable result, and what is presented as scientific fact or 'lab reality', all brings normativity to the table. Science shapes and reproduces societal values and, through that, influences behaviour, aspirations, and imaginations. (This is one reason not to consider IVG as a 'mere matter of biology'.)

At the same time, personal experience has taught that non-profit organizations working on science communication about new medical technologies regularly receive questions about whether 'there is more news' about IVG yet, which mostly come from people hoping for an ultimate possibility to fulfil reproductive aspirations in a (sometimes) turbulent fertility process, and who want to know when the technology will be clinically available. To the extent that science is a public good, accurate and non-overselling science reporting is important for more than one reason. It generally connects to responsibilities for the maintenance of trust, and it is vindicated by respect for personal autonomy, which encompasses access to information and knowledge as grounds on which people make life decisions and deal with reality. More specifically, discussing and communicating on IVG in the abstract from scientific state of the art would pay a disservice to those who hope to benefit from this technology because it would engender hope about opportunities that might never materialize. In part, this seems to circle back to the previous remark—what are the justificatory grounds to state that something will never materialize? Both the early pessimistic reading

[40] This point may have to be nuanced, since research reports like those of Li and colleagues and Murakami and colleagues belie the claim that there is no research into generating gametes from the opposite sex. Both studies have rekindled the debate about the possibility of same-sex reproduction in humans, but it is important not to jump to conclusions.

[41] Seppe Segers and Dorian Accoe, 'Mortality, Morbidity and Speculation: Ethical Considerations about the Lever for Ectogestation', in Christopher Stratman and Joona Räsänen, eds., *Ectogestation and the Future of Reproductive Rights: Technology, Justice and Abortion* (forthcoming).

[42] Seppe Segers, 'The IVG "Relatedness Paradox": Researchers Should Mind Speculation', *Trends in Biotechnology*, 41 (2023), 1220–1222.

à la Hinxton Group and the careful enthusiastic reading à la Adashi trigger questions about how to interpret 'potential use' of IVG from scientific findings, and, in its wake, discussions about 'false hope'.

Value Judgments about Hope and False Hope

Whether or not certain anticipated uses of IVG will become possible, partly[43] depends on what is found to be technically feasible. Understandings of technical feasibility, in turn, depend on scientific findings in lab settings, usually as an outcome of what is regarded as worthwhile to investigate in the first place. Depending on such understandings, imagined outlooks may become the subject of hope, or contrarily, of 'false hope'. There is moral relevance to paying considerate attention to what has been described as 'hoped for possibilities that may never materialize', and to pause on the associated notion of 'false hope'. The general part that hope has in moral thought and behaviour may already speak for this moral relevance. That someone hopes for X, and that this hope is judged to be false, can both be descriptive observations. However, the reasons why someone hopes for X, and the basis on which that hope is judged to be false, both boil down to value judgments—which, admittedly, have a place in moral considerations. In the context of assisted reproduction, we would say that there may be even more to this, illustrated by the way in which technical means to cater to reproductive aspirations are sometimes called 'hope technologies'.[44] For IVG too, this may add to the moral relevance of contemplating hope and false hope.

With IVG, the object of hope is twofold. Someone can hope for the successful development of the technology, driven by an underlying hope to have or share genetically related offspring. The value placed on genetic parenthood is shaped by societal norms. Indeed, hope is not only an individual state of mind, but is influenced by social, political, and economic positions held in systems of privilege and oppression.[45] By affecting what is seen as desirable or worth pursuing, the context shapes the state of affairs hoped for and the degree to which someone is hopeful that a desired outcome is probable.

The concept of *false* hope can be said to arise when an individual holds an 'inaccurate' belief regarding the possibility of a desired outcome. Because an individual cannot hope for an outcome known to be impossible, judgments of hope

[43] Technical feasibility is not all there is to whether or not possibilities materialize: something can be technically feasible, but legally prohibited for instance, which can stand in the way of the actual possibility.

[44] Cathy Herbrand and Rebecca Dimond, 'Mitochondrial Donation, Patient Engagement and Narratives of Pope', *Sociology of Health & Illness*, 40 (2018), 623–638.

[45] Katie Stockdale, *Hope under Oppression* (New York: Oxford University Press, 2021).

to be 'false' derive, by definition, from an external assessment.[46] While the probability of the outcome is not a determining factor for invocations of false hope—as long as the hoped-for outcome is possible—probability can become relevant when the likelihood of the outcome is judged to be too low. When a third party assesses the probability of a particular state of affairs (i.e., the object of hope) to be too low, it can result in a dynamic where the outcome is rendered impossible due to a lack of (further) effort and investment in making it possible (or an active effort of holding it back). Both the early scepticism about IVG as a facilitator for equal and shared genetic parenthood for LGBTQ+ persons, and the more 'optimistic' reading that this 'still nascent [technology] opens up for the first time the possibility that same-sex couples might achieve something that most opposite-sex couples readily pursue',[47] rely on value-laden assumptions. Yet, whether IVG is taken seriously as a scientific endeavour and its perceived likelihood of success can affect the chances of the technology becoming viable. The determination of which probability is considered 'too low' involves a complex interplay of value-relative understandings: a value, with practical (moral) consequences, is attributed to a statistical calculation.

These value-relative understandings should not be cloaked in seeming objectivity or neutrality, immune to reasoned disagreement and re-negotiation. The third-person judgment that hope about a desired prospect is false evokes questions about the authority by which this prospect can be said to be 'false', and about the normativity of what it means to earmark something as impossible in the first place.[48] In societies structured to award privileges to some people at the expense of others, outcomes desired by privileged groups are more likely to be seen as to be pursued and fulfilled than the hopes of members of oppressed groups.[49] Yet, if society deems the value of shared genetic parenthood for opposite-sex couples significant enough to warrant substantial investment in reproductive technologies, then the same consideration should be extended to the possibility of deriving gametes from the opposite sex for same-sex couples.[50] While the mere desire or hope for genetic parenthood through IVG does not automatically justify all efforts to achieve it, it would be inconsistent to suggest that only same-sex couples should redirect their hopes and values away from genetic parenthood while continuing to support the reproductive desires of infertile

[46] Dorian Accoe and Seppe Segers, '"False Hope" in Assisted Reproduction: The Normative Significance of the External Outlook and Moral Negotiation', *Journal of Medical Ethics*, 50 (2024), 181–184.

[47] Adashi, Hayashi, and Cohen, 'Ethical and Legal Challenges in Assisted Same-Sex Conception through In Vitro Gametogenesis', 322–323.

[48] Accoe and Segers, '"False Hope" in Assisted Reproduction', 181–184.

[49] Stockdale, *Hope under Oppression*.

[50] Seppe Segers, Guido Pennings, and Heidi Mertes, 'Ethical Reflections on Stem Cell-Derived Gametes', *Médecine de la reproduction*, 19 (2017), 298–306.

heterosexual couples. Reproductive technologies that are now well established were once speculative. Infertile heterosexual couples were considered legitimate recipients of efforts to develop these technologies, raising normative questions about why the same consideration should not be extended to other (same-sex or queer) couples. Some authors have brought these questions onto considerations of compensatory justice, which would point to rectifying (past and present) injustices by actively supporting and investing in the development of reproductive technologies enabling queer couples to achieve genetic parenthood.[51]

> Yet, importantly, similar to questions relating to the authority by which the prospect of certain IVG usages is judged 'possible' or 'impossible', and whether associated hope is false or not, generalizing statements about IVG as a 'hoped for' satisfier to the reproductive 'needs' of 'same-sex couples' (and 'opposite-sex couples' alike) should be tested. That is, on whose authority can these latter statements be made? We explore this in the next section.

How Empirical Data Can Expand Ethical Thought on IVG

The 'widely assumed [view] by scientists, medical professionals, companies, and the media' that IVG is 'a technological opportunity that directly addresses the needs of medically and socially infertile people' rests upon what we would call non-empirical gap-fillers.[52] More specifically, statements like these should make one wonder on whose authority such views are upheld, and in relation to which aspects of reality they can be said to be true, false, or doubtful. We agree with Le Goff and colleagues that '[t]his assumption needs to be tested empirically',[53] as it yet again illustrates the fact-value minuet in bioethics, which links back to our opening questions concerning why IVG should be a bioethical topic, and what it means to consider a theme as a bioethical one in the first place.

Le Goff and colleagues make an important point in stating that '[the views of] LGBTQ+ people on IVG have not been explored, reflecting the well-documented underrepresentation of such voices in matters of reproductive care and bioethics'.[54] They continue that tropes about the potential of IVG to 'democratize' reproduction, call for 'further in-depth engagement with a broad range of potential users of IVG'.[55] On pain of superfluity, respecting people's agency

[51] Timothy Murphy, 'Pathways to Genetic Parenthood for Same-Sex Couples', *Journal of Medical Ethics*, 44 (2018), 823–824. Seppe Segers et al., 'Using Stem Cell-Derived Gametes for Same-Sex Reproduction: An Alternative Scenario', *Journal of Medical Ethics*, 43 (2017), 688–691.

[52] Le Goff et al., 'Anticipating In Vitro Gametogenesis', 933–945.

[53] Le Goff et al., 'Anticipating In Vitro Gametogenesis', 933–945.

[54] Le Goff et al., 'Anticipating In Vitro Gametogenesis', 933–945.

[55] Le Goff et al., 'Anticipating In Vitro Gametogenesis', 933–945.

means that one should not suppress the other's voice when raising one's own.[56] In this case: broad claims about the supposed 'needs' of medically and situationally infertile people will have to be weighed against the voices of both those whose experiences do, and those whose experiences do not match generalized postulations about what one assumes those concerns or desires to be.

We do not wish to deny that people, regardless of sexual orientation or gender, can look upon IVG as a potential avenue towards genetic parenthood as something that is important to them. Moreover: there are moral reasons to provide help to satisfy such desires.[57] Yet, legitimizing *non*-genetic parenthood may be as much a matter of justice and respect for autonomy as securing equal access to reproductive services regardless of sexual orientation or gender. For one thing, in discussing IVG for non-hetero couples, it is often neglected that many within queer communities do not wish to assimilate to traditional genetic notions of family building (and some of those who want to become parents may rather wish to queer traditional parenthood conceptions, for which IVG could also be a vehicle, without being a necessary condition for it).[58]

As Katherine Kraschel points out, 'the queer community is not a monolith and discussions within it about the value of offspring that are genetically connected to both intended parents are nuanced'.[59] Given the heterogeneity of preferences and values in general, it is arrogant to assume otherwise and to project a monoculture of preferences onto LGBTQ+ people where everyone wants to have offspring 'just like' opposite-sex couple counterparts (or, vice versa, where everyone wants to 'tear down' traditional notions of family).[60] The empirical account of Le Goff and colleagues supports the view that queer family building can take the shape of 'challenging all types of biological conventions' (as one participant has it), while their study also showed support for IVG as 'another option [for] creating [a] family' enabling 'queer people to make their own reproductive choices'.[61] Importantly, their findings also showed that, among those who are all too often generalized as 'beneficiaries' of IVG, concerns exist that the advent of this technology might 'impact the validity of same-sex couples that chose to have a child with a donor' and that the state might recognize IVG procreators 'as more of a parent' than someone that chose to have a donor.[62]

[56] Eva F. Kittay, *Learning from My Daughter: The Value and Care of Disabled Minds* (New York: Oxford University Press, 2019).

[57] Segers, Pennings, and Mertes, 'Getting What You Desire', 487–495.

[58] National Academies of Sciences, Engineering, and Medicine, *In Vitro-Derived Human Gametes as a Reproductive Technology*. Le Goff et al., 'Anticipating in vitro gametogenesis', 933–45.

[59] National Academies of Sciences, Engineering, and Medicine, *In Vitro-Derived Human Gametes as a Reproductive Technology*.

[60] National Academies of Sciences, Engineering, and Medicine, *In Vitro-Derived Human Gametes as a Reproductive Technology*.

[61] Le Goff et al., 'Anticipating In Vitro Gametogenesis', 933–945. .

[62] Le Goff et al., 'Anticipating In Vitro Gametogenesis', 933–945.

Voiced concerns about how IVG might further entrench genetic parenthood ideals while devaluing constellations that do not live up to that presumed ideal, illustrate Erica Haimes' point on how empirical grounding sheds light not only on certain socio-political background theories and conditions, but also on how these can be enabling to some and constraining to others.[63] Moreover, feeding back to our central question about how empirical investigation can adjust 'our repertoire' of what counts as a matter of bioethical import, the above illustrates how such empirical data can 'expand' the content of bioethical debates about IVG by alerting us to the possibility of multiple perspectives on what is taken for granted, what is passed over, and which perspectives have hardly figured in what is talked and written about in 'IVG ethics'.[64]

There are other examples that illustrate this bioethical dynamic in the context of IVG. One that also flows from the study of Le Goff and colleagues is the voiced concern by participants from underserved communities 'that they could not devote mental, emotional, or organizational resources to thinking about IVG when they faced much more pressing concerns such as high rates of pregnancy-related death among Black and Indigenous birthing people . . . or the difficulty many trans people have in simply accessing basic needs such as food, housing, and health care'.[65] Here again, critical bioethics should ask why IVG as a technical means towards shared and equal genetic parenthood for queer persons should be 'bioethical' and the structural oppression of certain groups in healthcare not. It is nothing new that bioethics has a tendency to focus on high-tech neon problems, but choices about which 'questions' deserve moral attention are all but neutral. Directing empirical attention to matters like the above can therefore be an important normative counterweight to truncated bioethical considerations and function as a reminder to 'situate' canonical bioethical discussions as a specific practice of intellectual authority.[66]

Also if one accepts that IVG merits bioethical attention, it should be noted that the 'repertoire' out of which this attention takes shape is truncated. It is remarkable that IVG can be so hotly discussed as a means towards shared and equal genetic parenthood for same-sex couples, including for couples where surrogacy would be a necessary 'medium', all the while this requirement for surrogacy is hardly ever morally thematized in this regard. Surrogacy is a highly moralized and politicized praxis, considered a 'bioethical topic' in its own right. Given that it is no minor detail in making male same-sex procreation possible, it is all the

[63] Erica Haimes, 'What Can the Social Sciences Contribute to the Study of Ethics? Theoretical, Empirical and Substantive Considerations', *Bioethics*, 16 (2002), 89–113.

[64] Margaret Urban Walker, *Moral Understandings: A Feminist Study in Ethics* (New York: Oxford University Press, 2007).

[65] Le Goff et al., 'Anticipating In Vitro Gametogenesis', 933–945.

[66] Walker, *Moral Understandings.*

more notable that IVG can be presented as a magic bullet to 'democratize reproduction' while the involvement of a third person gestating and birthing a future child is either omitted from the solution or simply never thematized as a moral factor. Again, critical thinking in tandem with empirical attention for the concrete social, political, psychological, and ideological contexts from which that thinking emerges and to which it feeds back may show that attitudes towards and legislation of surrogacy can be ambiguous, oftentimes precarious, for both the surrogate and the intended parent(s), and obstructive to LGBTQ parenthood.[67]

We wish to close on that last note by tying in with our point about the normativity of what is labelled as '(im)possible' or '(un)feasible' and how there too the sphere of what 'is' and that of what 'ought' form an interconnected knot. Just like critical bioethics should interrogate the normative conditionals in claims earmarking certain applications of biotechnology as 'impossible' (with its manifold variations: 'false hope', 'science fiction', 'pipe dreams', 'speculations' . . .), it should interrogate the normative conditionals that are brushed aside when biotechnology—here IVG—is presented as a nascent technological fix to overcome limitations to people's reproductive strivings. In presenting IVG as a nascent technological fix to democratize reproduction, one should not silently pass over reflections indicating the material fact that the reason why some people cannot access certain reproductive services may have more to do with socio-political restrictions than with technological ones. For instance, fertility preservation procedures can presently already enable post-menopausal persons to reproduce, so limitations in this direction are not technical.[68] Our point here is that bioethics should not one-sidedly focus on the potential to overcome technical limitations since societal forces too can dampen the potential of IVG to enable genetic parenthood 'for everyone' (leaving it open whether that would be desirable or not). Even if it is accepted that technical obstacles could theoretically be overcome, one may still observe that political and socio-cultural factors can restrict which applications of IVG become available to whom. Systems-level bias, state-level legislation, and heterosexist views in public opinion can constitute a hostile environment for certain people's reproductive rights, which can be an important factor restricting IVG's potential to facilitate genetic parenthood 'for everyone'.[69]

[67] Tiia Sudenkaarne, 'Queering Bioethics: A Queer Bioethical Inventory of Surrogacy', *Ethics, Medicine and Public Health*, 6 (2018), 117–125. Sigrid Vertommen and Camille Barbagallo, 'The In/visible Wombs of the Market: The Dialectics of Waged and Unwaged Reproductive Labour in the Global Surrogacy Industry', *Review of International Political Economy*, 29 (2021), 1945–1966.

[68] And, if we take public opinion at face value, the advent of IVG might do little to change this, given empirical indication that lay opinions mostly oppose the willingness to allow IVG for post-menopausal persons. See: Heidi Mertes et al., 'Enthusiasm, Concern and Ambivalence in the Belgian Public's Attitude towards In-Vitro Gametogenesis', 156–168.

[69] Abbie E. Goldberg, 'LGBTQ-Parent Families: Diversity, Intersectionality, and Social Context', *Current Opinion in Psychology*, 49 (2023), 101517. Salvatore Monaco and Urban Nothdurfter,

Only citing technical, and not socio-political, obstacles as reasons why an imagined application of IVG may be very 'difficult if not impossible', and, vice versa, making it seem as if a technological fix is human reproduction's 'final frontier', is making bioethical analysis not only less politically engaged than it ought to be, but also less descriptively accurate. 'A variety of shiny objects',[70] like anticipated IVG, can distract from financial, familial, cultural, ideological, and other factors that shape (/limit) people's reproductive options and broader life plans, if it is not empirically tethered to material contexts, where what is presented as fact is also simultaneously and relentlessly questioned in terms of normative assumptions.

Concluding Thoughts

We have explored the fact-value interrelation in bioethics through the lens of debates about IVG. On one level, this interconnection is displayed in considerations of which anticipated uses of IVG, and which respective reproductive aspirations, are taken seriously. We have considered this by looking at invocations of 'false hope', and how normative content is brought to bear on certain empirical benchmarks—a scientific state-of-affairs—which, likewise, points back to normative interventions. Telling 'hope' from 'false hope' on the basis of empirical facts is not a value-neutral matter. Opening up to the value-ladenness of selection and interpretation of empirical facts brings into view how such decisions can be re-negotiated and how political factors can facilitate, hinder, and obscure such negotiations.

On a second level, we have considered how ethical questions about IVG can be re-calibrated under the influence of empirical insights. Such empirical enrichment can bring into view moral perspectives and moral experiences formerly excluded from bioethical attention. Such exclusionary factors contain multitudes. A constant openness to re-shifting the empirical lens to a different segment of interpersonal interactions, formerly excluded from ethical attention, is basic to this view, just as the recognition that pausing at a given selection of empirical 'facts' is never neutral.

'Discovered, Made Visible, Constructed, and Left Out: LGBT+ Parenting in the Italian Sociological Debate', *Journal of Family Studies*, 29 (2021), 471–488. Samantha Yee et al., 'Parenthood Desire, Perceived Parenthood Stigma, and Barriers to Achieving Parenthood in Childless Sexual Minority Men', *Journal of Assisted Reproduction and Genetics*, 41 (2024), 1739–1753.

[70] Josephine Johnston and Rachel L. Zacharias, 'The Future of Reproductive Autonomy', *Hastings Center Report*, 47 (2017), S6–11.

For IVG this means that critical bioethics should stimulate imagination of how things can be otherwise—for better and worse—through a multitude of means, spanning both the local and the more secluded spheres of privilege and prestige where policy is decided. Practical measures should be thought of as pointing out pivot points for change. Which and whose language is used to discuss IVG? Which and whose concerns? Which and whose imaginations? Critical bioethics can operate at these joints on all levels (scientific consortia, schools, public debates, the list goes on). This surely casts a very wide net, but it seems to us that it cannot be otherwise. The value-related questions (about IVG or otherwise) that are marked off as 'bioethical' do not emanate from an abstracted realm, but are seated in actual social, cultural, political contexts. Critical bioethics operates within advisory bodies, ethical committees, councils, public policy, and academe, just as much as through local science communication, education, speculative fiction, activism, or local social work. The realm of bioethics is not to be confused with or restricted to what is found in textbooks, but should be thought of as an active search for blind spots, exclusions, and assumptions in bioethical questions and answers. This calls for a wealth of empirical insights and experiences. It is in this sense that we advocate for a critical bioethics that keeps 'the knot in motion', by cultivating a 'double vision' in which the category of seemingly descriptive propositions are continuously tested for value-leadenness, while the category of prescriptive, normative valorization is tested in terms of empirical grounding.

Bibliography

Accoe, Dorian, and Segers, Seppe, '"False Hope"' in Assisted Reproduction: The Normative Significance of the External Outlook and Moral Negotiation', *Journal of Medical Ethics*, 50 (2024), 181–184.

Adashi, Eli Y., Hayashi, Katsuhiko, and Cohen, I. Glenn, 'Ethical and Legal Challenges in Assisted Same-Sex Conception Through In Vitro Gametogenesis', *Nature Medicine*, 30 (2024), 322–323.

Akatsuka, Kyoko, Hatta, Taichi, Sawai, Tsutomo, and Fujita, Misao, 'Public Attitudes in Japan Toward the Reproductive Use of Gametes Derived from Human-Induced Pluripotent Stem Cells', *Future Science OA*, 7 (2021), 1–13.

Bredenoord, Annelien L., and Hyun, Insoo, 'Ethics of Stem Cell-Derived Gametes Made in A Dish: Fertility for Everyone?', *EMBO Molecular Medicine*, 9/4 (2017), 396–398.

Clark, Amander T., Brivanlou, Ali, Fu, Jianping, Kato, Kazuto, Mathews, Debra, Niakan Kathy, et al., 'Human Embryo Research, Stem Cell-Derived Embryo Models and In Vitro Gametogenesis: Considerations Leading to the Revised ISSCR Guidelines', *Stem Cell Reports*, 16/6 (2021), 1416–1424.

Cyranoski, David, Contreras, Jorge L., and Carrington, Victoria T., 'Intellectual Property and Assisted Reproductive Technology', *Nature Biotechnology*, 41 (2023), 14–20.

Eichenlaub-Ritter, Ursula, 'Female Meiosis and Beyond: More Questions Than Answers?', *Reproductive Biomedicine Online*, 24 (2012), 589–590.

Goldberg, Abbie E., 'LGBTQ-Parent Families: Diversity, Intersectionality, And Social Context', *Current Opinion in Psychology*, 49 (2023), 1–6.

Haimes, Erica, 'What Can the Social Sciences Contribute To The Study Of Ethics? Theoretical, Empirical and Substantive Considerations', *Bioethics*, 16 (2002), 89–113.

Haraway, Donna J., *The Companion Species Manifesto: Dogs, People, and Significant Otherness* (Chicago: Prickly Paradigm Press, 2003).

Hendriks, Saskia, Dancet, Eline, Vliegenthart, Rens, and Repping, Sjoerd, 'The Acceptability of Stem Cell-Based Fertility Treatments for Different Indications', *MHR: Basic Science of Reproductive Medicine*, 23 (2017), 855–863.

Herbrand, Cathy, and Dimond, Rebecca, 'Mitochondrial Donation, Patient Engagement and Narratives of Hope', *Sociology of Health & Illness*, 40 (2018), 623–638.

Hinxton Group, 'Consensus Statement: Science, Ethics and Policy Challenges of Pluripotent Stem Cell-Derived Gametes', 2008, http://www.hinxtongroup.org/consensus_hg08_final.pdf.

Johnston, Josephine, and Zacharias, Rachel L., 'The Future of Reproductive Autonomy', *Hastings Center Report*, 47 (2017), S6–11.

Kittay, Eva. F., *Learning from My Daughter: The Value and Care of Disabled Minds* (New York: Oxford University Press, 2019).

Kruithof, Jaap, 'De plaats van de moraalwetenschap in de universiteit van morgen', *Ethiek & Maatschappij*, 6 (2003), 145–188.

Kushnir, Vitaly. A., Smith, Gary D., and Adashi, Eli Y., 'The Future of IVF: The New Normal in Human Reproduction', *Reproductive Sciences*, 29/3 (2022), 849–856.

Le Goff, Anne, Jeffries Hein, Robbin, Hart, Ariel, Roberson, Isaias, and Landecker, Hannah, 'Anticipating In Vitro Gametogenesis: Hopes and Concerns for IVG Among Diverse Stakeholders', *Stem Cell Reports*, 19 (2024), 933–945.

Li, Zhi-Kun, Wang, Le-Yun, Wang, Li-Bin, Feng, Gui-Hai, Yuan, Xue-Wei, Liu, Chao, et al., 'Generation of Bimaternal and Bipaternal Mice from Hypomethylated Haploid ESCs with Imprinting Region Deletions', *Cell Stem Cell*, 23/5 (2018), 665–676.

Master, Zubin, 'Embryonic Stem-Cell Gametes: The New Frontier in Human Reproduction', *Human Reproduction*, 21 (2006), 857–863.

Mathews, Debra J. H., Donovan, Peter J., Harris, John, Lovell-Badge, Robin, Savulesca, Julian, and Faden, Ruth, 'Pluripotent Stem Cell-Derived Gametes: Truth and (Potential) Consequences', *Cell Stem Cell*, 5 (2009), 11–14.

Mertes, Heidi, 'Gamete Derivation from Stem Cells: Revisiting the Concept of Genetic Parenthood', *Journal of Medical Ethics*, 40 (2014), 744–747.

Mertes, Heidi, Goethals, Tina, Segers, Seppe, Huysentruyt, Marie, Pennings, Guido, and Provoost, Veerle, 'Enthusiasm, Concern and Ambivalence in the Belgian Public's Attitude Towards In-Vitro Gametogenesis', *Reproductive Biomedicine & Society Online*, 14 (2022), 156–168.

Monaco, Salvatore, and Nothdurfter, Urban, 'Discovered, Made Visible, Constructed, and Left Out: LGBT+ Parenting in the Italian Sociological Debate', *Journal of Family Studies*, 29 (2021), 471–488.

Moreno, Immaculada, Míguez-Forjan, Jose Manuel, and Simón, Carlos, 'Artificial Gametes from Stem Cells', *Clinical and Experimental Reproductive Medicine*, 42 (2015), 33–44.

Murakami, Kenta, Hamazaki, Nobuhiko, Hamada, Norio, Nagamatsu, Go, Okamoto, Ikuhiro, Ohta, Hiroshi, et al., 'Generation of Functional Oocytes from Male Mice In Vitro', *Nature*, 615/7954 (2023), 900–906.

Murphy, Timothy, 'Pathways to Genetic Parenthood for Same-Sex Couples', *Journal of Medical Ethics*, 44 (2018), 823–824.

National Academies of Sciences, Engineering, and Medicine, *In Vitro–Derived Human Gametes as a Reproductive Technology: Scientific, Ethical, and Regulatory Implications: Proceedings of a Workshop, Washington, D.C, USA* (Washington: The National Academies Press, 2023).

Pennings, Guido, 'Why We Need Stem-Cell Derived Gametes', *Reproductive Biomedicine Online*, 47 (2023), 1–3.

Robinson, Meghan, Sparanese, Sydney, Witherspoon, Luke, and Flannigan, Ryan, 'Human In Vitro Spermatogenesis as a Regenerative Therapy—Where Do We Stand?', *Nature Reviews Urology*, 20 (2023), 461–479.

Rose, Nikolas, *The Politics of Life Itself: Biomedicine, Power, and Subjectivity in the Twenty-First Century* (Princeton: Princeton University Press, 2007).

Segers, Seppe, 'The IVG "Relatedness Paradox": Researchers Should Mind Speculation', *Trends in Biotechnology*, 41 (2023), 1220–1222.

Segers, Seppe, and Accoe, Dorian, 'Mortality, Morbidity and Speculation: Ethical Considerations about the Lever for Ectogestation', in Christopher Stratman and Joona Räsänen, J., eds., *Ectogestation and the Future of Reproductive Rights: Technology, Justice and Abortion* (forthcoming).

Segers, Seppe, Mertes, Heidi, de Wert, Guido, Dondorp, Wybo, and Pennings, Guido, 'Balancing Ethical Pros and Cons of Stem Cell Derived Gametes', *Annals of Biomedical Engineering*, 45/7 (2017), 1620–1632.

Segers, Seppe, Mertes, Heidi, Pennings, Guido, de Wert, Guido, and Dondorp, Wybo, 'Using Stem Cell-Derived Gametes for Same-Sex Reproduction: An Alternative Scenario', *Journal of Medical Ethics*, 43 (2017), 688–691.

Segers, Seppe, Pennings, Guido, and Mertes, Heidi, 'Ethical Reflections on Stem Cell-Derived Gametes', *Médecine de la Reproduction*, 19 (2017), 298–306.

Segers, Seppe, Pennings, Guido, and Mertes, Heidi, 'Getting What You Desire: The Normative Significance of Genetic Relatedness in Parent–Child Relationships', *Medicine, Health Care and Philosophy*, 22 (2019), 487–495.

Smajdor, Anna, and Cutas, Daniela, 'Will Artificial Gametes End Infertility?', *Health Care Analysis*, 23/2 (2015), 134–147.

Stockdale, Katie, *Hope Under Oppression* (New York: Oxford University Press, 2021).

Sudenkaarne, Tiia, 'Queering Bioethics: A Queer Bioethical Inventory of Surrogacy', *Ethics, Medicine and Public Health*, 6 (2018), 117–125.

Taelman, Jasin, Mishra, Swati, Van der Jeught, Margot, and Heindryckx, Björn, 'Stem Cell-Derived Spermatozoa', in Jaime A. Horcajadas and José Gosálvez, eds., *Reproductomics* (London: Academic Press, 2018), 315–345.

Testa, Guiseppe, and Harris, John, 'Ethical Aspects of ES Cell-Derived Gametes', *Science*, 305 (2004), 1719.

Testa, Giuseppe, and Harris, John, 'Ethics and Synthetic Gametes', *Bioethics*, 19 (2005), 146–166.

Urban Walker, Margaret, *Moral Understandings: A Feminist Study in Ethics* (New York: Oxford University Press, 2007).

Vertommen, Sigrid, and Barbagallo, Camille, 'The In/Visible Wombs of the Market: The Dialectics of Waged and Unwaged Reproductive Labour in the Global Surrogacy Industry', *Review of International Political Economy*, 29 (2021), 1945–1966.

Wilson, Duncan, *The Making of British Bioethics* (Manchester: Manchester University Press, 2014).

Wu, Genie M. J., Chen, Andy C. H., Yeung, William, S. B., and Lee, Yin L., 'Current Progress on In Vitro Differentiation of Ovarian Follicles from Pluripotent Stem Cells', *Frontiers in Cell and Developmental Biology*, 11 (2023), 1–16.

Yee, Samantha, Mamone, Amar A., Fatima, Maryam, and Sharon-Weiner, Maya, 'Parenthood Desire, Perceived Parenthood Stigma, and Barriers to Achieving Parenthood in Childless Sexual Minority Men', *Journal of Assisted Reproduction and Genetics*, 41 (2024), 1739–1753.

Author Biographies

Dorian Accoe is a PhD researcher at the Department of Philosophy and Moral Sciences, Ghent University, and member of the Bioethics Institute Ghent (BIG) and METAMEDICA. He obtained a master's degree in moral sciences at Ghent University in 2021. His research focusses on the ethics of reproductive medicine. Particularly, he addresses ethical issues concerning the management of (suspected) genetic conditions in gamete donation. Dorian Accoe is secretary of the Bioethics Institute Ghent (BIG).

Kyoko Akatsuka is a researcher in the Uehiro Research Division for iPS Cell Ethics at Kyoto University's Center for iPS Cell Research and Application. She earned her PhD in human and environmental studies from Kyoto University in 2017. Akatsuka's recent research focusses on the ethical implications of human genome editing technologies, examining them through both empirical (sociological) and normative lenses.

Hilary Bowman-Smart is an empirical bioethicist with a background in genetics and philosophy. She is a Postdoctoral Research Fellow at the University of South Australia. Prior to this, she was a researcher at the Ethox Centre, University of Oxford, as part of a research project on non-invasive prenatal testing. Hilary's research focusses on reproduction, genomics, and the philosophy of health.

Misao Fujita is a Professor of Bioethics at the Center for iPS Cell Research and Application and the Institute for the Advanced Study of Human Biology at Kyoto University. She is a Member of the UNESCO International Bioethics Committee, the ISSCR Ethics Committee, and the Expert Panel on Bioethics of the Council for Science, Technology, and Innovation, Cabinet Office. Her research group is committed to investigating issues concerning unproven cell-based interventions, the use of human iPS cells to create chimeric animals and germ cells, and related questions about domestic and international ethical regulation and government policy, public attitudes, and current practice.

Christopher Gyngell is a Team Leader in Biomedical Ethics at the Murdoch Children's Research Institute and an Associate Professor at the University of Melbourne in the Department of Paediatrics. He has over seventy peer-reviewed publications in a diverse range of ethics, philosophy, science, and medical journals. Chris is committed to public engagement, frequently contributing to public discussions and policy debates on the ethical aspects of biomedical advancements.

Taichi Hatta is a Senior Assistant Professor at Shizuoka Graduate University of Public Health. His research interests lie in health science, medical sociology, medical ethics, cultural psychology, quantitative and qualitative research, and mixed methods. His doctoral paper, published online in 2018 in the *Journal of Mixed Methods Research*, demonstrated a new way of analysing temporal changes of moods in physician-patient dialogues for cancer treatment at hospitals in Japan using a Japanese narrative framework. Following that, he was awarded a PhD from Kyoto University in 2019 and also received the Honorable Mention award for the Mixed Methods International Research Association's Dissertation Award in 2022. He served as an Executive MMIRA Board Member at Large from 2019 to 2022.

Katsuhiko Hayashi, born on December 2, 1971, holds the position of Full Professor in the Department of Genome Biology, Graduate School of Medicine, Osaka University. Throughout his career, he has been dedicated to the study of germ-cell development and its *in vitro* reconstitution. His academic career began with the completion of the MS course at Meiji University from 1994 to 1996. Subsequently, he served as an Assistant Professor at Tokyo University of Science from 1996 to 2002 and later worked as a staff researcher at Osaka Medical Center from 2002 to 2005. During this time, in 2004, he obtained his PhD from Tokyo University of Science. From 2005 to 2009, Hayashi engaged in post-doctoral research at Azim Surani's laboratory in the Gurdon Institute, University of Cambridge, and subsequently held the position of Associate Professor at Kyoto University from 2009 to 2014. From 2014 to 2021, he served as a Full Professor at Kyushu University. Since 2021, Katsuhiko Hayashi has been in his current role at Osaka University. The research focus of the Hayashi group centres on comprehending the genetic and epigenetic regulation of mammalian germ-cell differentiation.

Anantharaman Muralidharan is a Research Fellow at the Centre for Biomedical Ethics, Yong Loo Lin School of Medicine, National University of Singapore. He has written on vaccine allocation, epistemic permissivism, political liberalism, AI ethics, trust, and IRB legitimacy.

Heidi Mertes is an Associate Professor in medical ethics at Ghent University and one of the founding members of the Bioethics Institute Ghent, Ghent University's METAMEDICA consortium, and Reproductive Health Education.be (RHE.be). Her academic research focusses on the ethical implications of innovations in healthcare, with a particular focus on reproductive medicine, genetics, embryonic stem-cell research, and more specifically ethical challenges at the intersection of these different domains.

Timothy F. Murphy, PhD, is Professor of Philosophy in the Biomedical Sciences at the University of Illinois College of Medicine in Chicago. He has published widely on ethical aspects of medicine as they involve sex and gender minorities, including ethical aspects of assisted reproductive treatments.

César Palacios-González is a Senior Research Fellow in Practical Ethics. His research interests include bioethics, philosophy of medicine, neuroethics, and applied philosophy. Dr Palacios-González graduated from the University of Manchester in 2016 with a PhD in science ethics, where he was part of the Wellcome Trust funded project 'The Human Body, its Scope, Limits and Future'. His doctoral work explored the ethics of human/non-human chimera research.

Seppe Segers is a Professor of ethics and moral science at Ghent University, where he is a member of the Bioethics Institute Ghent. He is the current Junior Deputy of the Special Interest Group Ethics and Law of ESHRE. His main research interests theoretical and substantive ethics.

Anna Smajdor is Professor of Practical Philosophy at the University of Oslo. She has written extensively on reproductive ethics. Her research includes questions about the nature of reproduction, the degree to which harm can be a feature of moral frameworks governing access to fertility treatments, and ethical issues in the development of artificial wombs.

Rob Sparrow is a philosopher who deals in the often uncomfortable real-world ethical implications of adopting new technologies. Advances being developed in all walks of science and engineering, from artificial intelligence to genetic engineering, raise fundamental issues of democracy and consent. Rob's research focusses on formulating ethical arguments that contribute to the public and political debate on these controversial topics.

Sonia M. Suter's scholarship focusses on issues at the intersection of law, medicine, and bioethics, with a particular focus on reproductive rights, emerging reproductive technologies, and ethical and legal issues in genetics. She has published widely in law reviews, peer-reviewed interdisciplinary journals, and science journals. One of her recent articles was selected for Editors' Choice 2020 by the *Journal of Law and the Biosciences*, one of the leading peer-reviewed journals in medical ethics and legal medicine. Another was chosen by the same journal's editorial team as one of its 'favorite widely read and cited articles' that has 'made a big impact'. An internationally recognized expert in genetics and the law and assisted reproductive technologies, Professor Suter is a co-author of the leading textbooks in those areas. She also participates in national working groups and advisory boards and as a consultant to policymakers on issues in her field of expertise. At GW Law, she teaches Torts, Law and Medicine, Genetics and the Law, and Assisted Reproductive Technologies.

Julian Savulescu is the Chen Su Lan Centennial Professor of Medical Ethics and Director of the Centre for Biomedical Ethics at the National University of Singapore Yong Loo Lin School of Medicine. He is an award-winning bioethicist and moral philosopher trained in neuroscience, medicine, and philosophy who has held the Uehiro Chair in Practical Ethics at the University of Oxford since 2002. He is a Visiting Professorial Fellow in Biomedical Ethics at Murdoch Children's Research institute and Distinguished International Visiting Professor in Law at the University of Melbourne, where he leads the Biomedical Ethics Research Group. He is also a Fellow of the Australian Academy of Health and Medical Sciences, is Fellow of the Australian Academy of the Humanities, and received an honorary doctorate from the University of Bucharest.

Tsutomu Sawai is a Professor at Hiroshima University, where he directs the Uehiro Division for Applied Ethics and the ELSI Unit of the Center for Collaborative Sciences. He also serves as Program Faculty for the Frontier Development Program for Genome

Editing and as a Visiting Professor at the Centre for Biomedical Ethics, National University of Singapore. He received his PhD from Kyoto University in 2016 and previously held academic appointments there. His research focusses on a wide range of ethical issues arising in advanced science, engineering, and medicine.

Helen Watt holds a PhD in philosophy from the University of Edinburgh and is a Research Fellow of Blackfriars Hall, Oxford. From 2010 to 2021 she was Senior Research Fellow at the Anscombe Bioethics Centre in Oxford, prior to which she was the Anscombe Centre Director. She joined the Centre in 1992 as Research Fellow and from 1993 to 1996 was also Senior Research Associate at Peterhouse, Cambridge. Her publications include the Routledge monographs *The Ethics of Pregnancy, Abortion and Childbirth* (2016) and *Life and Death in Healthcare Ethics* (2000), together with the edited collections *Cooperation, Complicity and Conscience* (2005), *Incapacity and Care* (2009), and *Fertility and Gender* (2011). She has also published internationally in journals including the *Journal of Medical Ethics* and *Clinical Ethics*. Her research interests include reproductive ethics, gender issues, action theory, and cooperation and conscientious objection.

Index

For the benefit of digital users, indexed terms that span two pages (e.g., 52–53) may, on occasion, appear on only one of those pages.